Organic Electronics Materials and Devices

Shuichiro Ogawa
Editor

Organic Electronics Materials and Devices

Second Edition

Editor
Shuichiro Ogawa
Asahi Kasei Corporation
Fuji, Japan

ISBN 978-4-431-56938-1   ISBN 978-4-431-56936-7   (eBook)
https://doi.org/10.1007/978-4-431-56936-7

© The Editor(s) (if applicable) and The Author(s), under exclusive license to Springer Nature Japan KK 2015, 2024

This work is subject to copyright. All rights are solely and exclusively licensed by the Publisher, whether the whole or part of the material is concerned, specifically the rights of translation, reprinting, reuse of illustrations, recitation, broadcasting, reproduction on microfilms or in any other physical way, and transmission or information storage and retrieval, electronic adaptation, computer software, or by similar or dissimilar methodology now known or hereafter developed.

The use of general descriptive names, registered names, trademarks, service marks, etc. in this publication does not imply, even in the absence of a specific statement, that such names are exempt from the relevant protective laws and regulations and therefore free for general use.

The publisher, the authors, and the editors are safe to assume that the advice and information in this book are believed to be true and accurate at the date of publication. Neither the publisher nor the authors or the editors give a warranty, expressed or implied, with respect to the material contained herein or for any errors or omissions that may have been made. The publisher remains neutral with regard to jurisdictional claims in published maps and institutional affiliations.

This Springer imprint is published by the registered company Springer Nature Japan KK
The registered company address is: Shiroyama Trust Tower, 4-3-1 Toranomon, Minato-ku, Tokyo 105-6005, Japan

Paper in this product is recyclable.

Collaborators

Prof. Jun Mizuno
National Cheng Kung University
Tainan City, Taiwan

Prof. Toshiyuki Watanabe
Tokyo University of Agriculture and Technology
Tokyo, Japan

Satoru Toguchi
NEC Corporation
Tokyo, Japan

Prof. Kazuaki Furukawa
Meisei University
Tokyo, Japan

Preface

In 2022, Springer Japan, the publisher proposed that we publish a second edition, which was discussed by the Japanese Research Association for Organic Electronics Materials (JOEM) and the Organizing Committee of the JOEM Academy. We are very pleased to publish this second edition after 8 years since the original publication.

Compared to inorganic semiconductors such as silicon, organic semiconductors are generally more flexible, thinner, lighter, manufactured using printing processes, and lower cost. However, organic semiconductors are slower in operating speed, due to a large degree of freedom in their structure from van der Waals forces with weak intermolecular bonds, and vulnerable to oxygen and moisture.

Despite its disadvantages, the global market for organic electronic devices is driven by glucose sensors including the large scale of small- and medium-sized Organic Light Emitting Diode (OLED) elements that are increasingly being adopted in smartphones, and its market size reached US$ 35 billion in 2021 and expected to grow to US$ 60 billion by 2035.

JOEM Academy has also continued to lecture and evolved with these emerging trends in the recent period while overcoming the COVID19 pandemic. In this revision, we have added the latest research results on organic electroluminescence (EL), organic thin film transistor (TFT), and perovskite solar cells and others. We have also asked new professors to write about applications to various sensors. Since the book also has an aspect of being a textbook, we have made some modifications to the second edition so that it can be used for basic and theoretical studies, as well as improving the understanding of the science and mechanism of topic covered.

What we regret a little is that we were not able to add cutting-edge research on Digital Transformation technologies (DX) and sustainability technologies in the field of organic electronics. We have decided that we still need more time to compile it into this kind of textbook-like book. If there is another opportunity in the future, we would like to add these technologies and researches to the book.

Once again, I am profoundly thankful to the members of the JOEM Academy committee who are also coeditors of this book: Prof. Jun Mizuno at National Cheng

Kung University, Taiwan, Prof. Toshiyuki Watanabe at the Tokyo University of Agriculture and Technology, Prof. Kazuaki Furukawa at Meisei University and Mr. Satoru Toguchi at the NEC Corporation. I am also grateful to Emeritus Prof. Yoshio Taniguchi at Shinshu University, who is the emeritus chairman of JOEM; Dr. Hiroyuki Suzuki, who is the president of JOEM, and executive directors Mr. Kei Fujinami and Dr. Ryuichi Nakamura for their help and encouragement. I am also thankful to Mr. Yasuo Sakai and Ms. Aya Tomita of the JOEM secretariat and Dr. Shinichi Koizumi at Springer Japan, the publisher for their help.

Again, some mistakes certainly remain because of my inability to amend and correct them. However, I sincerely hope that this second edition will be of some help to graduate students and researchers who are doing research in the field of organic electronics, as well as to researchers who are struggling daily to commercialize their work in the industry.

Fuji, Shizuoka, Japan Shuichiro Ogawa
Summer 2023

Contents

1 **Physics of Organic Field-Effect Transistors and the Materials** 1
Tatsuo Hasegawa

2 **Organic Light-Emitting Diodes (OLEDs): Materials, Photophysics, and Device Physics** 73
Ryo Nagata, Kenichi Goushi, Hajime Nakanotani, and Chihaya Adachi

3 **Organic Solar Cells** 119
Shuzi Hayase

4 **Printed Organic Thin-Film Transistors and Integrated Circuits** 147
Hiroyuki Matsui, Kenjiro Fukuda, and Shizuo Tokito

5 **Ultra-Flexible Organic Electronics** 185
Tomoyuki Yokota

6 **Polymer Nanosheets with Printed Electronics for Wearable and Implantable Devices** 221
Tatsuhiro Horii and Toshinori Fujie

7 **Solution-Processed Organic LEDs and Perovskite LEDs** 239
Hinako Ebe, Takayuki Chiba, Yong-Jin Pu, and Junji Kido

8 **Transient Properties and Analysis of Organic Photonic Devices** 283
Hirotake Kajii

9 **Microfluidic Self-Emissive Devices** 317
Takashi Kasahara and Jun Mizuno

Chapter 1
Physics of Organic Field-Effect Transistors and the Materials

Tatsuo Hasegawa

Abstract Organic semiconductors that were discovered more than half century ago in Japan (H. Inokuchi, Org. Electron. **7**, 62 (2006)) are now transfigured into the practicable electronic materials by the recent concentrated studies of the materials, thin-film processing, and device fabrication technologies. In this chapter, we first present and discuss fundamental aspects of electronic phenomena in organic semiconductors as the bases to understand and study the organic electronics technologies. Then we discuss how to understand the charge carrier transport and the carrier states in organic field-effect transistors (OFETs or also frequently referred as organic thin-film transistors, or OTFTs). Finally, we introduce some new concepts to fabricate OFETs by print production technologies, and also present some recent studies to develop materials and processing for realizing high performance printed OFETs.

Keywords Organic semiconductor · Organic field-effect transistor · π-electron · Carrier dynamics · Printed electronics · Layered crystallinity

1.1 Fundamentals for Crystalline Organic Semiconductors

1.1.1 Semiconductors with Hierarchical Structure

Organic semiconductors are a class of semiconducting organic materials composed mainly of carbon elements. The rigorous definition of semiconductors—i.e., the filled electronic states and the empty electronic states are divided energetically by a moderate width of forbidden band or energy gap—is naturally satisfied by all the organic semiconductors, as is similar to other inorganic semiconducting materials. In fact, the most basic (or crude) characteristics of the materials and the devices based on the organic semiconductors are typical of semiconductors, whereas they exhibit

T. Hasegawa (✉)
Department of Applied Physics, The University of Tokyo, Bunkyo-ku, Tokyo, Japan
e-mail: t-hasegawa@ap.t.u-tokyo.ac.jp

© The Author(s), under exclusive license to Springer Nature Japan KK 2024
S. Ogawa (ed.), *Organic Electronics Materials and Devices*,
https://doi.org/10.1007/978-4-431-56936-7_1

specific characteristics unique to this whole class of the materials. The organic semiconductors may be defined, for a rather practical reason, as the semiconductors composed of organic molecules that are synthesized by the techniques of organic synthetic chemistry. Along with this feature, however, the organic semiconductors are quite unique in that the whole solid-state properties are ascribed to a hierarchical nature of [atom–molecule–solid], where the molecules are composed of atoms held together by covalent bonds, and the solids are formed by discrete molecules held together by van der Waals interactions. The key player to bridge this hierarchy is the π-electrons that are the source for all the functional electronic properties of the organic semiconductors. In this section, we outline the electronic structure and the origin of fundamental and specific characteristics of the organic semiconductors, with specially focusing on the roles of the π-electrons. Then we briefly outline the basic architecture of the organic field-effect transistors.

1.1.2 π-Electrons as Source of Mobile Carriers

A major source for an enormous number of organic materials is the unique nature of carbon that can form chains, rings, or branches by stable covalent σ- or π-bonds. The σ-bonds are formed by the 2s-2p hybrid orbitals (sp^1, sp^2, or sp^3) between the adjacent atoms, whereas the π-bonds are formed by the overlap between 2p orbitals of adjacent atoms that do not participate in the formation of σ-bonds (Fig. 1.1). The terms of σ and π are originally associated with the symmetry of the bonds with respect to the rotation along the interatomic axes, although the term of π is now frequently utilized to refer to the electrons in the π-bonds. The σ-bonds are relatively strong, and the electrons in the σ bonds are likely to be localized. In contrast, the π-electrons are widely delocalized when the 2p orbitals of respective atoms along the connected linear chains or rings are all aligned in parallel, as presented in Fig. 1.1.

Fig. 1.1 Schematic for π bond and σ bond between carbon atoms

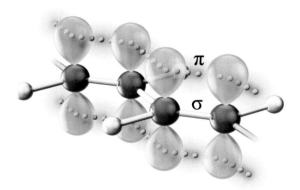

1 Physics of Organic Field-Effect Transistors and the Materials

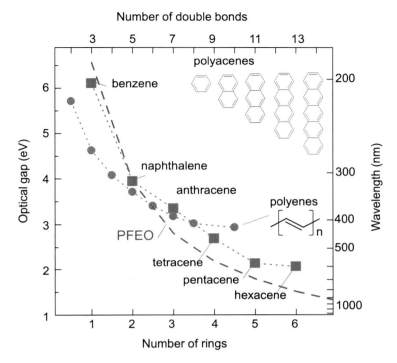

Fig. 1.2 Optical gap energy of polyacenes as a function of number of fused rings or double bonds (red squares); values calculated by PFEO model (blue dashed line); optical gap energy in polyenes as a function of double bonds (gray circles)

The effect of the delocalized π-electrons most obviously appears in the color of the materials. The usual organic or plastic materials that are formed only by the σ-bonds do not have colors or are transparent. This is because the σ-bonds are so strong that the electronic excitation energy becomes high and the optical gap energy is much larger than the visible photon energy range (1.6–3.3 eV). In contrast, when the π-electrons are delocalized over the molecule, the electronic excitation energy considerably decreases, and the materials become colored. Figure 1.2 presents the optical gap energies of polyacenes (and polyenes), plotted as a function of the number of fused benzene rings and double bonds. When the molecules become larger and the delocalized π-electrons are more extended, the excitation energy becomes considerably lowered and becomes colored due to the absorption of visible light.

In both the polyenes and polyacenes, each carbon has one π-electron along the alternating sequence of single and double bonds in the chemical notation. Actually, however, these π-electrons do not belong to each double bond but rather to a group of atoms along the alternating sequence of single and double bonds. The sequence is

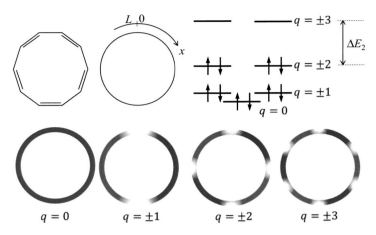

Fig. 1.3 Energy levels and wave functions in perimeter-free electron orbital (PFEO) model

often called as conjugated double bonds, which allow a delocalization of π-electrons across all the adjacent aligned p-orbitals.

Here we present the most intuitive picture for the delocalized π-electrons of polyacene by a perimeter-free electron orbital (PFEO) model [1]. We assume naphthalene, composed of two fused benzene rings as an example, that has ten delocalized π-electrons along the circle, as presented in Fig. 1.3. For the simplicity, it is considered that the ten π-electrons can move freely ($V = 0$) along the circle with a length L, but infinite potential ($V = \infty$) outside the circle. The wave function ϕ^{PFEO} is the simple plane wave as a free electron, and the energy E can be written as a solution of the Schrödinger equation by the following form:

$$\phi_q^{PFEO} = \frac{1}{\sqrt{L}} \exp\left[-i\left(\frac{E_q}{\hbar}t - \frac{2\pi q}{L}x\right)\right], \tag{1.1}$$

$$E_q = \frac{\hbar^2 k^2}{2m} = \frac{\hbar^2}{2m}\left(\frac{2\pi}{L}\right)^2 q^2. \tag{1.2}$$

Here, $\hbar$ is the Planck constant, k is the wave number, and q ($= 0, \pm 1, \pm 2$) is the quantum number. The energy depends on the number of nodes in the wave functions, which is equal to $2q$. The energy diagram of the systems is depicted in Fig. 1.3. The circle length can be represented as $L = Na = (4n + 2)a$, where N is the number of atoms ($N = 10$ in the case of naphthalene), a is the interatom distance, and n is the ring number ($n = 2$ in the case of naphthalene). As two electrons are filled at each levels, the highest filled level will have $q = n$, and the lowest unfilled level have $q = n + 1$, so that the energy difference between the levels can be written (with using typical interatomic distance value of $a = 0.138$ nm) as

$$\Delta E_n = \frac{\pi^2 \hbar^2}{2ma^2} \frac{1}{2n+1} = \frac{19.7 \text{ eV}}{2n+1}, \quad (1.3)$$

The dashed line in Fig. 1.2 presents the result of the calculation. In spite of such simplicity, it is surprising to find the overall consistency as to the trend and rough values. Another important result is also obtained in terms of the stability of the molecules. As the electrons are filled from the lower levels, these ring-like molecules become stable, if there are (4n + 2) π-electrons per molecule (one electron per each carbon atom). This is the origin of the stability of aromatic compounds and is often called as Hückel's rule. These features demonstrate that the delocalized nature of the π-electrons where the weak linkage between the 2p orbitals can form the nearly free electrons within the molecules. The unique nature of organic molecular materials is the designability of materials in terms of the shape and size of such free electron system within the molecules.

1.1.3 Molecular Orbitals

The molecular orbital (MO) theory is used to determine the π-electronic states in the molecules [2]. The MO theory is based on the concept that the electrons are not assigned to the individual bonds between atoms but to the molecular orbital that is extended to the whole molecule. This concept is in contrast to the valence bond (VB) or Heitler-London theory. The simplest model is based on linear combinations of atomic orbitals (LCAO) to present the molecular orbital ϕ as follows:

$$\phi = \frac{1}{\sqrt{N}} \sum_{n=1}^{N} c_n^a \chi_n^a, \quad (1.4)$$

where N is the number of atoms and c_n^a and χ_n^a are the coefficient and atomic orbitals of the nth atom, respectively. The linear combination should compose the eigenfunction of one-electron molecular Hamiltonian h with eigen energy ε as

$$h\phi = \varepsilon\phi. \quad (1.5)$$

By multiplying $\chi_{n'}^{a*}$ (complex conjugate of $\chi_{n'}^a$) and integrating both sides as $\int \chi_{n'}^{a*} h\phi d\tau = \varepsilon \int \chi_{n'}^{a*} \phi \, d\tau$, and by substituting Eq. (1.4) into Eq. (1.5), simultaneous equations for coefficients c_n^a are obtained. In the Hückel theory for π-electrons, the following simplifications are assumed

Fig. 1.4 Molecular orbital and energy diagram of benzene

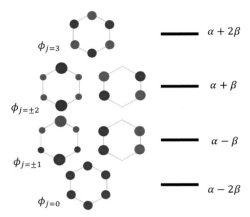

$$h_{nn'} = \int \chi_{n'}^{a*} h \chi_n^a d\tau = \begin{cases} \alpha & (n'=n) \\ \beta & (n'=n+1) \\ 0 & (n' \neq n, n+1) \end{cases} \quad (1.6)$$

$$s_{ij} = \int \chi_i^{a*} \chi_j^a d\tau = \begin{cases} 1 & (i=j) \\ 0 & (i \neq j) \end{cases}. \quad (1.7)$$

Here α and β are called as Coulomb integral and resonance integral, respectively. The s_{ij} is called as overlap integral. For example, the simultaneous equations in the case of benzene ($i = 1, 2, \ldots, 6$) can be simply represented by the matrix formula as

$$\begin{pmatrix} \alpha - & \beta & 0 & 0 & 0 & \beta \\ \beta & \alpha - & \beta & 0 & 0 & 0 \\ 0 & \beta & \alpha - & \beta & 0 & 0 \\ 0 & 0 & \beta & \alpha - & \beta & 0 \\ 0 & 0 & 0 & \beta & \alpha - & \beta \\ \beta & 0 & 0 & 0 & \beta & \alpha - \end{pmatrix} \begin{pmatrix} c_1^a \\ c_2^a \\ c_3^a \\ c_4^a \\ c_5^a \\ c_6^a \end{pmatrix} = 0. \quad (1.8)$$

By solving the equation, eigen energy and eigenfunction can be obtained. The eigen energy is obtained as $= \alpha - 2\beta, \alpha - \beta, \alpha + \beta, \alpha + 2\beta$, as presented in Fig. 1.4.

It is also important to understand, for solving the equation, that the coefficients c_n^a obey some relations due to the molecular symmetry. In the case of benzene, the $\pi/3$ rotation of the eigenfunction as to the axis perpendicular to the molecular plane at the center of the molecule still affords the eigenfunction with the same eigen energy. It means that the $\phi = c\phi'$, where c is complex number with absolute value 1 and ϕ' is the molecular orbital after the rotation. Furthermore, the six times repetition of the $\pi/3$ rotation should give it back to the original eigenfunction. Therefore, the eigenfunction of benzene can be represented as

$$\phi = \frac{1}{\sqrt{6}} \sum_{n=1}^{6} e^{i\frac{2\pi n}{6}j} \chi_n^a \quad (j = 0, \pm 1, \pm 2, \pm 3). \quad (1.9)$$

The wave functions of molecular orbitals of benzene are depicted in Fig. 1.4.

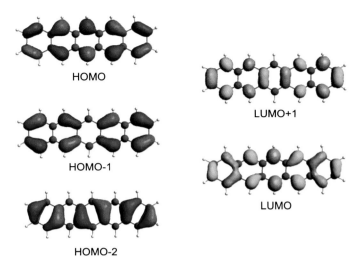

Fig. 1.5 Wavefunctions of molecular orbitals in a pentacene molecule calculated by ADF program [5]

The cyclic compounds can also include non-carbon elements within the cycles (which is called as hetero atoms). For example, five-membered ring becomes stable due to the Hückel's rule ($N = 4n + 2$), in case that one sulfur atom is included, because two electrons in the lone pair of 3p orbitals contribute to the molecular orbitals. In particular, the thiophene ring is known to be an extremely important unit to obtain high-performance organic semiconductors. This is associated with the characteristics of the 3p orbitals of sulfur, which is more extended outside the molecules than 2p orbitals of carbon, and is effective to increase the intermolecular interactions when the semiconducting molecular crystals are formed.

In the simplest MO calculations, empirical values are used for α and β to describe the one-electronic states in the molecules. Electrons in the molecules fill the states from lower levels with satisfying the Pauli principle. Ground states of the molecules that are formed by many electrons are approximated by using the Slater determinant, the treatment of which is called as the Hartree-Fock approximation. The detailed calculations for the many-electron system are conducted by ab initio (or first principles) calculations. In the states of many electrons, Coulomb and exchange interactions between electrons should be considered and are treated by configuration interactions that hybridize many excited electronic states to minimize the total energy and to obtain the more reasonable ground states of the molecule.

Recently, calculations using density functional theory (DFT) is more frequently used to calculate ground states of many-electron system; the method is based on the calculation of distribution function of electron density, $n(r)$, and the effective potential. These methods are now familiarized, owing to the rapid development of programs and calculation speed of the computers. Today, DFT calculations are widely conducted using Gaussian [3] or ADF [4] programs with the use of personal computers. Figure 1.5 illustrates the wave function of molecular orbitals of pentacene as obtained by the ADF calculations.

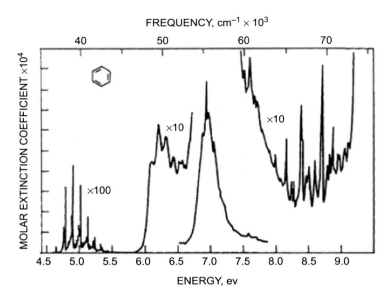

Fig. 1.6 Molecular absorption spectrum of benzene. Source: Ref. [5]

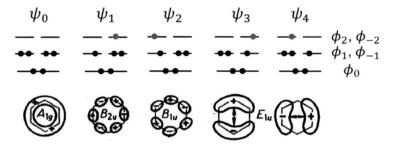

Fig. 1.7 Schematic of ground and excited states of benzene. Source: Ref. [6]

1.1.4 Optical Transition of Molecule

UV/vis light irradiation on the π-conjugated molecules causes excitation of one electron into an unoccupied state. Figure 1.6 presents molecular absorption spectrum of benzene [5]. A strong absorption band appears at 7.0 eV, whereas two weak absorption bands are observed at lower energy of around 5.0 eV and 6.2 eV. The energies are roughly as expected from Fig. 1.2, while the difference in absorption intensity of the bands should reflect the symmetry of molecules and thus of electronic states. The ground and excited states in benzene are schematically depicted in Fig. 1.7 [6]. The first exited states are degenerated fourfold but are split due to electron-electron interaction. The benzene molecule forms the point group D_{6h}, and

Fig. 1.8 Adiabatic potential and Frank-Condon transition in molecules

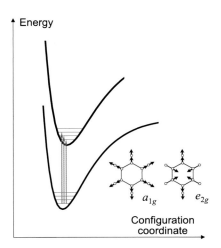

the ground and first excited states have symmetry represented by the irreducible representations as A_{1g}, B_{2u}, B_{1u}, and E_{1u}, respectively. As the optical transitions take place if the integral $M = e \int r\psi_n \psi_0^* dr$ does not vanish, the transition into E_{1u} states only becomes quite large.

As seen in Fig. 1.7, the respective electronic absorption bands are also composed of a series of discrete transitions, which is called as vibronic transitions as associated with the coupling with molecular vibrations. It is important to understand that atomic mass is much larger than electron mass, so that the motion of nuclei is always accompanied by instantaneous variation of electronic states. This is called as Born-Oppenheimer approximation. According to the approximation, atoms of molecules move along potential (called as adiabatic potential), as shown in Fig. 1.8. The potential curves are usually plotted one-dimensionally as a function of configuration coordinate, to represent the group motion of atoms or the molecular vibration. The coordinate that gives energy minimum represents the equilibrium position of the molecular shape, and the curve determines the molecular vibration which is quantized into discrete states. Similar curves can be drawn for the electronic excited states.

The time required for the optical transition to the electronic excited states is in the order of $\sim 10^{-15}$ s due to electric field oscillation of light, which is much faster than atomic motions ($\sim 10^{-13}$ s). Thus, the optical transition occurs vertically in Fig. 1.8, which is called as Frank-Condon transition. The transitions into respective vibrational states of the electronic excited states are called as vibronic transition. The vibrational coupling leads to the optically forbidden electronic excited states into weakly optically allowed states, as observed in the B_{2u} states at 5.0 eV of benzene in Fig. 1.6.

1.1.5 Electronic Band Formation and Carrier Transport

When same kinds of organic molecules are gathered, they are self-organized to form crystals, if the molecular shapes have relatively high symmetry and the molecules can be packed densely without opening within the molecular arrangement that holds the translational symmetry. In the crystals composed of densely packed molecules, molecular orbitals of π-electrons, composed of spatially extended carbon 2p orbitals, are overlapped and interacted with those of the adjacent molecules. Thus, the electronic states of the π-electrons can be extended widely over the crystals. In the crystals, the intermolecular interactions between the combination of adjacent molecules should become equivalent with others between a crystallographically equivalent combinations. Such a formation of crystals that have translational symmetry is essentially important for the (translational) charge-carrier motion in the semiconductors.

Here we discuss the electronic wave function in the crystals with certain intermolecular interactions. In the case of organic semiconductors in which the molecules are bound by relatively weak van der Waals interactions, it is quite effective, as a first approximation, to use the tight-binding model in which the wave function is formulated by linear combination of molecular orbitals for obtaining wave functions in solids. This is analogous to the LCAO for the formation of molecular orbitals by the linear combination of atomic orbitals. The solid-state wave function $\varphi^s(\mathbf{r})$ is represented by the following form [7]:

$$\varphi^s(\mathbf{r}) = \frac{1}{\sqrt{N}} \sum_{n=0}^{N} c_n^m \phi_n^m (\mathbf{r} - \mathbf{R}_n), \tag{1.10}$$

where N is the number of molecules and c_n^m and $\phi_n^m(\mathbf{r} - \mathbf{R}_n)$ are the coefficient and molecular orbitals of the nth molecule at the position $\mathbf{R}_n$, respectively. Because of the translational symmetry of the crystals, the wave function can be represented by the following form (by obtaining the same procedure as Eq. (1.9)):

$$\varphi_k^s(\mathbf{r}) = \frac{1}{\sqrt{N}} \sum_{n=1}^{N} \exp(i\mathbf{k} \cdot \mathbf{R}_n) \phi_n^m (\mathbf{r} - \mathbf{R}_n). \tag{1.11}$$

This kind of formula is called as the Bloch function and satisfies the Bloch's theorem.

Electronic energy, $E(k)$, which is plotted as a function of wave number k is the electronic band structure. The tight-binding method affords the trigonometric function (or called as "cosine band"). Because the intermolecular interaction is relatively weak in organic semiconductors, the highest occupied valence band (HOVB) is mainly composed of HOMOs of the molecules, and the lowest unoccupied conduction band (LUCB) is mainly composed of LUMOs of the molecules. Therefore, the solids composed of closed-shell molecules should be the semiconductors (each band is filled) and are divided by the energy gap. Effective mass of each band is approximated by the following equations:

$$\frac{1}{m^*} = \frac{1}{\hbar^2} \frac{\partial^2 E(k)}{\partial k^2}. \tag{1.12}$$

Thus, the effective mass m^* is proportional to the intermolecular transfer integrals. Band transport usually means the free-carrier motion with the effective mass as is determined by the band curvature. Under an electric field, free carriers are not infinitely accelerated but are frequently scattered such as by phonons in the crystals at finite temperature, so that they have average velocity. In other words, the mean free path of free carriers is finite at room temperature in the crystals, so that the transport of charge carriers becomes "diffusive" motion. In the diffusive motion, the average velocity of diffusive drift motion of carriers under an electric field is proportional to the applied electric field, whose proportional coefficient is defined as the drift mobility (or simply mobility) μ of carriers as

$$v = \mu E. \tag{1.13}$$

By using the mobility, the electrical conductivity σ is described by the following equation:

$$\sigma = ne\mu. \tag{1.14}$$

Here n is number of carriers per unit volume and e is the elementary charge. In the diffusive motions of carriers, the following Einstein relation generally holds with assuming the diffusion constant D (defined as the coefficient in diffusion equation):

$$\mu = \frac{eD}{k_B T}. \tag{1.15}$$

Intrinsic mobility in the semiconductor single crystals should be determined by such a mechanism.

1.1.6 Crystal Structures

For achieving efficient carrier transport in organic semiconductors, it is necessary to design organic molecules which can form highly crystalline solids with large intermolecular interactions and translational symmetry. The molecular orbital calculations using the DFT are also utilized to calculate intermolecular interactions between the molecules by using the atomic coordinates as is obtained by the crystal structure analysis. Figure 1.9 shows a result of the intermolecular interactions in pentacene calculated by the ADF method. The pentacene is known to crystallize

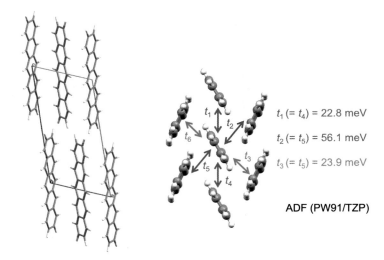

Fig. 1.9 Crystal structure and intralayer transfer integrals of thin-film pentacene

Fig. 1.10 Herringbone-type molecular packing and atomic contacts shown by space-filling model in pentacene crystals

layered-crystal structures where the molecules are packed by herringbone-type packing motif, as presented in Fig. 1.10. It is empirically known that a number of materials with this type of packing motif afford high-performance OFETs.

Crystal structure analysis [8] is an indispensable tool for investigating the molecular packing motif and thus for properly evaluating the potential of organic semiconductors if they afford high-performance OFETs. The structure analysis is done by irradiating monochromatic X-ray beam on single crystals and collecting data of a number of Bragg reflections of respective indexes. By the Fourier transformation of the intensity distributions of each index, distribution of electron density in the unit cells can be obtained. As is different from inorganic materials, it is difficult to conduct the crystal structure analysis of organic semiconductors by means of x-ray analysis for polycrystalline films or powders, because much more atoms are involved in the unit cell. It is indispensable to use full x-ray single-crystal structure analysis to obtain the reliable crystal structures.

It is often difficult to conduct full crystal structure analysis for organic semiconductors, which often burdens the evaluation and development of materials. For the further development of new materials, it is desirable to be able to predict in advance, at the molecular design stage, what type of crystal structure will be formed. So far it is not possible to predict correctly what kind of crystal structure is formed by a molecule before it is synthesized and crystallized. The prediction of the crystal structure of molecular materials is renowned as a computationally difficult problem [9]. Thus, it is still uncertain whether the designed molecules on the desk are promising until they are synthesized. Probably from many actual examples, however, we could empirically predict what kind of crystal structures can be obtained in the designed molecules. Further studies are now being carried out to accelerate the development of materials through crystal structure prediction targeted at highly layered-crystalline molecular materials.

1.1.7 Optical Properties and Excitonic Effects

The UV/vis absorption spectrum of organic semiconductors is usually quite similar to that of isolated molecules, though some energy shift is observed. The feature is in contrast to the case of covalent-bonded inorganic semiconductors in which strong band-to-band optical transitions are ordinarily observed. Some people may wonder, "where is the band-to-band transition in this semiconductor?". Origin of the spectral similarity is that the electron-hole relative motion is confined to the respective molecules because of relatively small intermolecular transfer integral and strong Coulomb interaction due to low dielectric constant of organic solids. In this sense, the strong molecular nature is apparent in the optical properties of organic semiconductors.

Thus it is natural to assume that transition dipoles of molecules should be simultaneously formed and arranged over the irradiated area of crystals. This is associated with the fact that the distance between molecules is much shorter than the wavelength of light, which means that the optical transition is allowed at wavevector $k = 0$. It leads to the effect derived from the electrostatic interactions between transition dipoles. These solid-state photo-excited states are called as molecular or Frenkel excitons.

Figure 1.11 presents the actual polarized absorption spectra of pentacene singlecrystal thin film, measured with polarization of incident light parallel to *a* and *b* axes, respectively. The absorption is stronger, and the peak photon energy is lower along *a* axis than along *b* axis. The former feature is attributed to the fact that the orientation of molecular short axis has larger component along *a* axis than along *b* axis. As is understood from the feature of HOMO and LUMO shown in Fig. 1.5, the transition dipole of the lowest electronic excitation of pentacene is polarized

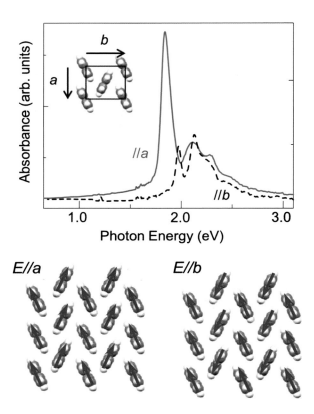

Fig. 1.11 Upper: Polarized absorption spectra of pentacene single-crystal thin film. Lower: Arrangement of transition dipole moments in crystals

parallel to the molecular short axis of pentacene molecule. The latter feature is ascribed to the fact that the unit cell contains two molecules with different molecular short-axis orientations. Mutual orientations of the transition dipoles formed in the crystals are distinct between the cases of incident light polarization parallel to the a and b axis, as presented in the lower panel of Fig. 1.11. The feature leads to the difference of intermolecular electrostatic interaction energy. The observed optical anisotropy in the peak photon energy is called as Davydov splitting.

1.1.8 Architecture of Organic Field-Effect Transistors

Figure 1.12 shows typical device structure of OFETs. The device is composed of semiconductor layer, gate dielectric, and gate/source/drain electrodes. In the device, drain current flows between source and drain electrodes, by applying drain voltage between the source and the drain electrodes. The drain current can be controlled by the gate voltage which is applied between the source and the gate electrodes. Nominally, there should be no current through the gate dielectric layer. Carriers

Fig. 1.12 Schematic for organic thin-film transistors

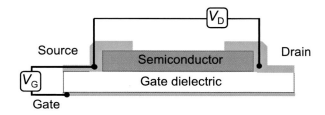

are accumulated both at the gate electrode and semiconductor layer as like a capacitor. These accumulated charges contribute to the drain current as a drift current.

The channel semiconductors are usually composed of intrinsic organic semiconductors without intentional doping. This feature of the device is much different from that of conventional FETs composed of inorganic semiconductors like covalent-bonded crystals of silicon [10]. The first reason of this feature is that the intentional doping in organic semiconductors is difficult, because the crystal lattice in which molecules are densely packed without opening and are bound by weak intermolecular interaction is very easily broken by the introduction of dopants with different shapes. The second is a rather positive reason that the surface states due to dangling bonds in covalent-bonded crystals are not formed in organic semiconductors, so that the carrier injection is possible without intentional doping.

Due to the feature of the OFETs as presented above, a type of the device is rather close to the enhancement-type Si-MOSFET [11]. Carriers are injected into the organic semiconductor layer, if the Fermi level of metal for source/drain electrode coincides with the band energy of semiconductors. Therefore, the organic semiconductors with higher HOVB energy (i.e., lower work function) usually show *p*-type operation, while those with lower LUCB energy (i.e., higher work function) usually show *n*-type operation. Indeed, even if the small number of charge carriers with different types (hole for *n*-type operation or electron for *p*-type operation) could be injected into the semiconductor layer by the direct semiconductor-metal junctions, they are usually trapped in some trap agents within the semiconductor layers and do not contribute to the drain current. We note that if the carriers are not accumulated under the application of the gate voltage, gate electric field should be penetrated into the channel semiconductor layers.

We give the expression for the drain current I_D as a function of gate voltage V_G and drain voltage V_D. When V_D is smaller than V_G, the charge accumulation covers the whole channel region between the source and drain electrodes. In this case, I_D is proportional to $(V_G - V_T)$ (V_T is the threshold voltage), which is called as the linear regime and can be described by the following equation:

$$I_D \cong \frac{Z}{L}\mu C_0 (V_G - V_T) V_D, \qquad (1.16)$$

where Z is channel widith, L is channel length, C_0 is capacitance of the gate dielectric layer per unit area, and V_T is the threshold voltage. The threshold voltage is associated with the number of traps within the semiconductor layer. When V_D is larger than V_G, the charge accumulation does not cover the whole channel region but is limited in the region close to the source electrodes. The location where the charge accumulation is depleted is called as the pinch-off point. In this case, I_D is proportional to $(V_G - V_T)^2$ and becomes independent of V_D. This is called as the saturation regime and can be described by the following equation:

$$I_D^{\text{sat}} \cong \left(\frac{Z\mu C_0}{2L}\right)(V_G - V_T)^2. \tag{1.17}$$

These equations are used to evaluate the mobility of the semiconductor layers in the experiment.

From the architecture of the field-effect transistors, layered-crystalline organic semiconductors are quite suitable to afford high-performance OFETs, as it is possible to produce highly-uniform two-dimensional semiconductor-insulator interfaces. It has been demonstrated that a high degree of layered crystallinity is essential for the production of single-crystalline or uniaxially oriented polycrystalline thin films in which high-mobility carrier transport occurs along the film planes.

1.2 Charge Carrier Dynamics in Organic Field-Effect Transistors

1.2.1 Overview

In the Sect. 1.1, we discussed that the electronic band structures with relatively narrow bandwidths are formed in crystalline organic semiconductors that feature periodic crystal lattices. This picture provides a primary fundamental basis for understanding the charge-carrier transport in organic semiconductor thin films of OFETs. However, we also have to know that the charge-carrier dynamics in actual OFETs is not that simple as is determined by the "free motion" of charge carrier with an effective mass that is prescribed by the electronic band structure. As a clue to address this issue, temperature dependence of carrier mobility (or conductivity) was frequently utilized to characterize the carrier transport in real devices as either "metallic type" or "activation type." Figure 1.13 presents an example of current-voltage characteristics of a pentacene OFET at various temperatures (gate voltage is fixed). When assuming free-carrier (i.e., metallic-type) transport, the mobility is expected to gradually increase by lowering the temperature due to the reduction of phonon scattering. In reality, however, almost all the OFETs including single-crystalline OFETs present activation-type characteristics at least at low temperature, even if the "metallic-type" behavior is observed at relatively high temperature [12–

Fig. 1.13 Transfer characteristics of pentacene OFETs at various temperature

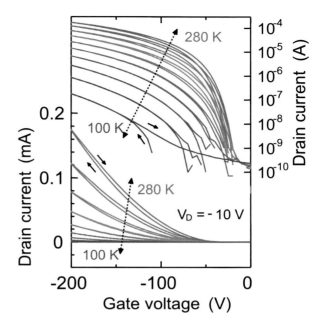

14]. In this respect, the charge transport has an intermediate character between the band transport and the charge localization at finite temperature.

Here we have to comment that the use of the "hopping" theory should not be justified in crystalline organic semiconductors that show relatively high carrier mobility, even if the activation-type characteristics are observed. In the hopping model, it is assumed that the charge transport is based on a hopping process between diabetic molecular states, where the molecules are energetically relaxed to localize the hopped charge on the respective molecule. The hopping frequency, $k_{hopping}$, is simply given by [15–17]:

$$k_{hopping} = \frac{t^2}{\hbar} \left(\frac{\pi}{\lambda k_B T}\right)^{1/2} \exp\left(-\frac{\lambda}{4k_B T}\right), \tag{1.18}$$

where t is transfer integral, $\hbar$ is Planck constant, λ is reorganization energy, k_B is Boltzmann constant, and T is temperature. The hopping model makes a basic assumption that the nuclear motion that "reorganizes" the molecule via ionization is much faster than the charge hopping rate between molecules. However, it is difficult to apply this picture into the crystalline organic semiconductors that involve strong intermolecular interactions in the range of 0.01–0.1 eV; the energy is comparable to the calculated reorganization energy for isolated molecules which is limited by a fast intramolecular vibrational mode, such as C=C bond stretching, and is less than 0.2 eV at most. Thus, the hopping picture may be only applicable to the disordered amorphous organic semiconductors, as used for organic light-emitting diodes, whose intermolecular hopping rate is very small due to the small intermolecular electronic coupling less than 1 meV. It was also pointed out that the hopping transport between the neighboring molecules is improbable by the Hall effect measurements for some OFETs [18].

We also briefly note here that it has been demonstrated that carrier transport becomes apparently "metallic type" in the similar but different types of π-conjugated organic molecular solids, if the number of carriers becomes large enough [19]: Two-component organic charge-transfer (CT) compounds composed of similar π-conjugated molecules have high enough carrier density (typically one carrier per two molecules) and show metallic behavior down to the lowest temperature [20]. Particularly, the effective nature of electronic band structure has been preponderantly demonstrated in the single crystals of these compounds; the coincidence of Fermi surface topology between the theory and experiments is investigated by the carrier transport studies under high magnetic field and at low temperature.

Back to the discussion on the temperature dependence of carrier mobility in OFETs, it is most probable that the thermally activated behavior should be ascribed to the existence of local potential (with either intrinsic or extrinsic origin) that disturbs the translational symmetry or lattice periodicity and forms carrier trap states within the semiconductor channel layers. This effect should be essentially important in the OFETs whose device operation is carried by a limited number of charge carriers: The number of carriers induced by gate voltages in the OFETs is roughly estimated as small as 10^{-12} cm^{-2} at most, which roughly corresponds to "one carrier per one thousand molecules," if we assume that the charges are accumulated at the semiconductor/gate dielectric interfaces for conventional organic semiconductor crystals. Such a tiny amount of charge carriers should be directly affected by the disordered potential in the crystals or in the gate dielectric layers. In other words, carrier transport in OFETs should be dominated by shallow or deep trap states that are formed in the vicinity of the bandedge states. The distribution of these trap states is also quite important in understanding the device operation of OFETs especially at the subthreshold regions, as discussed later.

Many theoretical reports have been reported, so far, to identify the origin of charge localization in organic semiconductors. Role of thermal fluctuation [21–23], vibration coupling [24, 25], and fluctuation in gate dielectric layer [26] have been discussed so far, as associated with the unique nature of organic semiconductors. A number of experimental studies have been also reported so far to investigate trap density of states in organic semiconductors, mainly on the basis of electrical device-characteristics measurements [27]. In spite of these works, the real picture has not been established due to the limited number of microscopic experimental studies. This is also related to the fact that the number of charge carriers is strictly limited in OFETs because the carrier transport takes place only at semiconductor-insulator interfaces, which makes it difficult to conduct these studies. In this section, we especially focus on microscopic charge-carrier transport in OFETs by providing experimental results by field-induced electron spin resonance (FESR) measurements. It is shown that the measurements can exceptionally probe microscopic

1.2.2 Field-Induced Electron Spin Resonance

1.2.2.1 Electron Spin Resonance

Figure 1.14 presents a schematic for the principle of electron spin resonance (ESR) technique. When holes (electrons) are accumulated in the OFETs by negative (positive) gate voltages, they are accommodated at the HOMO (LUMO) levels and can move from molecule to molecule through the intermolecular interactions or the electronic band states. When we view the carrier states in terms of electronic spins, only charge carriers are unpaired and have finite magnetic moments with spin quantum number $S = 1/2$. The ESR technique probes the response of electronic magnetic moment in terms of the magnetic resonance absorption of microwaves under static magnetic field [28].

In the presence of an external static magnetic field, the electronic magnetic moment aligns itself either parallel ($m_s = -1/2$) or antiparallel ($m_s = +1/2$) to the field. They have different magnetic energies whose separation is called as Zeeman splitting, as shown in Fig. 1.14. The split is proportional to the applied static magnetic field strength B and is given by $g\mu_B B$. Here g is a dimensionless constant called as g factor, and μ_B ($= 9.3 \times 10^{-24}$ J/T) is the Bohr magneton. When we irradiate the spin system with microwave at frequency ν, magnetic resonant absorption takes place at the resonance condition:

$$h\nu = g\mu_B B, \tag{1.19}$$

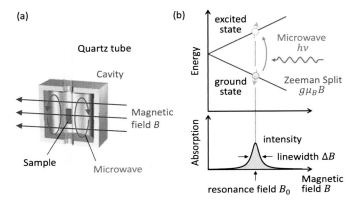

Fig. 1.14 Schematic of electron spin resonance measurement

by which the spin direction is converted between the parallel and antiparallel alignment. In usual cases, the ESR signal cannot be detected with the channel organic semiconductors under no gate bias due to the intrinsic semiconductor nature. The ESR measurements allow us to probe, sensitively and selectively, the carriers that are accumulated at the semiconductor-insulator interface in the OFETs by the gate voltages.

Because of the simple and general features of ESR measurements as presented above, it is expected that similar experiments should be possible for the devices based on inorganic semiconductors. However, there are no examples, except for the OFETs, to detect the charge carriers in field-effect transistors by ESR experiments. This is associated with the fact that the number of carriers induced by the gate voltages is considerably limited in the device, whose detection is not feasible. An essential point is that extremely highly sensitive detection of electronic spins in the OFETs is possible, because the relaxation time of excited spin state is fairly long in organic materials due to the small spin-orbit interactions of light atomic elements, so that much narrow ESR spectrum is observed.

ESR measurements of field-induced carriers in OFETs are first reported by Marumoto and Kuroda in 2004 [29]. They successfully demonstrated that the detected FESR signal from the OFETs is proportional to the applied gate voltages and is surely ascribable to the charges accumulated at the organic semiconductor interfaces. They also claimed that the carrier states are spatially extended over several molecules in pentacene OFETs by the fact that the observed FESR linewidth is narrower than that of the isolated cationic molecule in solution [30]. Then Matsui and Hasegawa observed the so-called motional narrowing effect in the FESR spectra of OFETs, from which the various aspects of charge-carrier dynamics that are directly connected to the device operations can be extracted [31].

1.2.2.2 Field-Induced Electron Spin Resonance Technique

The FESR measurements could be done with conventional X-band ESR apparatus equipped with a cavity with high Q value (4000–6000 in TE_{011} mode). Figure 1.15a schematically illustrates the device structure of an OFET as used for the FESR measurements. A 100-μm-thick poly(ethylene naphthalate) film was used as the nonmagnetic substrate, and an 800-nm-thick parylene C film was used as the gate dielectric layer. The capacitance C_i is estimated at 4.5 nF cm^{-2} by AC method at 1 mHz. The semiconductor layer of pentacene was fabricated by vacuum deposition to form a total area 2.5 mm × 20 mm and a thickness of 50 nm on top of the gate dielectric layer. As the gate, source, and drain electrodes, vacuum-deposited gold films with a thickness of 30 nm were used. The thickness is much smaller than the skin depth of gold at the X-band microwave. By using these devices, it is possible to eliminate the ESR signals from the device components. It is also useful to use a stack of sheet devices for the high-precision FESR measurements.

The ESR experiment detects magnetic resonance absorption at the condition: $B = h\nu/g\mu_B$, by sweeping static magnetic field at a constant microwave frequency.

1 Physics of Organic Field-Effect Transistors and the Materials

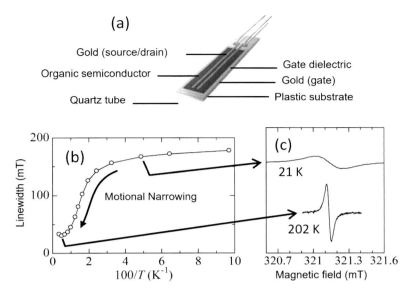

Fig. 1.15 (a) Schematic of device structure of OFET for FESR measurement, (b) temperature dependence of FESR linewidth, (c) typical FESR spectrum. Source: Ref. [31]

The spectrum is obtained in the form of a first derivative curve of the resonance absorption as a function of magnetic field due to the use of lock-in detection technique. An example of the FESR spectrum is presented in Fig. 1.15c for the pentacene OFETs. The spectra usually exhibit symmetric, single-shaped, and very narrow resonance absorption line with width of ten to several hundred µT. The resonance field allows to evaluate the g factor which is usually close to that of free electron (=2.002319) in organic semiconductors. The g factor depends on the direction of applied static magnetic field as to the semiconductor crystals to afford anisotropic g tensor, which is mainly originated from the anisotropy of spin-orbit interaction of the component molecules.

1.2.2.3 Motional Narrowing Effects

Here the "motional narrowing effect" is briefly outlined as a core concept to analyze the FESR experiments [32, 33]. First, we presume that charge carriers with electronic spins do not move and continue to stay at respective sites. In this case, the origin of finite linewidth in the ESR spectrum is classified into two fundamentally different cases. The one comes from a decay of the excited spin state, i.e., lifetime width, which leads to the Lorentzian line shape. This effect becomes more crucial at higher temperature because of the increased phonon scattering. The other is a result of inhomogeneity in local magnetic field, ΔB_{local}, at respective sites, i.e.,

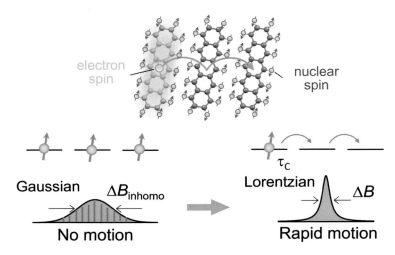

Fig. 1.16 An electron spin moving through randomly oriented nuclear spins

inhomogeneous width. Especially, an important origin of the ΔB_{local} is the interaction of electronic spins with nuclear spins which is known as hyperfine interaction. The hyperfine interaction with proton nuclear spin in π-conjugated molecules reaches as high as 0.1–1 mT. As a result of the independent nature of respective nuclear spin orientations, the ESR spectrum is inhomogeneously spread, as schematically shown in Fig. 1.16, because of the probability distribution of nuclear spin moments as given by

$$B = h\nu/g\mu_B - \Delta B_{\text{local}}. \tag{1.20}$$

The motional narrowing takes place in the latter case. Let us assume that the electronic spins of charge carriers move rapidly within the space that involves nonuniform distribution of ΔB_{local}. The electronic spins should feel the ΔB_{local} that rapidly varies with time, but they effectively feel the average magnetic field within a certain period of time, in terms of the magnetic resonance absorption. Because the fluctuation width of the averaged magnetic field, ΔB_{local}, is smaller than that of local magnetic field, the spectral width of the obtained ESR spectrum becomes narrower than that of the ESR spectrum without motion (see Fig. 1.15). This is the motional narrowing effect. The motional narrowing is observed only when the motion velocity exceeds over a threshold value. If we define k as the motion frequency between sites per unit time, the condition is given by the relation

$$k > \gamma \left(\langle \Delta B_{\text{local}}^2 \rangle\right)^{1/2}, \tag{1.21}$$

where $\gamma = 1.8 \times 10^{11}$ T^{-1} s^{-1} is gyromagnetic ratio. Considering that the ΔB_{local} is in the range of 0.1–1 mT, the threshold motion frequency is estimated at about 10^7–10^8 s^{-1}. Motional narrowing does not appear when the motion frequency is lower than the threshold value. The ESR spectrum begins to be narrowed when the motion frequency becomes higher than the threshold value, and the linewidth is inversely proportional to k (see Eq. (1.21)). Therefore, the motion frequency of charge carriers can be estimated by the measurement of motionally narrowed linewidth.

1.2.3 Carrier Transport Inside Microcrystal Domains

Based on the backgrounds as presented in the preceding subsections, we focus on various aspects of microscopic charge-carrier dynamics in OFETs as experimentally revealed by the FESR measurements in the subsequent three subsections. The first two are based on the experiments where the static magnetic field is applied perpendicular to the plane of polycrystalline organic semiconductor thin films. In the OFETs showing relatively high mobility, polycrystalline films are usually composed of layered microcrystals (as discussed in Sect. 1.1) that are uniaxially oriented with the layer parallel to the substrate plane. Therefore, the obtained FESR spectra are equal for all the microcrystals in terms of the measured direction of g tensor, where

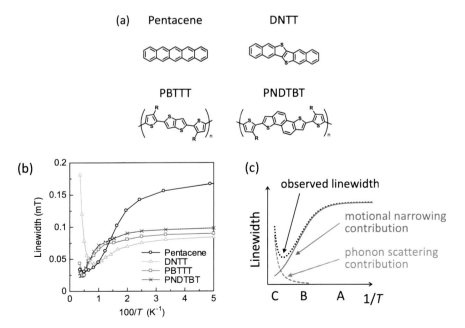

Fig. 1.17 (**a**) Molecular structures of pentacene, DNTT, PBTTT, and PNDTBT. (**b**) Temperature dependence of ESR linewidth. (**c**) Schematics of motional narrowing and phonon scattering contributions to linewidth. Source: Ref. [34]

the magnetic field is applied parallel to the crystal axes perpendicular to the layer. In contrast, the last section is based on the experiment where the static magnetic field is applied parallel to the film plane.

Figure 1.17 presents the temperature dependence of the ESR linewidth for four kinds of OFETs (pentacene, DNTT, PBTTT, and PNDTBT) measured at the magnetic field perpendicular to the film plane [34]. The plots present unique temperature variation which can be commonly divided into the three regions for all the OFETs: (1) low-temperature range where the linewidth is independent of temperature, (2) mid-temperature range where the linewidth decreases with increasing temperature, and (3) high-temperature range where the linewidth increases with increasing temperature.

The polycrystalline OFETs do not operate in the low-temperature range (see Fig. 1.13). It means that the motion of charge carriers should be "frozen," which is consistent with the temperature-independent nature of the ESR linewidth where no motional narrowing is observed. In sharp contrast, the temperature-dependent feature in the mid-temperature range should be attributed to the motional narrowing effect, where the carriers are thermally detrapped (or released) from the trap sites and begin to move in the crystals. This interpretation of the temperature dependence is also demonstrated by the dependence of the ESR linewidth on the input microwave power at various temperatures; the results indicate that the spectrum in the low-temperature range is inhomogeneous, while the spectrum character becomes more homogeneous with the increase of temperature in the mid-temperature range. In the high-temperature range, the linewidth turns to increase with the increase of temperature, where the temperature dependence should be dominated by the fast spin-lattice relaxation. Actually, this feature becomes more apparent in DNTT than in pentacene, most probably because the DNTT are composed of larger element of sulfur, leading to the larger spin-orbit coupling.

When the ESR spectrum is narrowed by the motional narrowing effects, motion frequency of charge carriers, denoted as k_1, can be estimated by the observed ESR linewidth at respective temperature, as is given by the following relation:

$$k_1 = \gamma \langle \Delta B_{\text{local}}^2 \rangle / \Delta B. \tag{1.22}$$

$\langle \Delta B_{\text{local}}^2 \rangle$ can be estimated by the original inhomogeneous linewidth at the low-temperature range, where the carrier motion is frozen. The estimated k_1, presented in Fig. 1.18a, is in the range of 10^7–10^9 s^{-1}, indicating that the residence time of charge carriers at respective trap sites is estimated at about 1–100 ns. The k_1 increases by the increase of temperature, but the temperature-dependent nature is relatively gradual and the activation energy is estimated at 2–21 meV. Here we must note that the estimated average residence time of charge carriers is extremely long, if the hopping theory is tentatively assumed; k_{Marcus} should be in the range of 10^{13}–10^{14} s^{-1} under the assumption of hopping length as the intermolecular distance.

Then we deliberate what the obtained motion frequency represents for, as is related to the charge-carrier motion. According to the multiple trap-and-release (MTR) model [35, 36], the average time at traps is much longer than that for

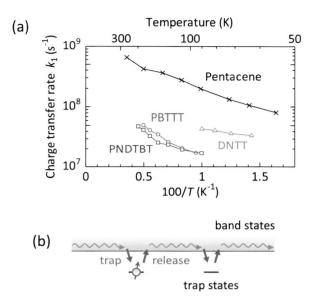

Fig. 1.18 (a) Intradomain charge transfer rate k_1. (b) Schematics of trap and release model

traveling from trap to trap. In this case, the average traveling length d between the traps is simply represented by the diffusive motion of charge carriers as [37]:

$$d = \sqrt{4D\tau_C} = \sqrt{\frac{4k_B T \mu' \tau_C}{e}}, \qquad (1.23)$$

where D is the diffusion constant, k_B the Boltzmann constant, μ' the effective charge-carrier mobility inside the microcrystals, and e is the elementary charge. The second derivation utilizes Einstein's relation, as presented in Eq. (1.15), which implies a statistical nature for the total stochastic carrier motion, including the trap-and-release processes. From Eq. (1.23), d is calculated to be about 10–50 nm at room temperature. In Fig. 1.18b, we show the schematic of the charge-carrier motion, as achieved by the FESR experiments. The charge carriers should move from trap to trap by thermally activated trap-and-release process. The activation energy for motion frequency, which is comparable to or smaller than the thermal energy at room temperature, should correspond to the averaged depth of these traps. The motional nature between the traps should be band-like carrier transport, while the feature is not clear from the FESR experiments.

1.2.4 Density of Trap States Inside Microcrystal Domains

As we discussed in the former sections, FESR linewidth becomes almost constant (or temperature independent) in the low-temperature range, typically below a few

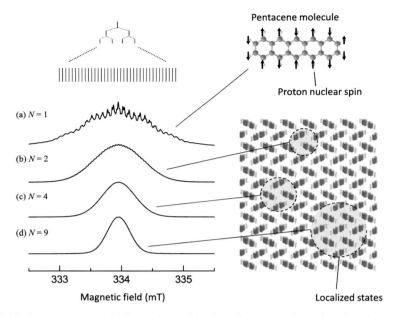

Fig. 1.19 Local magnetic field due to the combination of proton nuclear spin orientations and its effect on ESR spectrum [38, 39]. (**a**) Measured ESR spectrum of cationized pentacene molecule in solution, and caluculated ESR spectra for the shallow or strongly localized electronic states with spatial extension over (**b**) $N = 2$, (**c**) $N = 4$, (**d**) $N = 9$ molecules. The right shows schematic for these localized states

tens K. In this temperature range, the charge carriers lose thermal energy, and become immobile, as they should be trapped by shallow or deep traps. It means that the FESR spectrum at the lowest temperature should include rich information about the trapped or weakly localized carrier states. Therefore, the analysis of the spectrum should allow the detailed investigation of these trap states that take crucial roles in the charge-carrier transport at elevated temperature [38, 39].

The low-temperature FESR spectrum is inhomogeneous, because the local magnetic field at respective sites is different with each other. As discussed above, the origin of the local magnetic field is ascribed to the hyperfine interaction with nuclear spin moments. As an example, we show the ESR spectrum of cationized pentacene molecule in solution in Fig. 1.19a [40]. The spectrum exhibits complicated hyperfine splitting. As the respective pentacene molecules are isolated with each other in solution, an unpaired electronic spin within the cationic pentacene molecule interacts with 14 proton nuclear spins. As the ESR line should split per each proton nuclear spins ($I = 1/2$), the total number of splits amounts to $2^{14} = 16{,}384$ (number of independent combinations is 372 from the molecular symmetry), where both ends are composed of either all up or all down spins. The ESR spectrum is composed of these resonance lines that have finite linewidth and are overlapped with each other. Although the actual spectrum is waved, the envelope of the resonance lines is close to the normal (or Gaussian) distribution function.

On the other hand, the electronic spins in solids should be extended over several molecules. Namely, the electronic wave function in solids is not restricted to a single molecule. When we assume that the wave function is extended over N molecules, the electronic spin interacts with nuclear spins included in all the N molecules, so that the split number exhibits exponential dependence on N. Eventually, the linewidth becomes narrowed by a factor of $\sqrt{1/N}$, as compared to the case of cationic single molecule, according to the central limit theorem. Based on the presumption, Marumoto and Kuroda pointed out that the wave function is extended over about ten molecules in pentacene OFETs, because the observed spectral linewidth is almost 1/3 of the linewidth of radical molecules in solution. But they did not take account of the fact that the obtained spectrum is not Gaussian and also of the fact that the linewidth is temperature dependent at least at room temperature.

In order to reveal the spatial extension of trapped charge states, spectral analyses could be conducted where the spectrum is decomposed into several Gaussian curves with different linewidths. Considering that there are several kinds of weakly localized states, the actual spectrum, denoted as $S(B)$, should be composed of the sum of several Gaussian curves, denoted as $G(B, N)$, with different widths for N, as is given by:

$$S(B) = \int_{1}^{+\infty} \frac{\partial G(B,N)}{\partial B} D(N) dN, \quad (1.24)$$

where $G(B, N)$ is given by:

$$G(B,N) = \sqrt{\frac{N}{2\pi\sigma_0^2}} \exp\left[-\frac{(B-B_0)^2}{2(\sigma_0^2/N)}\right]. \quad (1.25)$$

By solving the integral equation numerically, we can obtain the density of localized states, $D(N)$, from the ESR spectrum. In Fig. 1.20, we show the $D(N)$ as obtained by the analyses of FESR spectrum of pentacene OFETs. As seen, the $D(N)$ is composed of two discrete peaks (denoted as A and B) and a broad structure (denoted as C) whose spatial extension is over 6–20 molecules. Such shallow localized states should originate from a weak attractive potential that may be ascribed to the possible slight deviation of molecular location from the equilibrium position or a chemical change in molecules. These localized states should take crucial roles in the device operation of pentacene OFETs at room temperature.

1.2.5 Carrier Transport Across Domain Boundaries

As discussed in the former subsections, channel semiconductor layers of the OFETs with showing relatively high mobility are composed of uniaxially oriented layered microcrystals whose layers are parallel to the substrate plane. When the magnetic

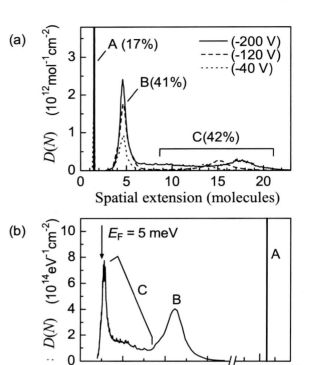

Fig. 1.20 Distribution of trap density of states, $D(N)$, in pentacene TFTs are plotted against (**a**) the spatial extent N and (**b**) the binding energy of the trap states, as obtained by stochastic optimization analysis of FESR spectrum at 20 K and at gate voltage of -200 V. Source: Refs. [38] and [39]

field is applied perpendicular to the substrate, the same ESR signal should be obtained for all the layered microcrystals where the magnetic field is parallel to the normal of the layer (which is denoted as c-axis). In striking contrast, when the magnetic field is applied parallel to the substrate, static magnetic field is applied with different angles on each microcrystal domains within the layers (crystal axes within the layer are denoted as a and b axes). Because of the anisotropy of g values, resonance magnetic field should be distributed in the latter measurements. Here we demonstrate that these measurements allow us to observe motional narrowing between microcrystals and to estimate grain boundary potential.

Figure 1.21 shows the temperature dependence of the FESR spectra obtained by the measurement with applying the magnetic field parallel to the film (or substrate) plane. Single monotonous peak is observed at room temperature, while the spectrum split into the two peaks by the decrease of the temperature. The respective peaks observed at low temperature should correspond to the resonance signals and at magnetic field applied parallel to the a and b axes, respectively. We note that such a clear peak splitting is observed in DNTT and PBTTT OFETs, but not in pentacene OFETs. This difference is associated with the relatively large g anisotropy in DNTT and PBTTT, due to the large spin-orbit interactions of sulfur.

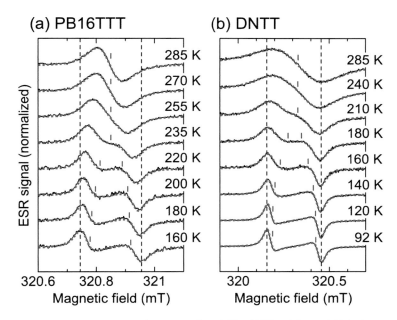

Fig. 1.21 Temperature-dependent ESR spectra for (**a**) PB16TTT and (**b**) DNTT, both at magnetic field perpendicular to the film plane. Fitting curves are indicated by dotted lines. Source: Ref. [34]

The peak of the spectrum obtained at room temperature is located almost at the center between the two peaks that are split at low temperature. It is clear that the spectrum at room temperature cannot be reproduced by the simple sum of the split peaks at low temperature. The temperature dependence of the spectrum can be understood in terms of the motional narrowing effect where the anisotropy of g values is averaged by the motion of charge carriers across the domain boundaries.

In order to analyze the motional narrowing effect due to the averaging of the two different local magnetic fields, it is necessary to make fitting analyses by the theoretical curves. Figure 1.21 also presents the result of fitting, which is found to reproduce well the experiment. In addition, it is possible to estimate the motion frequency moving across the domain boundaries. Figure 1.22 presents the temperature dependence of motion frequency k_2 across the boundaries for the case of DNTT and PBTTT. The k_2 is about an order of magnitude smaller than the k_1 which is the motion frequency between the traps as described in the former subsection. In addition, activation energy estimated by the temperature dependence of k_2 is 42 meV for DNTT and 86 meV for PBTTT. The result indicates that there are large energy barriers at the grain boundaries.

It is thus concluded that the effective mobility values in polycrystalline OFETs are rate determined by the motion across the grain boundaries. Although the PBTTT films are known to show high degree of crystallinity as polymer semiconductor, observation of surface morphology by AFM measurements is not effective to

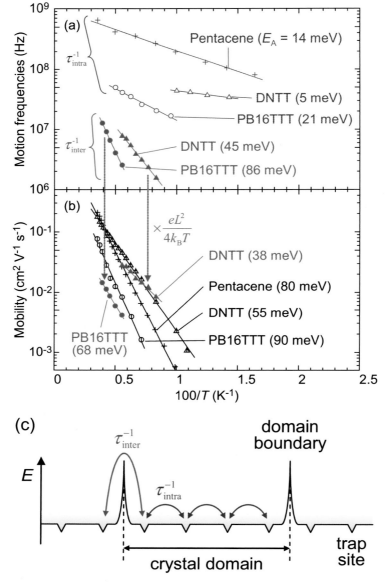

Fig. 1.22 (a) Intra- (k_1) and interdomain (k_2) motion frequencies and (b) field-effect mobility for the PB16TTT, DNTT, and pentacene OTFTs. The mobility calculated from interdomain motion frequencies is also shown. (c) Diffusion model in polycrystalline films. Source: Ref. [34]

identify the domain structures. Nonetheless, it was clearly demonstrated in this measurement that the grain boundary is the rate-determining process for the PBTTT.

1.2.6 Short Summary and Outlook

A microscopic charge-carrier dynamics that is crucial to understand device operation of OFETs is discussed, by the experimental results using FESR experiments that probe charge carriers accumulated at the organic semiconductor interfaces. By the analyses of motional narrowing effects in the FESR spectrum, it was shown that the charge-carrier transport can be understood in terms of the two aspects: the trap-and-release transport within the microcrystal domain and the transport across the domain boundaries with a relatively high barrier potential (a few times larger than $k_\mathrm{B}T$).

1.3 Gate Modulation Spectroscopy and Imaging

1.3.1 Introduction

It is possible to control the amount of charge carriers accumulated in channel organic semiconductors with the use of the OFET device structures. When charge carriers are injected and accommodated within the channel organic semiconductors, the excitonic optical spectra undergo a unique variation. Charge modulation (CM) or gate-modulation (GM) spectroscopy is a unique technique for probing the change in exciton absorption spectra of organic semiconductors caused by the gate-induced charge carrier accumulation (e.g., [41]). The technique is based on a well-established methodology of electroreflectance spectroscopy which is used to unravel the optical properties of semiconductors through the application of electric fields [42].

In organic FETs, hole density of up to about 5×10^{12} cm^{-2} can be accumulated with the use of conventional gate dielectric layers. The change of absorbance, $\Delta \alpha d = -\Delta T/T$ (d is film thickness and T is transmittance), by the charge accumulation reaches as large as 10^{-4}. To observe the signal, rectangular voltage pulses (V_ac) at frequencies ranging from several Hz to kHz with a certain DC bias are applied to the gate, and the modulation signal is processed by a lock-in technique in the modulation measurements.

1.3.2 Separation of Charge and Field Effects

Figure 1.23 presents the typical GM spectra of a polycrystalline pentacene OFET [43]. It is clearly seen that the observed GM spectra are much different between under the negative and the positive gate bias modulation. Such a difference is clearly

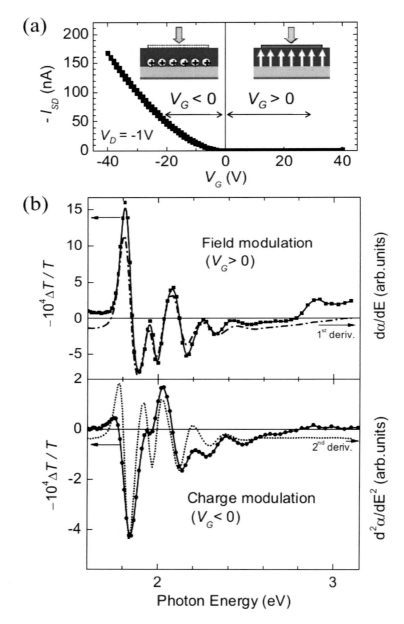

Fig. 1.23 (a) Transfer characteristic of pentacene thin film OFET. The inset shows schematics for negative and positive gate bias, corresponding to charge and field modulation, respectively. (b) Field and charge modulation spectra of pentacene thin film OFET. Source: Ref. [43]

ascribed to the unipolar nature of pentacene OFETs. The charge density should be modulated by applying a negative gate bias, whereas field modulation should be achieved by applying a positive gate bias.

Under the positive gate-bias modulation, electric field effects are clearly observed in the absorption spectrum. It is seen that the observed GM spectrum is well fitted by the first derivative curve of the original absorption spectrum, as seen in Fig.1.23b. The feature indicates that the excitonic absorption presents a homogeneous quadratic Stark effect with an energy shift of ΔE as $\Delta \alpha = (d\alpha/dE) \Delta E$, over an entire range between 1.8 eV and 2.6 eV. The result is in contrast to earlier studies of electromodulation spectroscopy [44]; the report indicated the appearance of some spectral features that do not simply follow the first derivative curve at higher energy. We conclude that the use of OFET devices allowed us to obtain pure electric-field modulation by eliminating the hole injection.

In contrast, charge accumulation effects are clearly observed in the absorption spectrum under the negative gate-bias modulation. Spectral features of the GM spectrum obtained with negative gate bias are essentially different from that obtained with positive gate bias and could be explained by the sum of bleaching (negative absorption)-like curve and second derivative curve of the original absorption spectrum. The second derivative-like feature indicates that the absorption spectrum is slightly broadened by charge accumulation. It is most probable that the original exciton absorption is narrowed by exchange narrowing, and the exciton coherence should be violated by the field-induced accumulated charges, which should eventually affords the second derivative-like feature [45].

1.3.3 Development of GM Imaging Technique

The GM spectroscopy is recently coupled with two-dimensional area-image sensor, such as charge-coupled device (CCD) or complementary metal-oxide-semiconductor (CMOS), which allows to develop the gate-modulation (GM) imaging technique [46, 47]. The technique is to measure different optical images of semiconducting channel layers between at gate-on and at gate-off states and thus visualizes active channel layers of operating OFETs.

Figure 1.24 presents an apparatus for the GM imaging technique, composed of optical microscope, area-image sensor, and image-subtraction module. The polycrystalline or single-crystal OFETs are illuminated by monochromatized light, and the transmitted (or reflected) light was captured at gate-on and at gate-off states at ~ several tens Hz as an area image at various photon energies. In the image processing, the difference in the images between those recorded at gate-on and those recorded at gate-off is integrated over $\sim 10^4$ times to obtain a GM image ($-\Delta R/R$ or $-\Delta T/T$) with enough high signal-to-noise ratio. The technique allows spatially resolved electro-optic responses of channel organic semiconductor layers as optical microscope images.

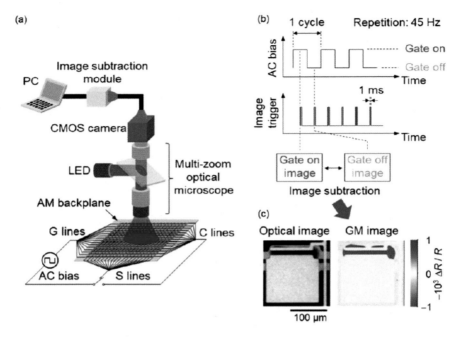

Fig. 1.24 Schematics for GM imaging of AM backplane; (**a**) experimental setup, (**b**) measurement principle, and (**c**) optical micrograph and corresponding GM image for a single AM pixel. Source: Ref. [48]

1.3.3.1 Fast Optical Inspection of Integrated Circuits

The important target of organic FETs is to produce active-matrix (AM) backplanes that facilitate versatile use in manufacturing displays, electronic papers, and two-dimensional sensor arrays. The AM backplane is basically composed of an array of many organic OFETs with pixel numbers as large as 7×10^6, as for the case of an AM backplane for an ordinary 20-inch display with a 200 pixel-per-inch (ppi) resolution. However, inspecting all the pixels using conventional electrical measurements would take an infinitely long time. Thus, it is quite important to develop a fast and effective inspection technique for the operation of a large number of AM pixels collectively.

It was demonstrated that the GM imaging technique is quite useful for a fast optical inspection of the operation of large-area AM backplanes. Figure 1.25 shows an example of the measurement for an AM backplane composed of one-transistor–one-capacitor cells arranged at a 150 pixel-per-inch resolution [48]. The use of a high-frame-rate CMOS image sensor and a high-speed image-subtraction module facilitates the bulk inspection of 30,000 pixels within 3 min. The technique allows not only to identify both defective transistors and capacitors quickly by examining a zoomed-out image but also to investigate the origin of defects within channel layers or capacitors by investigating an expanded image.

1   Physics of Organic Field-Effect Transistors and the Materials   35

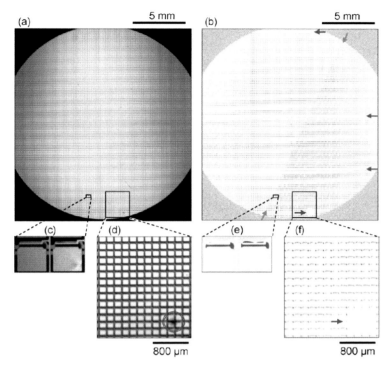

Fig. 1.25 (**a**) Large-area optical micrograph and (**b**) corresponding GM image for an AM backplane. The red and blue arrows indicate the line and point defects, respectively, and the green arrows indicates the border between the strong and weak GM signal areas. (**c, d**) Expanded optical micrographs and (**e, f**) GM images. Source: Ref. [48]

1.3.3.2   High-Resolution Microscopic GM Imaging

It is possible to conduct microscopic GM imaging measurements with diffraction limit resolution through the use of an objective lens that has high magnifying power and high numerical aperture [49]. The technique allows visualizing both the channel charge density and electric field distribution at the microcrystal level in polycrystalline pentacene thin-film OFETs.

Figure 1.26 shows the optical microscopic image and the corresponding microscopic GM image of the pentacene thin films, recorded at the probe photon energy of 1.85 eV with high spatial resolution less than 500 nm. The obtained microscopic GM image presents unexpectedly inhomogeneous distribution of positive (red-colored) and negative (blue-colored) signals over the intragrain and grain boundaries. The boundary of a monocrystalline grain is denoted by a thick white curve, as identified by the crossed-Nicols polarized micrograph. The comparison of the images indicates that the microscopic GM signal distribution depends clearly on the grain-shape

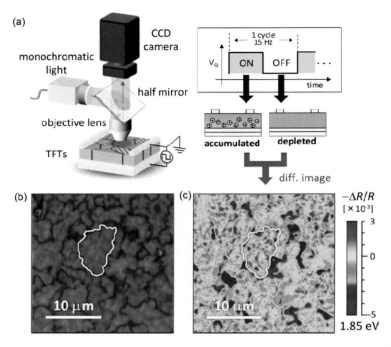

Fig. 1.26 (a) Schematics of the experimental setup for microscopic GM imaging. (b) Optical microscope image and (c) the corresponding GM image for a polycrystalline pentacene thin film probed both at photon energy of 1.85 eV. White curves indicate the boundary of a specific monocrystalline grain. Source: Ref. [49]

profile. The positive GM signal areas are mostly located inside the grains, whereas the negative GM signal areas are mostly located near the grain boundaries.

Figure 1.27 presents the result of spectroscopic analyses, measured at a dark-blue-colored area denoted as B and at a red-colored area denoted as A. The respective GM spectra were achieved by a series of the GM image measurements at various photon energies. The results show that two distinct effects appear, simultaneously, within the polycrystalline pentacene channel layers: The negative GM signals at 1.85 eV originate from the second derivative-like GM spectrum caused by the effect of charge accumulation, whereas the positive GM signals originate from the first derivative-like GM spectrum caused by the effect of leaked gate fields. Comparisons with polycrystalline morphologies indicate that grain centers are predominated by areas with high leaked gate fields due to the low charge density, whereas grain edges are predominantly high-charge-density areas with a certain spatial extension as associated with the concentrated carrier traps.

All the results indicate that the accumulated charge distribution is much more inhomogeneous than we expected in polycrystalline pentacene thin-film OFETs, where microscopic local areas with fairly high and fairly low charge carrier density coexist within the channels. To probe the charge current through the polycrystalline

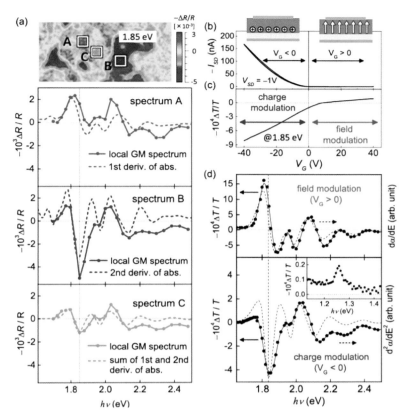

Fig. 1.27 (**a**) GM spectra at local areas (by each 4 pixel detections with a size of 0.012 μm²) showing positive [A], negative [B], and intermediate [C] signals in the GM images of pentacene polycrystalline thin film measured at 1.85 eV [49]. (**b**) Transfer characteristic, (**c**) GM signal intensity, and (**d**) field- (upper) and charge-modulation (lower) spectra. Source: Ref. [43]

films, a time-resolved microscopic GM imaging was also conducted by using a fast image-intensifier system that could amplify a series of instantaneous optical microscopic images acquired at various time intervals after the stepped gate bias is switched on [50]. The temporal evolution of the channel current and accumulated charges was successfully observed with a time resolution of as high as 50 ns.

1.3.4 Observation of Giant Excitonic Electro-Optic Response

The OFETs can exhibit excellent switching characteristics when they involve trap-eliminated semiconductor interfaces with highly hydrophobic gate dielectric layers, as we show in the Sect. 1.5. It was recently reported that the GM imaging for trap-eliminated single-crystal FETs presents a unique enhancement of electro-optic response under the application of drain voltages [51].

The device for the GM imaging measurements is schematically illustrated in Fig. 1.28. In the device fabrication, high-quality flake-like pentacene single crystals were laminated on top of a gate dielectric layer of the most hydrophobic amorphous perfluoropolymer, Cytop. Figure 1.29 presents the obtained GM image. It is seen that the drain-unbiased GM image exhibits a uniform spatial distribution, which is consistent with the accumulated carrier density in the channel. In sharp contrast, the drain-biased GM image presents a peculiar spatial distribution with sharp increases around the edges of the source and drain electrodes, as seen in Fig. 1.29c. The following intriguing features are observed: (1) The sharp increase in the GM signal distribution around the electrode edge is like the lateral electric field distribution as measured by Kelvin probe force microscopy, and (2) the GM spectra, extracted from the respective GM images measured at different wavelengths, present a second derivative-like shape that implies the broadening of exciton absorption. The observed unique excitonic electro-optic response is associated with the delocalized carrier accumulation at trap-eliminated interfaces of pentacene single-crystal FETs, where the gate-induced holes that are weakly bound to shallow traps should be detrapped by lateral electric fields. These effects should induce the enhanced violation of exciton coherence by delocalized carrier accumulations under drain bias, which eventually enhance the electro-optic response, as observed in Fig. 1.29c.

1.4 Print Production of Electronic Materials: Its Challenges and Technologies

1.4.1 Introduction

Studies of "organic electronics" have had a new outlook in the past decades, in responding to a strict requirement from industrial circles to demonstrate a clear

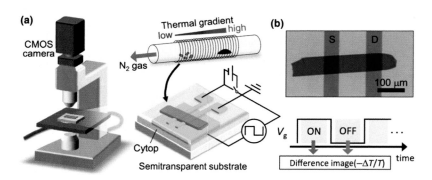

Fig. 1.28 (a) Schematic for a setup of GM imaging measurements of pentacene single-crystal OFET composed of a Cytop gate dielectric. (b) An optical micrograph of the device with source and drain electrodes. Source: Ref. [51]

1 Physics of Organic Field-Effect Transistors and the Materials

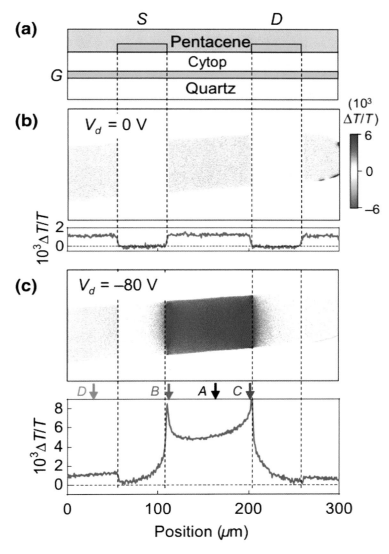

Fig. 1.29 (a) Cross-section schematic of the measured device. (**b**, **c**) Microscope GM images and profiles of GM signal distribution along the channel layer, obtained (**b**) with a modulated gate bias at $V_{g\_mod} = -80$ V without drain bias, and (**c**) with dc drain bias at $V_{d\_dc} = -80$ V. Source: Ref. [51]

comparative advantage of the organic semiconductors over the other semiconducting materials. The concept of the "printed electronics" was born against such a background [52]. In the printed electronics, printing technologies that are used to produce documents or photo images on papers are used to manufacture electronic devices at ambient conditions. Particularly, it is expected that the printed electronic technology should replace the vacuum, lithography, and heat-treatment technologies that have been indispensable for the production of all the electronic devices thus far. By these replacements, massive vacuum facilities become unnecessary, and/or the use of flexible plastic sheets as base plates becomes possible, allowing us to realize flexible electronic products that have lightweight, thin, and impact-resistant characteristics. Among other semiconducting materials, π-conjugated organic molecular materials should take key roles in the innovation of the electronic device productions by the use of printing technologies. This is because the organic materials are soluble in kinds of solvents and also because the convenient thin-film formation is possible for the materials by convenient solution processes such as spin coating or drop casting at ambient conditions.

A possible future role of the printed OFETs in the printed electronic technologies can be considered as that any plastic surfaces are decorated electronically by arraying a large number of OFETs by the printing technologies to function as active backplanes for displays or sensors to realize what is called "ubiquitous electronics" society (Fig. 1.30). Because it is now possible to utilize commercially available inkjet printers, as an example, to print documents or photo images on papers with a spatial resolution higher than 1200 × 1200 dpi, it appears auspiciously easy to manufacture electronic devices with a similar high pattern resolution by the printing technologies. However, it is also true that there is a crucial difference between document printing and production of electronic devices. In order to power electronic devices, it is absolutely necessary to form and stack patterned layers of electronic functional materials that are uniform at atomic or molecular scales on flat substrate surfaces. In this section, we first discuss various problems to encounter for producing semiconductor devices by the use of conventional printing technologies. Then we outline some particular printing technologies for printing patterned semiconductor layers and metal wiring [53–55].

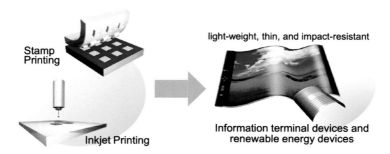

Fig. 1.30 Future image of the printed electronics

1.4.2 Printing Semiconductor Devices?

The printing technologies generally cover a wide range of methods and techniques which include the stamp printing that transfers ink by convex part of resin stamp or by concave part of metal stamp (the former is called as flexography and the latter as rotogravure), the screen printing that transfers inks through minute opening in the stencil, the inverse printing that removes unnecessary part of deposited inks by a stamp (above all are so-called stamp printings), and the inkjet printing that directly deposits microdroplet inks without stamps (which is the plateless printing). These technologies are now, respectively, utilized, depending on their costs and uses in a wide variety of fields. It might be possible for us to define and characterize all these printing technologies as to freely allocate and deposit a tiny amount of fluidic medium including coloring matters (i.e., ink) on such as papers. The inks as used also have a wide variety, whose viscosity ranges from high to low, and are used depending on the respective printing techniques. In the conventional printing technologies, crystallinity, grain size, or grain boundaries of coloring matters are not a matter of concern after the deposited inks are dried out, as long as the coloring matter is precisely positioned. However, the most important subject for "printing semiconductor devices" is to obtain the semiconductor layers and metal wirings with high uniformity, after the microdroplet ink is positioned.

It is necessary, for manufacturing high-performance OFETs by a printing technology, to use solution of layered-crystalline organic semiconductors as ink, to deposit the solution ink on flat substrate surface, and to produce uniform patterned semiconductor films composed of the layered crystals whose layers are parallel to the substrate surface. In order to uniformly crystallize the materials from the solution, it should be much more advantageous to utilize low-viscous solution fluid of semiconducting materials possibly with no additives. On the other hand, it is known to be quite difficult to form uniform thin solid layers from a tiny volume of low-viscous droplets—volume of printed microdroplets produced by such as inkjet printing technique is in the range of 1–100 pL—by precipitation within the droplet through solvent evaporation. In particular, according to the fluid science, solvent evaporation occurs efficiently at around the (solid-liquid-air) contact line of the sessile droplets, so that the resultant outward capillary flow carries solutes toward the contact line of microdroplets and form ring-like deposits at around the contact line [56, 57]. (This is the so-called "coffee-ring effect," see Fig. 1.31.) In the conventional printing, papers are used as printed medium where the surfaces are highly uneven in mesoscopic level including opening or mesoscopic pore that efficiently absorbs solvents, so that the colored matters are dispersed and adhered to the fabrics in papers by solvent absorption, where these nonuniformity problems are not apparently exposed. However, in order to manufacture uniform semiconductor layers for "printing semiconductor devices," it should be necessary to control the convection flow and solvent evaporation within the microdroplets.

In order to find out a clue to resolve this issue, it would be meaningful to look back on the birth course of the concept of printed electronics. The concept has an

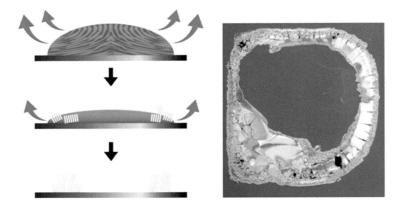

Fig. 1.31 Schematic for the formation of coffee-ring-like deposits by solvent evaporation and outward capillary flow within the deposited microdroplet. The right image shows an example

origin in the spin coating of solution processable π-conjugated polymer semiconductors to manufacture organic light-emitting diodes, organic solar cells, and OFETs [58–60]. It was later discovered that the spin coating is also applicable to fabricate high-quality thin films of soluble small-molecule semiconductors that shows layered crystallinity and thus to manufacture high-performance OFETs [61, 62]. Due to these progresses, the spin coating is now widely accepted as a standard technique for manufacturing thin films in the studies of developing soluble organic semiconductor materials [63, 64]. Spin coating is the process to produce thin solution layer on top of the substrate surfaces by centrifugal force as a result of spinning of substrates and to subsequently produce thin solid films by uniform solvent evaporation from the entire liquid-air interfaces [65]. However, the usability of the spin coating is limited as an industrial production process of semiconductor layers, both due to the enormous loss of raw materials during the process and to the difficulty in scaling up the device area which is strongly demanded in the application of flexible electronic devices. Nonetheless, the film formation mechanism may be quite reasonable to realize uniform thin films for the materials showing high degree of layered crystallinity. In the subsequent sections, we illustrate some printing technologies that are designated to produce uniform thin films with layered-crystalline organic semiconductors and ultrafine metal wirings.

1.4.3 Double-Shot Inkjet Printing Technique

The double-shot inkjet printing (IJP) is a concept to use two inkjet print head and to deposit two kinds of microdroplets at the same positions on the substrate surfaces. It was reported that when antisolvent crystallization, i.e., a binary liquid mixture of a material solution and an antisolvent, is incorporated into an IJP-based microdroplet process, it becomes possible to manufacture highly uniform thin films as precipitates, where the coffee-ring effect can be eliminated. The scheme of double-shot inkjet printing technique is presented in Fig. 1.32. In the process, an antisolvent microdroplet is first deposited by the inkjet head on substrates, and then the semiconductor solution microdroplet is over-deposited at the same position by the other inkjet head, to form mixed sessile droplets on the substrate surfaces. In the mixed droplets, semiconductor layer is first formed and then is dried by evaporating solvent after several minutes. Finally, uniform crystalline semiconductor layer with uniform thickness of about 200 nm is formed over the area, if the suitable semiconductor material is utilized.

It was expected that the use of antisolvent crystallization could separate the time of occurrence of the solute crystallization and the solvent evaporation [66]. However, as it will be shown later, the semiconductor thin films grew at the whole area over the liquid-air interfaces of the mixed droplets. This feature is in striking contrast to conventional macroscopic antisolvent crystallization that produces a large mass of microcrystals due to rapid turbulent mixing inside the liquids. It is thus quite likely that the chemically different binary microdroplet will exhibit unique mixing phenomena essentially different from macroscale fluids [67], as described in the next section. Furthermore, the use of drop-on-demand process which is unique to the

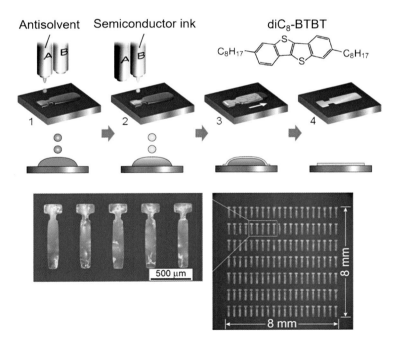

Fig. 1.32 Upper: Schematic for producing semiconductor single-crystal thin films by the double-shot inkjet printing technique. Molecular structure of diC$_8$BTBT is also shown. Lower: Micrographs for single-crystal thin film arrays produced by the double-shot IJP technique. Source: Ref. [53]

inkjet printing process allows us to control the crystal growth and to form single crystalline thin films.

1.4.3.1 Mixing Process of Chemically Different Microdroplets

It has been frequently pointed out that microfluids present distinct dynamical characteristics that are different from those of macroscale fluids. For example, it is difficult to mix two kinds of microfluids rapidly inside microchannels [68, 69]. This feature has been discussed in terms of the low Reynolds number of the fluid flow in microchannels, which causes laminar flow to dominate over turbulent flow. Another unique characteristic of microfluids is that the surface (or interfacial) tension becomes predominant because it is primarily attributed to the high surface-area-to-volume ratio of microfluids. Specifically, microdroplets that have free liquid–gas interfaces may exhibit more diversified dynamics in terms of the surface transformation than those confined within microchannels. It is thus expected that the mixing dynamics of the microdroplets should be considerably affected by the difference in surface tension between the mixed liquids. However, the dynamic nature of mixing binary microdroplets on solid surfaces has not been extensively studied so far.

We found that the mixing process between microdroplets is predominated by surface tension, which is much different from the mixing of macroscale fluids that is accompanied by turbulent flow. Particularly, if the microdroplets whose surface tension is lower than the sessile droplet, the over-deposited microdroplet rapidly covers the entire surface of the sessile droplet, as presented in Fig. 1.33. This is a type of Marangoni effect whose surface flow along the droplet surface is driven by the surface tension gradient. Therefore, it is possible to form thin semiconductor solution layer on top of the antisolvent droplet surface by the use of suitable combination of solvents. Then the solvent molecules are slowly diffused and mixed in the droplet, which contribute to the layered self-organization of the semiconductor molecules within the thin solution layer and thus to form the uniform crystalline films.

1.4.3.2 Single-Domain Formation

The "drop-on-demand (DOD)" feature, which means to allocate and deposit required volume of functional ink at a predefined position, is unique to the inkjet printing technology. If the nucleus is randomly generated at the surface of mixed microdroplets, the films are obtained as polycrystalline films. In contrast, it is found that the DOD function is also useful to form suitable concentration gradient within the droplet, which allows to control the nucleus generation and thus to grow single-domain crystal. Particularly, the single-domain formation should be advantageous to improve the device characteristics, because the domain boundaries between the grains are eliminated. We found that the single-domain formation is possible

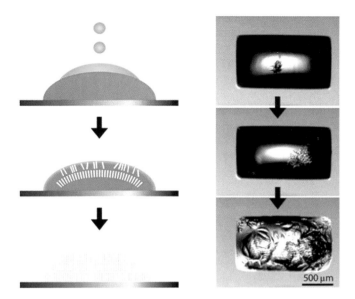

Fig. 1.33 Left: Schematic for the microdroplet mixing process in the double-shot IJP process. Right: Time-lapse micrographs of the droplets after the deposition of semiconductor solution microdroplet on top of the antisolvent sessile droplet. It is observed that the over-deposited microdroplet rapidly covers the entire surface of the sessile droplet, and the thin films are grown. Source: Ref. [67]

over the films by a tactic design of the droplet shape of the deposited sessile droplet by the surface modification of the substrate.

For the single-domain formation, we first controlled the droplet shape by hydrophilic/hydrophobic surface modification of substrates. The surface modification on silicon dioxide surface is conducted by the combination of the hydrophobization (formation of $SiO-Si(CH_2)_5CH_3$) with hexamethyldisilazane (HMDS) and the partial silanol (SiOH) formation by UV/ozone treatment. It was found that the droplet shape formation of rectangular hydrophilic area with the necked region, as depicted in Fig. 1.32, is effective to produce concentration gradient and to control the growth direction of the films: When we deposit semiconductor ink at the head region beyond the neck part, a part with semiconductor ink with high concentration is formed which accelerates the formation of nucleus.

1.4.3.3 Thin-Film Characteristics

Here we present examples for the double-shot IJP process. We used 2,7-dioctyl[1] benzothieno[3,2-b][1] benzothiophene (denoted as diC_8BTBT) as the solute organic semiconductors that has high degree of layered crystallinity [70–72]. We used 1 mM solution of diC_8BTBT in 1,2-dichlorobenzene (DCB) as the over-deposited microdroplet and N,N-dimethylformamide (DMF) as the antisolvent sessile droplet. Note that all the process temperature for producing the single-crystal thin films and the subsequent devices is below 30 °C.

The obtained thin film is 30–100 nm in thickness, depending on the printing condition such as ink concentration, and is quite high uniform to exhibit molecularly flat surfaces. The synchrotron X-ray study was performed for the films, and it was found that all the diffractions were observed as clear spots (Fig. 1.34). The result clearly indicates the high crystallinity of the films. The refined unit cell based on the analyses of the observed Bragg diffractions is also consistent with that of diC_8BTBT. Additionally, the films were also investigated by a crossed-Nicols microscope that is suitable for the observation of anisotropic crystals. It was found that the color of almost the entire film changes from bright to dark, simultaneously, on rotating the film about an axis perpendicular to the substrate (Fig. 1.35). The results imply the single-domain nature of the whole semiconductor thin films. Furthermore, stripe-like features with intervals of several micrometers to several tens of micrometers were observed in the ordinary optical microscope images of the films (Fig. 1.36). It is also seen in the atomic-force microscope image that the features correspond to the step-and-terrace structure with step height of 2.3–2.8 nm. The stripes are associated with the height of the molecular step in diC_8BTBT and are characteristic of the semiconductor single-crystal thin films.

As presented above, the double-shot inkjet printing technique is quite useful for manufacturing organic semiconductor thin films with high uniformity, having been the main challenge in the printable electronic technology.

1.4.4 Push Coating Technique

As discussed in Sec. 1.4.2, the spin coating technique is widely utilized for the production of plain thin films of organic semiconductors. There are many problems, however, in the spin coating technique, with regard to the large material loss and

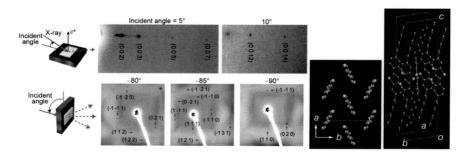

Fig. 1.34 Left: X-ray oscillation photographs of the organic semiconductor single-crystal thin films. Out-of-plane (upper) and in-plane diffractions (lower). Right: Crystal structure of diC_8BTBT. Source: Ref. [53]

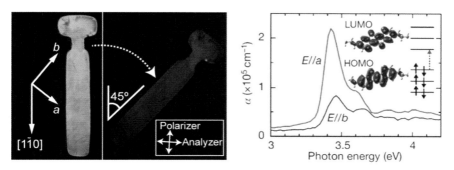

Fig. 1.35 Crossed-Nicols polarized micrographs (left) and polarized optical absorption spectra with polarization parallel to the a and b axes (right) in the inkjet printed single-crystal film of diC$_8$BTBT. Source: Ref. [53]

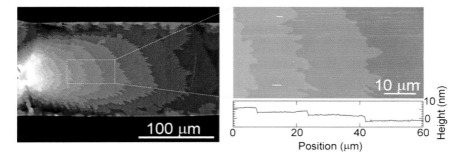

Fig. 1.36 Optical micrograph (left) and atomic-force microscopy image (right) showing the step-and-terrace structure on the organic semiconductor single-crystal thin film of diC$_8$BTBT. Source: Ref. [53]

limited controllability for the thin-film growth. For example, it is known that crystallinity of organic semiconductors can be improved by the use of highly hydrophobic substrates or the use of high-boiling point solvent, whereby the device characteristics can be improved [73–75]. However, it is quite difficult by spin coating to form uniform thin films with the use of such a substrate or solvent, as most of the material will be lost.

We have developed a "push coating technique", as an alternative, which uses viscoelastic silicone stamp using polydimethylsiloxane (PDMS) to spread polymer solution on the substrate and subsequently to absorb solvent slowly from the thin solution layer. Figure 1.37 shows the schematic of the process. First, we deposit a tiny amount of polymer semiconductor solution on the substrate and then press a viscoelastic stamp on to spread the solution, by which uniform thin solution layer is formed between the stamp and the substrate. Subsequently, thin solid film is formed by extracting solvent with the stamp. Finally, stamp is peeled off from the film. The

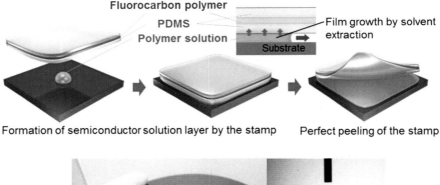

Fig. 1.37 Schematic of the film production process by the push coating technique (upper). P3HT film produced by the push coating technique on highly hydrophobic surface (lower left), and water contact angle on the substrate surface (lower right). Source: Ref. [54]

uniform thin-film formation process is quite similar with the spin coating technique, in terms of the thin solution layer formation and subsequent extraction of solvent. The feature is quite fitted to the polymeric organic semiconductors that have high degree of the layered crystallinity.

In the lower part of Fig. 1.37, we present an example of the obtained plain organic semiconductor films by the push coating technique; the typical polymer semiconductor, poly-3-hexylthiophene (P3HT), is used to form films on highly hydrophobic silicon substrate whose surface (water contact angle is 110°) is treated with silane-coupling agent. Although the high-boiling point solvent (1,2,4-trichlorobenzene with boiling point at 214 °C) is used, uniform thin films can be manufactured with eliminating the material loss. It takes about a few minutes to absorb solvent from the solution layer. As the absorbing velocity of the solvent is slower than the spreading velocity of the solution, it was easy to form larger-area films.

It was advantageous to use the stamp with trilayer structure which is composed of PDMS both-sided surface layer and solvent-resistant fluorinated silicone layer. There are two advantages in the use of the stamp with the trilayer structure. The first one is the shape stability of the stamp. If the single-layer PDMS is used, the stamp will be easily deformed by the solvent absorption, so that it is not easy to form uniform thin solution layer. In contrast, the stamp with trilayer structure is not

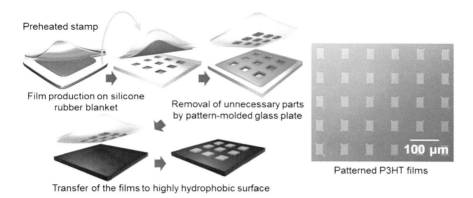

Fig. 1.38 Negative image patterning process for a push-coated film. Source: Ref. [54]

deformed against the solvent absorption. In addition, the repeated use of the stamp is also possible after the solvent extraction due to such shape stability. The second advantage is that the stamp can be easily detached from the film. As the solvent retains within the PDMS stamp for a long period of time, semiwet nature is retained at the stamp surface during the film formation. As a result, adhering force between the film and stamp is always weaker than that between the film and substrate. Therefore, it is possible to detach the stamp with all the films left on the substrate.

Various types of patterning method are applicable in the push coating technique, as it can form thin films on any substrate surfaces. In Fig. 1.38, we show an example for the thin-film patterning with the use of the inverse printing method. First, we fabricated polymer semiconductor films on a silicone blanket. When we heat the stamp at 35 °C, by which the solvent absorbing capability of the stamp is higher than that of the blanket, all the films are left on the blanket, after the stamp is peeled off. Next, we remove the unnecessary part of the film by attaching the molded glass plate. The films left on the blanket can be then transferred even to the substrate with any highly hydrophobic surface, and as a result, high-resolution pattern film is obtained.

Additionally, it is possible to use high-boiling point solvent in the push coating technique, so that the temperature and duration for the film growth can be widely controlled. We fabricated P3HT films at various temperature with using 1,2,4-trichlorobenzene, and examined the intensity profile of (100) diffraction peak. By the increase of temperature for the film growth, it is seen that the diffraction profile is narrowed. From the analyses of the peak profile, interlayer distance is distributed (1.34–1.39 nm) in the spin-coated films, while the interlayer distance of the films fabricated at 150 °C is uniform at 1.34 nm. It means that the high-temperature process is quite effective to improve the structural order between the polymer chains. With the use of the push-coated films, we fabricated bottom gate/bottom contact OFETs. We used silicon dioxide as gate dielectric layer and gold film as the source/drain electrode (with channel length at 5–100 μm and width at 5 μm). As a result, the

OFETs based on the push-coated films exhibit mobility (0.5 cm^2 V^{-1} s^{-1}) about one order higher than the OFETs based on the spin-coated films.

1.4.5 Printing of Ultrafine Metal Wiring

1.4.5.1 Materials and Techniques for Printing Electrodes and Wiring

Any electronic circuits and devices, including the AM backplanes and organic TFTs, require fine metal electrodes and wirings as indispensable building blocks. Many efforts have been dedicated, so far, to develop conductive inks and printing techniques for the print production of fine metal patterns [76, 77]. One of the most promising materials is the concentrated metal nanocolloids, or nanometal inks, composed of highly concentrated metal nanoparticles dispersed in water or organic solvents. A great variety of metal nanoparticles and nanocolloids has been investigated and developed, so far, with various metal species, particle size and distribution, various encapsulating agents for colloidal dispersion, and dispersant. Two approaches have been adopted to prepare the metal nanoparticles: The top-down approach is to prepare nanoparticles by breaking bulk metal into small particles, by using such as mechanical grinding, laser ablation, rapid condensation of metal vapor, or plasma excitation of bulk. The bottom-up approach utilizes the so-called synthetic chemistry, where the metal nanoparticles are obtained by reducing metal ions or decomposing precursor metal-complex molecules. The latter bottom-up approach seems more advantageous, in terms of the cost and mass productions, than the former top-down approach.

The metal nanoparticles are dispersed in appropriate dispersant to obtain stable metal nanocolloids. An example of the dense silver nanocolloids is presented in Fig. 1.39. The dispersed metal nanoparticles should keep to exhibit rapid Brownian motion in the stable nanocolloids. Active or unstable bare metal surfaces should be encapsulated with insulating layers, such as polymeric material or various kinds of surfactants, to prevent aggregation or precipitation through the collision between metal nanoparticles.

Currently, silver is the most reported material for the conductive ink. Although the inkjet printing is usually used to obtain metal wiring patterning, the obtained resolution is not enough for producing the active-matrix backplane with resolution higher than 100 ppi or transparent conductive electrodes. The printed deposits also often suffer from nonuniform layer thickness due to the coffee-ring phenomenon. It has also to be mentioned that the encapsulating insulating layer, which is necessary for stable colloid dispersion, must be removed after the printing deposition to restore the metal conductivity. So, a post-printing process is required for the formation of continuous metallic contacts. This is usually conducted for the printed patterns by thermal annealing, light pulse irradiation, microwave irradiation, plasma sintering, or chemical sintering, while these processes have to be done under relatively mild conditions, at which the properties of the substrate material are not affected. Due to

1 Physics of Organic Field-Effect Transistors and the Materials

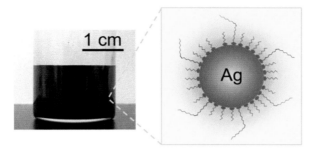

Fig. 1.39 Appearance of dense silver nanoink and schematic of the dispersed silver nanoparticle. The silver nanoparticles, obtained by thermal decomposition of oxalate-bridging silver alkylamine complexes, are encapsulated by alklamine/alkylacid and can be dispersed in octane/butanol at 40–60 wt%. Source: Ref. [78]

these difficulties, printed patterns (except for those described below) have not yet offered the material quality, pattern resolution, substrate compatibility, substrate adhesion, or throughput for mass production, required for the industrial application of high-resolution active-matrix backplanes, as long as conventional printing technologies are utilized.

1.4.5.2 Nanoparticle Chemisorption Printing Technique for Ultrafine Metal Wiring

A useful printing technique was reported by taking advantage of the unique nature of specific silver nanocolloids that show self-sintering characteristics [55]. The technique, called as surface photoreactive nanometal printing (SuPR-NaP), is based on unique chemisorption phenomena of silver nanoparticles on activated patterned polymer surfaces, as schematically presented in Fig. 1.40. The technique is only composed of simple two-step processes: The first step is the fabrication of a patterned activated surface by masked vacuum ultraviolet (VUV) irradiation of the perfluoro-polymer layer surface, and the second step is blade coating to expose the silver nanocolloids on the patterned activated surface for a short period of time (less than 1 s) at room temperature. It allows easy, high-speed, and large-area manufacturing of ultrafine metal wiring with a minimum linewidth of 0.8 μm that is conductive without any post-heating treatment and strongly adheres on the substrate surface, as presented in Fig. 1.41. The thickness profile of the printed silver lines presents flat distribution against the line cross section with no area-size dependence, clearly indicating that they do not suffer from the coffee-ring effect. These features are in striking contrast to the conventional printing technique which suffer from the limitations of the physisorption phenomena of the fluidic ink.

The technique utilizes the specific characteristics of a class of silver nanocolloids, obtained by thermal decomposition of oxalate-bridging silver alkylamine complexes

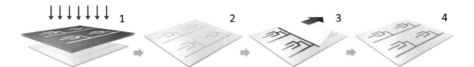

Fig. 1.40 Schematic of the surface photo-reactive nanometal printing (SuPR-NaP) technique. Source: Ref. [55]

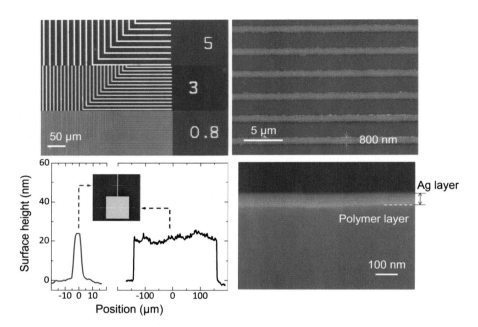

Fig. 1.41 Upper: Optical micrographs of printed patterns of parallel lines by the SuPR-NaP technique. Lower left: Thickness profiles of the printed silver lines at different line widths. Lower right: Enlarged SEM images of cross-section of the printed silver layer. Source: Ref. [55]

[78]. The peculiar nature of these silver nanocolloids is that the high dispersion stability is preserved for several months at room temperature, whereas the silver nanoparticles are readily self-fused with each other, eventually exhibiting metallic conductivity at room temperature, if the metal nanocolloids are dried out (i.e., the dispersant evaporated). According to the recent study based on the confocal dynamic light scattering, it is demonstrated that these silver nanocolloids are composed of a unique balance of ligand formulation and dispersant composition and that the unique balance enables the rapid silver nanoparticle chemisorption with maintaining the high dispersion stability of the silver nanocolloids, eventually leading to the ultrahigh-resolution patterning [79]. The technique is applicable to the production

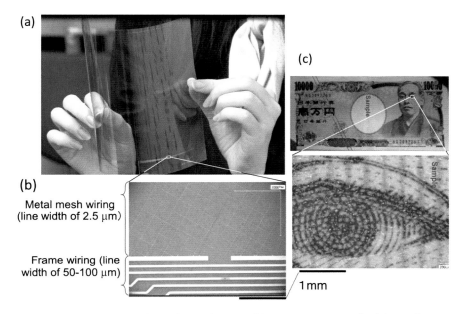

Fig. 1.42 (**a**) Photograph of an 18-cm-wide capacitive-type transparent flexible touch-screen sensor sheet, fabricated by the SuPR-NaP technique with a PET substrate. (**b**) Expanded microscope image of the sensor sheet, and the comparison with (**c**) an expanded microscope image of ten-thousand yen bill. The resolution of SuPR-NaP technique is more than ten times higher than the traditional high definition printing technique

of flexible and transparent touch screen sensor sheet composed of printed ultrafine metal mesh, as presented in Fig. 1.42. The technique should be also quite useful in the production of organic TFTs and high-resolution active-matrix backplanes [80, 81].

1.4.6 Short Summary and Outlook

Solution processability of both organic materials and metal nanoparticles is expected to provide essential advantages to the production of large-area and flexible electronic devices. To realize the scale-up production of these devices, the use and development of printing technologies is also quite promising. Owing to the intensive studies, some novel printing techniques for the formation of high mobility organic semiconductor layers as well as of ultrafine metal wiring have been recently developed. The feature of these printing methods is to take fully advantage of the unique characteristics of the electronic materials, such as the high layered crystallinity of semiconducting organic molecules and the chemisorption effect of metal nanoparticles.

1.5 Extremely Sharp Switching Operation of Printed Transistors Using Highly Layered-Crystalline Organic Semiconductors

1.5.1 Introduction

Particularly useful ways to realize high-performance OFETs are the integrated development of materials and processes. To this end, it is necessary to focus on developing organic semiconductor molecules that present highly layered-crystsalline characteristics and are suitable for the formation of a high-quality interface with the insulating layer and also on linking these layered crystallizations to the construction of device structures. Among them, molecular ordering is a decisive factor in the development of OFETs, which is in sharp contrast to the case of other organic semiconductors such as for organic light-emitting diodes or photoelectric converters in which the electronic function of the molecule itself is directly linked to the device function.

Recently, the development of a class of organic semiconductors showing a very high layered crystallinity prompted a rapid and considerable progress to develop the materials and processes as well as the device performances. In this section, we first discuss that a high layered crystallinity and excellent carrier transport characteristics can be obtained in some organic semiconductors that consist of simple organic molecules composed of rod-shaped π-electron skeletons (π-cores) linked with normal alkyl chain groups and then illustrate the molecular interpretation of these properties [82–88]. These molecules exhibit a property of facile self-ordering into molecular layers in a solution in which the molecules are dissolved in an organic solvent. By skillfully promoting and utilizing this self-film formability in the solution process, an ultrathin and highly uniform crystalline layer of organic semiconductors can be achieved [89–93]. Recently, techniques for manufacturing an extremely clean semiconductor-insulator interface have been devised, which enable OFETs to achieve practical device performances showing a low voltage and highly sharp switching with high stability, which has been challenging for many years [94–96]. These coating- or printing-based device manufacturing technologies are referred to as printed electronics. In addition to reducing the environmental load during thin-film device production, this approach is suitable for manufacturing of large-area devices using flexible plastic substrates and is expected to be applied in various manners to meet the demands for lightweight, wearable, and resource-saving devices [97, 98]. In the following sections, we introduce the status of developments in materials, processes, and devices, with mainly focusing on our research achievements.

1 Physics of Organic Field-Effect Transistors and the Materials 55

1.5.2 Development of Highly Layered-Crystalline Organic Semiconductors

1.5.2.1 Unsymmetrically Alkylated Organic Semiconductors and Origin of Its High Layered Crystallinity

A series of organic semiconductors showing a very high layered crystallinity can be obtained with various organic molecules that simply link rod-shaped π-cores with straight-chain alkyl groups, as illustrated in Figs. 1.43, 1.44, 1.45, and 1.46. The π-core is a rigid molecular unit obtained by linking (in a fused ring) cyclic conjugated systems such as five- or six-membered rings, which provides highly active π-electrons, which are the source of various electronic functions. Numerous organic semiconductors based on π-cores (thiophenes and others) containing sulfur atoms and having various fused-ring structures have been developed, and their excellent carrier transport properties have been reported [99–101]. Chemical modification by linking (or substitution, for hydrogen replacement) electronically inert alkyl groups to the π-core has often been used in organic semiconductors to impart solubility and modify solid-state properties, whereas the effectiveness and rationale for these modifications were unclear. We thoroughly studied variations in solubility and crystal structure depending on the chain length in several molecular systems composed of straight-chain alkyl groups ($-C_nH_{2n+1}$) linked with a benzothienobenzothiophene (BTBT) skeleton that affords excellent OFET properties. The substitution with an alkyl group longer than a certain length largely enhances the layered crystallinity [82, 83].

Figure 1.43 shows the crystal structure of Ph-BTBT-C_n at each chain length. Molecules substituted with alkyl groups at $n = 5$ or higher form a layered structure in

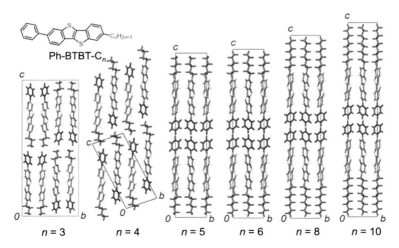

Fig. 1.43 Crystal structures of Ph-BTBT-C_n depending on the number of carbon atoms of alkyl group. The b-LHB packing is formed at $n = 5$ or higher. Source: Ref. [82]

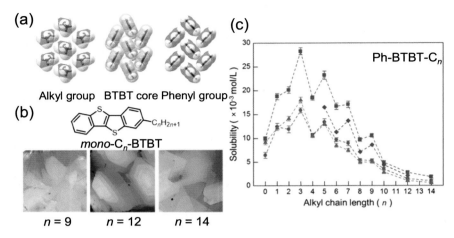

Fig. 1.44 (a) Intralayer molecular packing motif of Ph-BTBT-C_{10} depicted by space-filling model [82]. (b) An optical micrograph of mono-C_n-BTBT crystals precipitated by recrystallization from a solution. Ultrathin crystals whose step-terrace structure is visible by eye are precipitated [83]. (c) Alkyl-chain length dependence of solubility of Ph-BTBT-C_n at room temperature in toluene (filled circle), anisole (filled triangle), chlorobenzene (filled square), and toluene (filled diamond, in which solubility is calculated by the concentration at which crystals begin to precipitate at room temperature) [82]

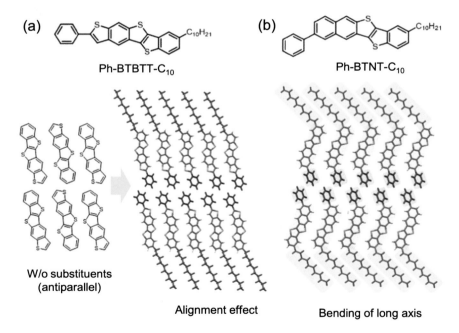

Fig. 1.45 Molecular structures and various layered molecular packing arrangements of (a) Ph-BTBTT-C_{10} [84] and (b) Ph-BTNT-C_{10} [85]

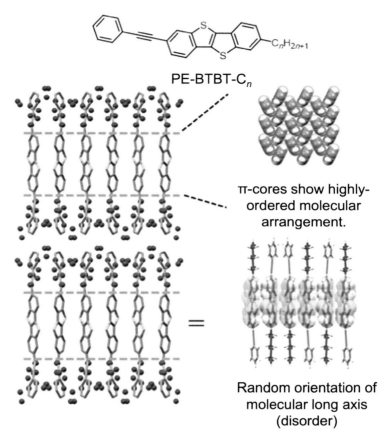

Fig. 1.46 Molecular structures and various layered molecular packing arrangements of PE-BTBT-C_n [87]

which adjacent molecules are aligned in the same direction of the long axis and are connected laterally to each other, which eventually increases the contact between the π-cores. In particular, the construction of a layered herringbone (LHB) structure (Fig. 1.44a), in which adjacent skeletal planes form a moderate dihedral angle (50–60°) within a layer, significantly improves two-dimensional carrier transport properties. The respective molecular layers with aligned molecular long axes are stacked in pairs to each other aligned in the opposite direction. These structural features are like those of bilayer lipid membranes that make up cell membranes and are referred to as bilayer-type herringbone (*b*-LHB) structures. A similar layered ordering by alkyl substitutions is observed in other π-core systems, as described later.

Regarding the origin of the attractive interaction between the molecules, high-precision quantum chemical calculations confirmed that dispersion forces dominate regardless of the π-core/alkyl moiety [83]. In particular, the dense molecular arrangement of straight-extended alkyl chains that form the all-trans configuration generates a strong cohesive force that further strengthens the layered structure between the π-cores, eventually affording the high layered crystallinity. Organic semiconductors with long-chain alkyl groups exhibit a remarkable tendency to precipitate ultrathin crystals with a molecular-level thickness from a solution (Fig. 1.44b [83]). Regarding the solubility of the molecules, an increase in solubility by alkyl substitution was observed for the shorter chains. However, once the chain length exceeds $n = 5$, the solubility decreases with the chain length (Fig. 1.44c [82]). This shows that the solubility of organic semiconductors is closely related to the stability of the layered crystal structure.

A series of asymmetric molecular systems in which one end of the rod-shaped π-core is replaced by an alkyl group tend to exhibit a higher layered crystallinity than that of symmetric molecular systems in which both ends of the π-core are replaced by alkyl groups with the same length. The feature is likely associated with the fact that asymmetric molecules can be more easily arranged with a high degree of freedom within the molecular layer, eventually increasing the independence of the layer. The phenyl group used as a counterpart to the alkyl group also contributes to the enhancement in layered crystallinity, as it mutually forms a strong herringbone-like arrangement within the molecular layer, as demonstrated in Ph-BTBT-C_n and Ph-BTNT-C_n.

1.5.2.2 Various Molecular Designs and the Effects for Layered Molecular Arrangements

One of the advantages of organic semiconductors is that materials can be designed and upgraded with a high degree of freedom by controlling the type, size, and substitution position of the π-core and substituents. We have been developing various π-core–straight-chain-alkyl linked molecular system for controlling layered crystallinity and various properties [84–86]. In benzothienobenzothienothiophene (BTBTT) (Fig. 1.45a) and benzothienonaphthothiophene (BTNT) (Fig. 1.45b), both with a fused ring number of 5 being one ring larger than BTBT, the skeleton has no inversion or mirror symmetry. Thus, for example, in the case of BTBTT, the orientation of adjacent π-cores becomes antiparallel in the crystal in the absence of the substitution, which results in poor carrier transport properties. However, when alkyl substitution is applied to one end, an alignment effect arises, in which the long axes of the asymmetric skeletons are oriented in the same direction, significantly improving its carrier transport properties (to 12 cm^2 V^{-1} s^{-1}) [84].

One of the aims of using stretched π-cores is to improve the thermal endurance, which is essential for practical use. For example, in Ph-BTNT-C_n, in which the π-core of Ph-BTBT-C_n is replaced by BTNT, the enhanced interactions between the cores provide a sufficient thermal durability, but the tradeoff for this is less than adequate solubility. Therefore, we developed a molecule that slightly weakens the linearity of the entire rod-shaped molecule by controlling the substitution position of the π-core to introduce a bending site. As a result, intermolecular forces between neighboring molecules were relaxed, and an organic semiconductor exhibiting both thermal durability and solubility was obtained (Fig. 1.45b) [85]. A similar effect of substitution position was also observed for mono-C_n-BTNT without phenyl groups [86].

1.5.2.3 Long-Axis Orientational Order and Liquid–Crystal Phase Transition

In general, rod-shaped organic molecules are typically able to take on a liquid crystalline state in which only the orientation of the molecular long axis is aligned in the aggregated state. When the layered organic semiconductors described above are in their crystalline phase, a long-range order is maintained between molecules both within and between layers, but their ability to form an order in the long- vs. short-axis direction of the molecule clearly differs. Many layered organic semiconductors such as Ph-BTBT-C_n and Ph-BTNT-C_n transition to the liquid crystal phase at high temperatures. As discussed below, we believe that this liquid crystallinity has a major role in enabling a highly uniform thin-film formation by solution processing. However, the actual liquid crystal phase has not been understood so far in detail in terms of emerging mechanism and molecular arrangement structure.

Therefore, we developed PE-BTBT-C_n, replacing the phenyl group in Ph-BTBT-C_n with a phenylethynyl (PE) group of certain length and rotational freedom (Fig. 1.46) [87]. The result was an emergence of unique molecular arrangement structure closely related to liquid crystallinity. In cases where the alkyl group of PE-BTBT-C_n is longer than the PE group ($n = 8, 10, 12$), we obtained a b-LHB structure in which the orientations of the molecular long axes in each layer are aligned at room temperature, and we observed a transition to the smectic liquid crystal (*SmE*) phase at a high temperature. However, a crystallographic analysis using electron diffraction showed that, when the alkyl and PE groups have approximately the same length ($n = 6$), the arrangement takes on a disordered LHB (*d*-LHB) structure with molecules randomly distributed in opposite directions within the layer (Fig. 1.46). In this case, the herringbone-type ordering between BTBT skeletons in the molecular layer is maintained despite the random orientation of the molecular long axis. As the transition to the *SmE* phase disappears at high temperatures, we deduce that the *d*-LHB phase is the frozen phase of the *SmE* phase. Furthermore, when small amounts of molecules with a longer alkyl chain length are mixed during the crystallization process of this system, the highly ordered *b*-LHB phase appears and improves the transport characteristics [88].

1.5.3 Meniscus-Guided Coating

Solution is an ideal environment for mildly grown solute crystallization. The high layered crystallinity of the organic semiconductors described above allows facile formation of ultrathin uniform crystalline layers with a molecular-level thickness over a wide area in a solution. Coating- or printing-based thin-film manufacturing that utilize this phenomenon include spin coating, drop casting, meniscus-guided coating, and double-shot inkjet printing [53, 102–104]. We describe the formation of ultrathin single-crystal layers mainly by the meniscus-guided coating technique. Single-crystal thin-film formation is particularly useful for elucidating layered-material-specific crystal growth and also intrinsic carrier transport mechanisms, which is essential to design new materials and processes and improve device performances.

1.5.3.1 Single-Crystal Thin-Film Formation and Interlayer Access Resistance

Figure 1.47a shows an optical microscopy image of the resulting Ph-BTBT-C_{10} thin film. By an appropriate selection of processing conditions, a single-crystal thin film with a size of several millimeters consisting of multiple molecular layers has been obtained under ambient temperature and pressure [89]. The film thickness is approximately 5–80 nm, which is equivalent to 1–16 times the thickness of a bilayer (approximately 5 nm). The atomic force microscopy image clearly shows a step-terrace structure with a step height equivalent to one bilayer. When these single-crystal thin films are formed on a silicon substrate with a SiO_2 surface layer, it is possible to distinguish differences in the number of layers of the films by each step at the molecular level, because of the optical interference in the multilayer structures including the SiO_2 layer (Fig. 1.47b). In the construction of a top-contact (staggered) device with an SiO_2 layer as a gate insulator using single-crystal thin films with different layer numbers (Fig. 1.47c), a small number of molecular bilayers (such as two layers) resulted in a mobility higher than 10 cm^2 V^{-1} s^{-1}, whereas stacking of several molecular bilayers increased the access resistance from the electrode to the channel region [89, 90]. However, as the SiO_2 surface layer generally involves many interface traps, the device exhibits high threshold gate voltage, and also it is difficult even to operate the bottom contact (inverse coplanar)-type device in most cases.

1.5.3.2 Construction of a Single Molecular Bilayer-Type Ultrathin Semiconductor Layer

The number of layers (film thickness) described above can be controlled to some extent by the solution concentration and film formation speed. The variability of the alkyl chain length can be utilized for a precise construction of ultrathin single

1 Physics of Organic Field-Effect Transistors and the Materials 61

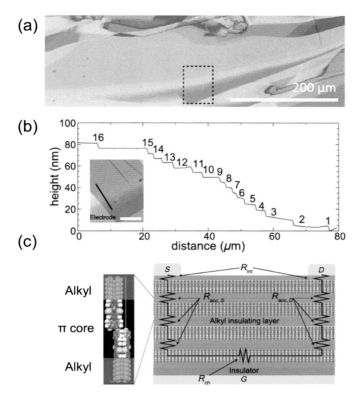

Fig. 1.47 (**a**) An optical micrograph of a Ph-BTBTT-C_{10} single-crystal thin film formed on an SiO$_2$/Si substrate. The layer number difference is visually recognized by the color variation. (**b**) A cross-sectional height profile obtained by AFM measurement for the area surrounded by dashed lines. Each step corresponds to bilayer thickness. (**c**) Schematic for top-contact-type device composed of single-crystal thin film. Source: Ref. [89]

molecular bilayers [91, 92]. The method utilizes the effect of suppressing interlayer stacking while maintaining intralayer molecular ordering through the mixing of molecules with different chain lengths. Figure 1.48 shows an optical microscopy image of a thin film formed using a solution in which both Ph-BTBT-C_6 and Ph-BTBT-C_{10} are dissolved. We obtained a thin film (single-crystal domain size of approximately 10 cm^2) with a uniform thickness of one bilayer over the whole wafer surface area (100 cm^2). An electron density distribution analysis using X-ray reflectometry revealed that single molecular bilayers are composed of interlayer stacking arrangement with head-to-head alignment of π-cores. These single molecular bilayers were obtained efficiently by mixing approximately 3–50% of long molecules. The single-layer formation as described above originates from the unevenness formed on the surface of the molecular layer due to the variation in chain length while maintaining intralayer molecular ordering.

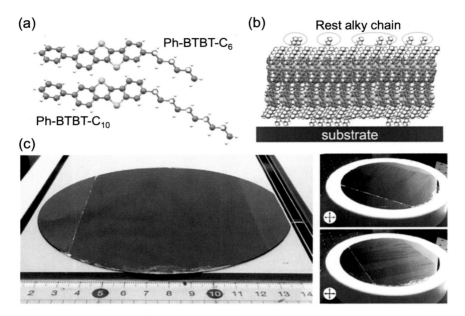

Fig. 1.48 (**a**) Molecular structures of Ph-BTBT-C_6 and Ph-BTBT-C_{10}. (**b**) Schematic for single molecular bilayer film obtained by the mixed solution of molecules. (**c**) Photographs of single molecular bilayer film manufactured on a 6-in. Si wafer (left) and of its crossed-Nicols polarized observation (right). Source: Ref. [91]

1.5.3.3 Molecular Orientation and Stepwise Ordering at the Air–Liquid Interface

In the meniscus-guided coating method, the solution is held between the substrate and blade tip by interfacial tension (originating from the attraction between the solid surface and liquid molecules), and a thin film is formed near the tip of the solution droplet that is spread wet on the substrate due to the hydrophilic nature of the substrate (the meniscus is the area where the air–liquid interface is curved by the surface tension of the solution). The ability to construct wafer-size single-crystal thin films by such simple film formation is a distinctive feature of layered organic semiconductors that is not observed for other semiconductors, which require a high vacuum, high temperature, epitaxial growth, etc. We consider that one background factor allowing to provide such highly uniform thin-film formation is a unique crystal-growth mechanism mediated via the liquid crystal state in the vicinity of the air–liquid interface, being characteristic of the layered organic semiconductors.

Indeed, observation of a solution droplet surface on substrate under optical microscope reveals the formation of ultrathin crystals along the droplet surface [87]. If the crystal nuclei of layered crystals were to form in a solution, the plane orientation of the precipitated layered crystals would become random, which would

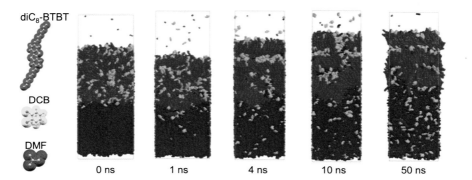

Fig. 1.49 Snapshots of the MD simulation for layered ordering process in diC$_8$BTBT solution of 1,2-dichlorobenzene (DCB) and dimethylformamide (DMF). Molecular layer formation parallel to the air-liquid interface is observed. Source: Ref. [93]

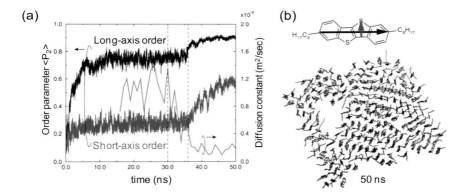

Fig. 1.50 (**a**) Temporal evolution of orientational order parameter along molecular long and short axes directions, respectively, and the corresponding in-layer diffusion constants (red line). (**b**) A snapshot ($t = 50$ ns) of molecular ordering process in liquid-crystal-like molecular layer formed at the air-liquid interface. Source: Ref. [93]

hinder a highly uniform film formation on the substrate. Molecular dynamics (MD) simulations of these behaviors in the solution show that, in the vicinity of the air–liquid interface, the rod-shaped molecules are aggregated with their molecular long axes perpendicular to the interface (Fig. 1.49) [93]. The result demonstrates that a molecular layer is first formed in which only the orientation of the molecular long axis is aligned, and then, over time, ordering proceeds in the direction of the molecular short axis (Fig. 1.50a). The state in which only the molecular long-axis direction is ordered is the liquid crystal state (lyotropic liquid crystal emerged by solvent mixing). We consider that the formation of a liquid crystalline precursor film, followed by stepwise crystal growth and subsequent sequential ordering, is crucial for obtaining highly uniform crystalline thin films over a large area.

The solution-based thin-film formation process not only provides an advantage in device manufacturing under ambient temperature and pressure but is also essential for achieving ultrathin, large-area, and highly uniform thin films, as described above. This understanding of the growth mechanism of layered crystals via the air–liquid interface is used as a basis for constructing an extremely clean insulator–semiconductor interface, discussed in the next section.

1.5.4 Clean Interface Construction and Very Sharp Switching

1.5.4.1 Realization of a Solution-Based Highly Uniform Thin-Film Formation on Highly Liquid-Repellent Surfaces

The highly uniform semiconductor crystal layers could be manufactured by taking advantage of the highly layered-crystalline characteristics of organic semiconductors, as described in the previous section. However, the obtained semiconductor layers were far from ideal in terms of the interfacial state between the semiconductor and insulating layer, so that their OFET characteristics were not sufficient for practical use. Recently, we developed a printing technique that allows to create an extremely clean semiconductor–insulator interface with minimal carrier traps and thus to operate a practical coplanar-type OFET showing very sharp switching at low voltages [94, 95].

An effective approach to produce an organic semiconductor with an extremely clean semiconductor–insulator interface is to construct an insulator interface that is lower in surface energy (i.e., less adhesive). Excellent switching characteristics can be obtained with an OFET fabricated by attaching a thin organic semiconductor monocrystal to the insulator using the amorphous perfluoropolymer Cytop [105]. However, the requirement that the surface energy is low also confers a highly liquid-repellent nature repelling the solution (Fig. 1.51). Thus, the applied solution does not spread wet on the surface, which has been in truth a significant obstacle to realize solution-based uniform thin-film processing on the highly liquid-repellent surfaces. We developed an "extended meniscus-guided (EMG) coating technique"

Fig. 1.51 Molecular structure of Cytop (perfluoropolymer resin) and the appearance of water droplet repellence on the surface of Cytop layer. Source: Ref. [94]

1 Physics of Organic Field-Effect Transistors and the Materials

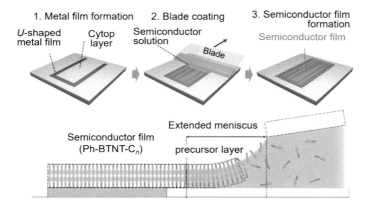

Fig. 1.52 Schematics for the film production by the EMG coating technique and the formation process of single molecular bilayer film. Source: Ref. [94]

that creates a state in which the semiconductor solution can be spread wet without curling up even on highly liquid-repellent insulator surfaces and thus enables to form an ultrathin, large-area, and highly uniform semiconductor crystal layer on the Cytop layer.

A conceptual diagram of the developed thin-film processing technique is shown in Fig. 1.52. As a mechanism for spreading the solution wet, we produced a region surrounded on three sides by a U-shaped metal film pattern on the highly liquid-repellent insulator surface and then applied the meniscus-guided coating to this region to produce the film. With this technique, semiconductor film formation first begins on the metal film with a high surface energy, and, even if the blade reaches over the highly liquid-repellent insulator surface, the semiconductor layer continues to grow and form on the insulator surface just as on the metal film surface with keeping the meniscus extended. Because the molecular film, formed at the air–liquid interface and covering the interface, is continuous with the thin solid film on the substrate, the condition of the thin liquid film that was spread wet (i.e., extended meniscus) is maintained stably without being affected by droplet surface tension. Figure 1.53 shows an optical microscopy image of the resulting Ph-BTNT-C_n thin film. The thin-film formation ends at the point where the U-shaped area surrounded on three sides by the metal film pattern ends. Real-time observation of crystal growth by optical microscopy shows that, when the droplet tip is covered by a semiconductor film, the extended meniscus is maintained. However, as soon as it leaves the U-shaped area, the extended meniscus ends and returns to its original state showing a large contact angle.

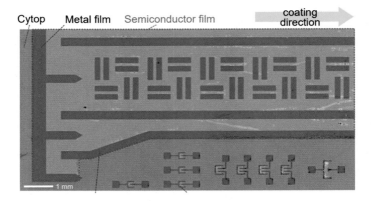

Fig. 1.53 An optical micrograph of printed Ph-BTNT-C_n crystalline films fabricated on Cytop surfaces by the EMG coating technique. Source: Ref. [94]

1.5.4.2 Very Sharp Switching Operation

The use of organic semiconductors with a high layered crystallinity enables to form a highly uniform thin film on a highly liquid-repellent insulating layer by the EMG coating technique. Suppression of multilayering, with the use of a solution in which molecules with different alkyl chain lengths are mixed, can yield even more effective thin-film formation. We used this technique to fabricate a Ph-BTNT-C_n thin film on the Cytop layer to produce a bottom-gate bottom-contact (inverse coplanar)-type OFET. Typical device characteristics are shown in Fig. 1.54. The obtained transfer characteristics show a sharp rise near 0 V, absence of hysteresis, and good mobility reaching 4.9 cm^2 V^{-1} s^{-1}. In addition, the subthreshold swing (SS), a measure of switching sharpness, had an average value of 67 mV dec^{-1}, which is close to the theoretical limit, as presented in Fig. 1.55. This excellent SS is attributed to the formation of an extremely clean semiconductor–insulator interface that minimizes the trap generations that inhibit carrier transport. A similar OFET construction is also effective when the source/drain electrodes are produced by the novel printing technique referred as the SuPR-NaP technique, as presented in Sec. 1.4.5 [55]; the technique can be used to print and form a high-definition silver electrode wiring with the use of a patterned irradiation of deep ultraviolet light on the Cytop layer. This enabled to fabricate an all-printed OFET exhibiting very sharp switching characteristics.

With regard to the mobility of the above devices, the behavior is dependent on the semiconductor material: Ph-BTNT-C_n exhibits a satisfactory performance, while Ph-BTBT-C_n presents a performance which is significantly lower than that obtained with the staggered-type device described above. In the latter case, the contact resistance at the source/drain electrodes is high. It indicates that stability near the three-way electrode–semiconductor–insulator interface constituting the coplanar-type OFET is significant.

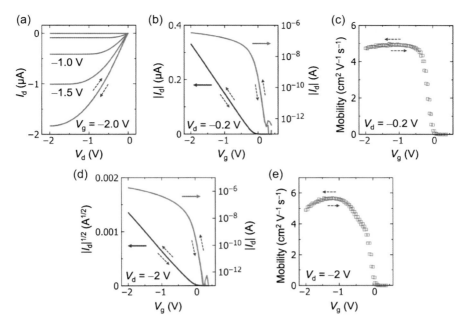

Fig. 1.54 Typical electrical characteristics of printed Ph-BTNT-C_n single-crystal TFTs. (**a**) Output characteristics. (**b, d**) Transfer characteristics in the linear regime ($V_d = -0.2$ V) and saturation regime ($V_d = -2$ V), respectively. (**c, e**) Gate voltage dependence of carrier mobility extracted from first derivative of transfer curve in the linear regime and saturation regime, respectively. Source: Ref. [94]

1.5.4.3 Spin Coating for a Very Sharp Switching Device

Organic semiconductors that exhibit a high layered crystallinity are also useful for the fabrication of highly uniform polycrystalline thin films by simple spin-coating or drop-casting methods. It was shown that OFETs that simultaneously exhibit high performances and very sharp switching have been obtained using a mixed solution of mono-BTNT-C_n with an insulating polymer such as polymethyl methacrylate (PMMA) or polystyrene [96]. Further performance improvements can be expected by the development of materials that further enhance the layered crystallinity.

1.5.5 Short Summary and Outlook

In this section, we described the status of the development of OFETs, which are still evolving by relying on the crucial concept that a certain class of organic semiconductors exhibits high layered crystallinity. We showed that the remarkably high layered crystallinity is realized by the synergistic effect of the π-cores and long-chain

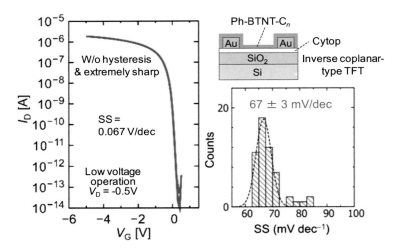

Fig. 1.55 Transfer characteristics (left) and distribution of SS values for 56 devices (right) of inverse-coplanar-type OFETs composed of Ph-BTNT-C_n single-crystal thin films produced on Cytop layers by the EMG coating technique. Source: Ref. [94]

alkyl groups, both of which can be ordered in layers, and this unique property enables formation of ultrathin, large-area, and highly uniform thin films by solution processing. Furthermore, by producing an extremely clean semiconductor interface with this property, it is possible to obtain a very sharp and stable device operation at low voltages. This remarkable property achieved by organic molecules provides a strong basis for further improvement of performances. Its application to various molecular systems is also expected to explore and provide new electronic and device functions.

For the development of such new materials, it is desirable to be able to predict in advance, at the molecular design stage, what type of crystal structure will be formed. However, the prediction of the crystal structure of molecular materials is renowned as a computationally difficult problem: It is still uncertain whether the designed molecules on the desk are promising until they are synthesized. Further studies are being carried out to accelerate this development through crystal structure prediction targeted at highly layered-crystalline molecular materials.

Acknowledgment The author is grateful to Dr. Satoru Inoue for his help in the preparation of this manuscript. I also would like to note that the later part of this chapter is based mainly on research findings that were obtained and published by our research group. The works were carried out in collaboration with many people including Prof. Hiroyuki Matsui, Dr. Jun'ya Tsutsumi, Dr. Satoshi Matsuoka, Dr. Hiromi Minemawari, Dr. Yuki Noda, Dr. Mitsuhiro Ikawa, Dr. Toshikazu Yamada, Dr. Satoru Inoue, Dr. Shunto Arai, Dr. Makoto Yoneya, Dr. Takamasa Hamai, Dr. Gyo Kitahara, and Dr. Seiji Tsuzuki.

References

1. M. Pope, C.E. Swenberg, *Electronic Processes in Organic Crystals and Polymers*, 2nd edn. (Oxford University Press, New York, 1999)
2. G.H. Wagniere, *Introduction to Elementary Molecular Orbital Theory and to Semiempirical Methods, Lecture Notes in Chemistry* (Springer, Berlin/Heidelberg, 1976)
3. Gaussian, *16, Revision C.01* (Gaussian, Inc, 2016)
4. Amsterdam Density Functional, ADF (2016), Scientific Computing & Modelling (SCM) N. V
5. M. Robin, *Higher Excited States of Polyatomic Molecules*, vol 3 (Academic, New York/London, 1985)
6. G. Varsányi, *Vibrational Spectra of Benzene Derivatives* (Academic, New York/London, 1969)
7. C. Kittel, *Introduction to Solid State Physics* (Wiley, New York, 2014)
8. W. Clegg (ed.), *Crystal Structure Analysis: Principles and Practice* (Oxford University Press, New York, 2009)
9. Methods and Applications of Crystal Structure Prediction, *Faraday Discussion 211* (Royal Society of Chemistry, 2018)
10. F. Seitz, N.G. Einspruch, *Electronic Genie: The Tangled History of Silicon* (UI Press, Illinois, 1998)
11. S. M. Sze, Semiconductor Devices: Physics and Technology, 2016
12. S.F. Nelson, Y.-Y. Lin, D.J. Gundlach, T.N. Jackson, Appl. Phys. Lett. **72**, 1854 (1998)
13. V. Podzorov, E. Menard, A. Borissov, V. Kiryukhin, J.A. Rogers, M.E. Gershenson, Phys. Rev. Lett. **93**, 086602 (2004)
14. M.E. Gershenson, V. Podzorov, A.F. Morpurgo, Rev. Mod. Phys. **78**, 973 (2006)
15. R.A. Marcus, Rev. Mod. Phys. **65**, 599 (1993)
16. J. Brédas, D. Beljonne, V. Coropceanu, J. Cornil, Chem. Rev. **104**, 4971 (2004)
17. J. Jortner, J. Chem. Phys. **64**, 4860 (1976)
18. J. Takeya, J. Kato, K. Hara, M. Yamagishi, R. Hirahara, K. Yamada, Y. Nakazawa, S. Ikehata, K. Tsukagoshi, Y. Aoyagi, T. Takenobu, Y. Iwasa, Phys. Rev. Lett. **98**, 196804 (2007)
19. J.S. Miller, J.S. Miller (eds.), *Extended Linear Chain Compounds*, vol 2 and 3 (Plenum Press, New York, 1983)
20. T. Ishiguro, K. Yamaji, G. Saito, *Organic Superconductors. Springer Series in Solid-State Sciences*, vol 88 (Springer, Berlin, 1998)
21. A. Troisi, G. Orlandi, Phys. Rev. Lett. **96**, 086601 (2006)
22. J.-D. Picon, M.N. Bussac, L. Zuppiroli, Phys. Rev. B **75**, 235106 (2007)
23. S. Fratini, S. Ciuchi, Phys. Rev. Lett. **103**, 266601 (2009)
24. V. Coropceanu, M. Malagoli, D.A. da Silva Filho, N.E. Gruhn, T.G. Bill, J.L. Bré´das, Phys. Rev. Lett. **89**, 275503 (2002)
25. N. Ueno, S. Kera, Prog. Surf. Sci. **83**, 490 (2008)
26. N. Hulea, S. Fratini, H. Xie, C.L. Mulder, N.N. Iossad, G. Rastelli, S. Ciuchi, A.F. Morpurgo, Nat. Mater. **5**, 982 (2006)
27. W.L. Kal, B. Batlogg, Phys. Rev. B **81**, 035327 (2010)
28. C.P. Slichter, *Principles of Magnetic Resonance, Springer Series in Solid-State Sciences*, vol 1 (Springer, Berlin, 1996)
29. K. Marumoto et al., J. Phys. Soc. Jpn. **73**, 1673 (2004)
30. K. Marumoto et al., Phys. Rev. Lett. **97**, 256603 (2006)
31. H. Matsui, T. Hasegawa, Y. Tokura, M. Hiraoka, T. Yamada, Phys. Rev. Lett. **100**, 126601 (2008)
32. R. Kubo, K. Tomita, J. Phys. Soc. Jpn. **9**, 888 (1954)
33. P.W. Anderson, J. Phys. Soc. Jpn. **9**, 316 (1954)
34. H. Matsui, D. Kumaki, E. Takahashi, K. Takimiya, S. Tokito, T. Hasegawa, Phys. Rev. B **85**, 035308 (2012)
35. G. Horowitz, P. Delannoy, J. Appl. Phys. **70**, 469 (1991)

36. G. Horowitz, R. Hajloui, P. Delannoy, J. Phys. III **5**, 355 (1995)
37. M.F. Calhoun, C. Hsieh, V. Podzorov, Phys. Rev. Lett. **98**, 096402 (2007)
38. H. Matsui, A.S. Mishchenko, T. Hasegawa, Phys. Rev. Lett. **104**, 056602 (2010)
39. S. Mishchenko, H. Matsui, T. Hasegawa, Phys. Rev. B **85**, 085211 (2012)
40. J.R. Bolton, J. Chem. Phys. **46**, 408 (1967)
41. J.H. Burroughes, C.A. Jones, R.H. Friend, Nature **335**, 137 (1988)
42. M.-F. Li, *Modern Semiconductor Quantum Physics* (World Scientific, Singapore, 1995)
43. S. Haas, H. Matsui, T. Hasegawa, Phys. Rev. B **82**, 161301 (2010)
44. L. Sebastian, G. Weiser, H. Bässler, Chem. Phys. **61**, 125 (1981)
45. J. Tsutsumi, S. Matsuoka, I. Osaka, R. Kumai, T. Hasegawa, Phys. Rev. B **95**, 115306 (2017)
46. R. Davis, L.N. Pye, N. Katz, J.A. Hudgings, K.R. Carter, Adv. Mater. **26**, 4539 (2014)
47. J. Tsutsumi, S. Matsuoka, T. Yamada, T. Hasegawa, Org. Electron. **25**, 289 (2015)
48. J. Tsutsumi, S. Matsuoka, T. Kamata, T. Hasegawa, Org. Electron. **55**, 187 (2018)
49. S. Matsuoka, J. Tsutsumi, T. Kamata, T. Hasegawa, J. Appl. Phys. **123**, 135301 (2018)
50. S. Matsuoka, J. Tsutsumi, H. Matsui, T. Kamata, T. Hasegawa, Phys. Rev. Appl. **9**, 024025 (2018)
51. S. Matsuoka, J. Tsutsumi, T. Hasegawa, Phys. Rev. Appl. **16**, 044043 (2021)
52. T. Kawase, T. Shimoda, C. Newsome, H. Sirringhaus, R.H. Friend, Thin Solid Films **279**, 438 (2003)
53. H. Minemawari, T. Yamada, H. Matsui, J. Tsutsumi, S. Haas, R. Chiba, R. Kumai, T. Hasegawa, Nature **475**, 364 (2011)
54. M. Ikawa, T. Yamada, H. Matsui, H. Minemawari, J. Tsutsumi, Y. Horii, M. Chikamatsu, R. Azumi, R. Kumai, T. Hasegawa, Nat. Commun. **3**, 1176 (2012)
55. T. Yamada, K. Fukuhara, K. Matsuoka, H. Minemawari, J. Tsutsumi, N. Fukuda, K. Aoshima, S. Arai, Y. Makita, H. Kubo, T. Togashi, M. Kurihara, T. Hasegawa, Nat. Commun. **7**, 11402 (2016)
56. J.C. Berg, *An Introduction to Interfaces & Colloids – the Bridge to Nanoscience* (World Scientific, 2010)
57. R.D. Deegan, O. Bakajin, T.F. Dupont, G. Huber, S.R. Nagel, T.A. Witten, Nature **389**, 827 (1997)
58. J.H. Burroughes, D.D.C. Bradley, A.R. Brown, R.N. Marks, K. Mackay, R.H. Friend, P.L. Burns, A.B. Holmes, Nature **347**, 539 (1990)
59. H. Sirringhaus, P.J. Brown, R.H. Friend, M.M. Nielsen, K. Bechgaard, B.M.W. Langeveld-Voss, A.J.H. Spiering, R.A.J. Janssen, E.W. Meijer, P. Herwig, D.M. de Leeuw, Nature **401**, 685 (1999)
60. H. Sirringhaus, T. Kawase, R.H. Friend, T. Shimoda, M. Inbasekaran, W. Wu, E.P. Woo, Science **290**, 2123 (2000)
61. J.E. Anthony, J.S. Brooks, D.L. Eaton, S.R. Parkin, J. Am. Chem. Soc. **123**, 9482 (2001)
62. H. Ebata, T. Izawa, E. Miyazaki, K. Takimiya, M. Ikeda, H. Kuwabara, T. Yui, J. Am. Chem. Soc. **129**, 15732 (2007)
63. H. Yan, Z. Chen, Y. Zheng, C. Newman, J.R. Quinn, F. Doetz, M. Kastler, A. Facchetti, Nature **475**, 679 (2009)
64. X. Zhang, S.D. Hudson, D.M. DeLongchamp, D.J. Gundlach, M. Heeney, I. McCulloch, Adv. Funct. Mater. **20**, 4098 (2010)
65. D.W. Schubert, T. Dunkel, Mat. Res. Innovat. **7**, 314 (2003)
66. M. Hiraoka, T. Hasegawa, T. Yamada, Y. Takahashi, S. Horiuchi, Y. Tokura, Adv. Mater. **19**, 3248 (2007)
67. Y. Noda, H. Minemawari, H. Matsui, T. Yamada, S. Arai, T. Kajiya, M. Doi, T. Hasegawa, Adv. Funct. Mater. **25**, 4022–4031 (2015). https://doi.org/10.1002/adfm.201500802
68. D. Stroock et al., Science **295**, 647 (2002)
69. J.M. Ottino, S. Wiggins, Phil. Trans. Roy. Soc. Lond. Math. Phys. Sci. **362**, 923 (2004)
70. T. Izawa, E. Miyazaki, K. Takimiya, Adv. Mater. **20**, 3388 (2008)
71. T. Uemura, Y. Hirose, M. Uno, K. Takimiya, J. Takeya, Appl. Phys. Express **2**, 111501 (2009)

72. T. Minari, C. Liu, M. Kano, K. Tsukagoshi, Adv. Mater. **24**, 299 (2012)
73. B. Meredig, A. Salleo, R. Gee, ACS Nano **3**, 2881 (2009)
74. T. Umeda, D. Kumaki, S. Tokito, J. Appl. Phys. **105**, 024516 (2009)
75. W.L. Kalb, T. Mathis, S. Haas, A.F. Stassen, B. Batlogg, Appl. Phys. Lett. **90**, 092104 (2007)
76. A. Kamyshny, S. Magdassi, Small **10**, 3515 (2014)
77. J.S. Kamyshny, S. Magdassi, O. Appl. Phys. J. **4**, 19 (2011)
78. M. Itoh, T. Kakuta, M. Nagaoka, Y. Koyama, M. Sakamoto, S. Kawasaki, N. Umeda, M. Kurihara, J. Nanosci. Nanotechnol. **9**, 1 (2009)
79. K. Aoshima, Y. Hirakawa, T. Togashi, M. Kurihara, S. Arai, T. Hasegawa, Sci. Rep. **8**, 6133 (2018)
80. K. Aoshima, S. Arai, K. Fukuhara, T. Yamada, T. Hasegawa, Org. Electron. **41**, 137 (2017)
81. G. Kitahara, K. Aoshima, J. Tsutsumi, H. Minemawari, S. Arai, T. Hasegawa, Org. Electron. **50**, 426 (2017)
82. S. Inoue, H. Minemawari, J. Tsutsumi, M. Chikamatsu, T. Yamada, S. Horiuchi, M. Tanaka, R. Kumai, T. Hasegawa, Chem. Mater. **27**, 3809 (2015)
83. H. Minemawari, M. Tanaka, S. Tsuzuki, S. Inoue, T. Yamada, R. Kumai, Y. Shimoi, T. Hasegawa, Chem. Mater. **29**, 1245 (2017)
84. T. Higashino, S. Inoue, S. Arai, H. Matsui, N. Toda, S. Horiuchi, R. Azumi, T. Hasegawa, Chem. Mater. **33**, 7379 (2021)
85. S. Inoue, S. Shinamura, Y. Sadamitsu, S. Arai, S. Horiuchi, M. Yoneya, K. Takimiya, T. Hasegawa, Chem. Mater. **30**, 5050 (2018)
86. S. Inoue, T. Higashino, S. Arai, R. Kumai, H. Matsui, S. Tsuzuki, S. Horiuchi, T. Hasegawa, Chem. Sci. **11**, 12493 (2020)
87. S. Inoue, K. Nikaido, T. Higashino, S. Arai, M. Tanaka, R. Kumai, S. Tsuzuki, S. Horiuchi, H. Sugiyama, Y. Segawa, K. Takaba, S. Maki-Yonekura, K. Yonekura, T. Hasegawa, Chem. Mater. **34**, 72 (2022)
88. K. Nikaido, S. Inoue, R. Kumai, T. Higashino, S. Matsuoka, S. Arai, T. Hasegawa, Adv. Mater. Interfaces **9**, 202201789 (2022)
89. T. Hamai, S. Arai, H. Minemawari, S. Inoue, R. Kumai, T. Hasegawa, Phys. Rev. Appl. **8**, 054011 (2017)
90. T. Hamai, S. Inoue, S. Arai, T. Hasegawa, Phys. Rev. Mater. **4**, 074601 (2020)
91. S. Arai, S. Inoue, T. Hamai, R. Kumai, T. Hasegawa, Adv. Mater. **30**, 1707256 (2018)
92. S. Arai, K. Morita, J. Tsutsumi, S. Inoue, M. Tanaka, T. Hasegawa, Adv. Funct. Mater. **30**, 1906406 (2020)
93. M. Yoneya, H. Minemawari, T. Yamada, T. Hasegawa, J. Phys. Chem. C **121**, 8796 (2017)
94. G. Kitahara, S. Inoue, T. Higashino, M. Ikawa, T. Hayashi, S. Matsuoka, S. Arai, T. Hasegawa, Sci. Adv. **6**, eabc8847 (2020)
95. G. Kitahara, M. Ikawa, S. Matsuoka, S. Arai, T. Hasegawa, Adv. Funct. Mater. **31**, 105933 (2021)
96. R. Miyata, S. Inoue, K. Nakajima, T. Hasegawa, A.C.S. Appl, Mater. Interfaces **14**, 17719 (2022)
97. M. Kaltenbrunner, T. Sekitani, J. Reeder, T. Yokota, K. Kuribara, T. Tokuhara, M. Drack, R. Schwödiauer, I. Graz, S. Bauer-Gogonea, S. Bauer, T. Someya, Nature **499**, 458 (2013)
98. R. Shiwaku, H. Matsui, K. Hayasaka, Y. Takeda, T. Fukuda, D. Kumaki, S. Tokito, Adv. Electron. Mater. **3**, 1600557 (2017)
99. K. Takimiya, S. Shinamura, I. Osaka, E. Miyazaki, Adv. Mater. **23**, 4347 (2011)
100. T. Okamoto, C.P. Yu, C. Mitsui, M. Yamagishi, H. Ishii, J. Takeya, J. Am. Chem. Soc. **142**, 9083 (2020)
101. S. Riera-Galindo, F. Leonardi, R. Pfattner, M. Mas-Torrent, Adv. Mater. Technol. **4**, 1900104 (2019)
102. S. Yamamura, M. Watanabe, M. Uno, C. Mitani, J. Mitsui, N. Tsurumi, Y. Isahaya, T.O. Kanoda, J. Takeya, Sci. Adv. **4**, eaao5758 (2018)

103. T. Rocha, K. Haase, Y. Zheng, M. Löffler, M. Hambsch, S.C.B. Mannsfeld, Adv. Electron. Mater. **4**, 1800141 (2018)
104. K. Bulgarevich, K. Sakamoto, T. Minari, T. Yasuda, K. Miki, A.C.S. Appl, Mater. Interfaces **9**, 6237 (2017)
105. B. Blülle, R. Häusermann, B. Batlogg, Phys. Rev. Appl. **1**, 034006 (2014)

Chapter 2
Organic Light-Emitting Diodes (OLEDs): Materials, Photophysics, and Device Physics

Ryo Nagata, Kenichi Goushi, Hajime Nakanotani, and Chihaya Adachi

Abstract Currently, organic light-emitting diodes (OLEDs) have reached the stage of commercialization, and there have been intense efforts to use them in various applications from small- and medium-sized mobile devices to illumination equipment and large TV screens. In particular, room-temperature phosphorescent materials have become core OLED materials as alternatives to conventionally used fluorescent materials because devices made with phosphorescent materials exhibit excellent light-emitting performance with internal electroluminescence efficiency (η_{int}) of nearly 100%. However, phosphorescent materials have several intrinsic problems, such as their structure being limited to organic metal compounds containing rare metals, for example, Ir, Pt, Au, and Os, and difficulty in realizing stable blue light emission. Therefore, the development of new materials has been anticipated. In this chapter, first, we shortly review the progress of OLED materials and device architectures mainly based on fluorescence (the first generation) and phosphorescence (the second generation) emitters. Then, we mention the third-generation OLED using a new light-emitting mechanism called thermally activated delayed fluorescence (TADF). Recently, highly efficient TADF, which had been difficult to realize with conventional molecular design, has been achieved by very sophisticated molecular structures, indicating unlimited freedom of molecular design in carbon-based materials. This has led to the realization of ultimate OLEDs that are made of common organic compounds without precious metals and can convert electricity to light at nearly $\eta_{int} = 100\%$. Further, we mention the recent progress of NIR-OLEDs.

Keywords OLED · TADF · Delayed fluorescence · RISC · Exciton · Annihilation

R. Nagata · K. Goushi · H. Nakanotani · C. Adachi (✉)
Center for Organic Photonics and Electronics Research (OPERA), Kyushu University, Fukuoka, Japan
e-mail: adachi@opera.kyushu-u.ac.jp

2.1 Introduction

Materials can be classified into three groups in terms of electrical conductivity: insulators, semiconductors, and metals. In general, organic molecules composed of carbon skeletons act as insulators, as exemplified by plastics. However, a very different property can be obtained by forming an ultrathin film of such insulating organic molecules with a thickness of approximately 0.1 μm. When two electrodes are equipped to both sides of the organic thin film and a voltage of approximately 10 V is applied, electrons and holes are injected from the anode and cathode into the film, respectively, by overcoming the energy barriers at the interfaces. These injected carriers hop toward the opposite electrode, following the electric potential gradient. When the electron and hole recombine at a molecule, an excited state, namely, a molecular exciton, is induced. Then, photons are emitted when the excited state returns to the ground state. This successive process is called as organic electroluminescence (EL). Based on the emission mechanism similar to inorganic light-emitting diodes (LED), then organic EL is widely called as organic LED (OLED). Because of the ultrathin film architecture, a high electric field of over 10^6 V/cm can be applied to the organic thin film. Under such an extreme condition, carriers are injected from the electrodes into the organic thin film and become capable of easily hopping between molecules. Thus, unprecedented high current injection and transport can be realized in this extreme case, while it behaves as a complete insulator under the low electrical field of $<10^3$ V/cm.

In 1950, Japanese scientists, H. Inokuchi and H. Akamatsu, who demonstrated the doping of donors and acceptors into organic solids based on a similar idea in inorganic semiconductors, discovered the first electrical conduction and established the category of organic semiconductors (Fig. 2.1) [1]. In 1965, a clear EL was confirmed from an anthracene single crystal by W. Helfrich and W. G. Schneider [2]. They also carried out research on liquid crystals in those days, when it was the dawn of liquid crystals. In fact, they also invented twisted nematic (TN) liquid crystals. At that time, scientists discussed whether OLEDs or liquid crystals were more suitable for display devices, as can be seen in the literature of those days. Eventually, liquid crystals were focused on as a research target for practical display devices, and the target of research on OLEDs shifted from single crystals to ultrathin films for realizing low-voltage operation. Since the OLED had used single crystals with a thickness of a few millimeters, it required a few thousand voltages, and it was supposed to be far from commercialization. After almost 50 years' development focusing on novel organic molecules and device architectures, OLEDs have now reached the stage of commercialization, albeit 20 years behind the commercialization of liquid crystals. In particular, the research and development of OLEDs have rapidly accelerated since 1990. Their practical application has leaned toward small display devices, such as mobile phones and MP3 players, and flat panel TV screens since 2000. Thus, it has been firmly confirmed that through sufficient π-electron orbitals of adjacent molecules, electric charges can transport between the molecules under a high electric field; consequently, thin films of organic molecules can be used

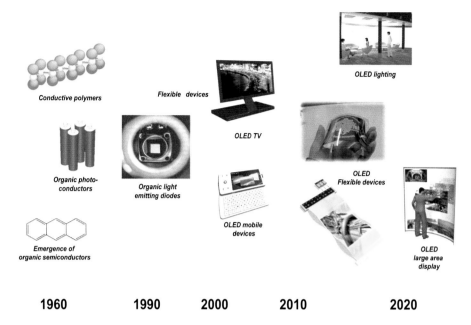

Fig. 2.1 History of research and development of organic semiconductors. The concept of organic semiconductors was established in the 1950s and 1960s. The research fields of conductive polymers and OPCs were established in the 1970s and 1980s. The full-scale research and development of organic electronic devices actively using a current density on the order of mA/cm^2 started from the 1980s. Afterward, the research and development of OLEDs, OSCs, organic transistors, and organic memories was carried out extensively, from basic research to practical application. Organic semiconductors are expected to be applied to bioelectronics in the future

as semiconductor thin films. Such a unique organic semiconductor behavior has been established through the development of OLEDs and has led to the emergence of novel semiconductors.

The organic photoconductor (OPC) unit in the xerographic process was the first commercialized electronic device using organic semiconductors and is the heart of copiers and laser printers used daily in homes and offices. When the charged layer formed on the surface of an OPC is irradiated with light, current flows through the organic semiconductor layer to form a latent image. OPCs have now become a main field of industry as a result of the dramatic spread of laser printers and copiers. In fact, the organic molecules currently used in OLEDs are an extension of the materials developed for OPCs. Also, we note that the progress of conducting polymers significantly influenced OLED R&D from both materials and device design. Since the early 1990s, various molecular skeletons for the carrier transport function of both holes and electrons were designed and synthesized on the basis of the molecular design for OPCs. For emitters in OLEDs, a wide variety of fluorescent molecules and the optimum device architectures were developed in the 1990s, and phosphorescent molecules and the device architectures have been in development since around 2000. In particular, we stress that electron transport materials are a key point in the development of OLED materials, since almost only hole transport materials had been used and few-electron transport materials had been identified in OPCs. Thus, the molecular design for organic materials that conduct electrons has been established through the research on OLEDs [3]. On the basis of research achievements with OLEDs, a wide range of research and development of next-generation organic devices, such as organic solar cells (OSCs), organic transistors, organic memories, and organic semiconductor lasers, is now ongoing, and these technologies will be connected to future bioelectronics.

We note that in the history of research on the abovementioned organic optoelectronics, OLEDs were the first devices based on organic thin films capable of being operated at current densities as low as the mA/cm^2 order and are considered to be the core organic optoelectronic devices realized by utilizing organic molecules as active semiconductors. Since 2000, organic electronics has become not only an independent academic field but also a newly established industry and is beginning to gain market value. The new organic semiconductor materials, device physics, and device engineering developed in relation to OLEDs have been applied to the creation of new next-generation devices. Thus, a new industrial field of electronics has evolved.

2.2 Basis of Organic Light-Emitting Diodes (OLEDs)

In this section, recent progress of OLED device architectures, organic fluorescence and phosphorescence materials, roll-off characteristics of η_{ext} in OLEDs, white OLED, and solution processing are introduced.

2.2.1 Progress of Device Structures

In OLEDs, excitons can be formed by the recombination of holes and electrons. The excitons lose their energy through the radiative decay process in an emitting layer (EML). In an EL process, recombined holes and electrons generate excitons with four different spin combinations of one singlet (antiparallel spins) and three triplets (parallel spins). Therefore, statistically, 25% of singlet excitons and 75% of triplets excitons are formed, resulting in the different radiative decay processes of fluorescence and phosphorescence, respectively. The fundamental structure of OLEDs with carrier flow is shown in Fig. 2.2. Holes are injected into the hole conduction level (E_v), and electrons are injected into the electron conduction level (E_e) from the anode and cathode layers. We call this phenomenon carrier injection. Successively, these carriers are transported through a hole-transport layer (HTL) and an electron-transport layer (ETL), respectively. In addition, on the anode side, a hole-injection layer (HIL) having a small energy gap to the work function of an anode layer and an HTL, which facilitates the injection of holes, has been widely introduced. In the

Fig. 2.2 Operating process in OLEDs EBL electron-blocking layer, EIL electron-injection layer, EML emitting layer, ETL electron-transport layer, HBL hole-blocking layer, HIL hole-injection layer, HTL hole-transport layer

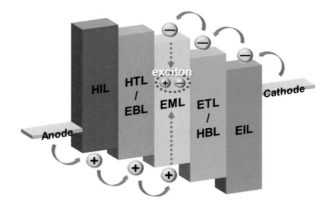

same manner, on the cathode side, an electron-injection layer (EIL) having an intermediate E_e between an ETL and a cathode layer has been introduced for effective electron injection and transport. This multilayer structure of OLEDs can improve carrier injection and transport efficiencies, resulting in the enhancement of the recombination of holes and electrons in an EML with the decrease of driving voltage. Further, electron- and hole-blocking layers (EBL and HBL) are also widely introduced for the charge balance of holes and electrons and the confinement of excitons, preventing the leakage of charge carriers and excitons from an EML to adjacent layers. Thus, present high-performance OLEDs are composed of multilayer structures.

The very first report on EL can be traced to Bernanaonose's observation of EL from organic dye-containing polymer thin films when applying a high alternating current (AC) voltage in 1950. Successively in 1965, emission of blue EL was observed from an anthracene single crystal by applying a rather high voltage over 1000 V [4]. It should be noted that the recombination of holes and electrons was found to result in the direct generation of both singlet and triplet excitons [5]. In the 1980s, a wide variety of thin film device architectures had been examined, and OLEDs based on a thin film of anthracene fabricated by vacuum deposition were shown to emit light at a low voltage of 12 V [6]. Further, advanced studies on multilayer OLEDs were reported by C. W. Tang et al. in 1987 [7]. The OLED using tris(8-quinolinolato)aluminum(III) (Alq$_3$) and aromatic amines as an ETL (EML) and an HTL, respectively, and Mg:Ag as a cathode, exhibited a high luminance of 1000 cd A^{-1} at 10 V and an external quantum efficiency (η_{ext}) of nearly 1% (Fig. 2.3). In the successive work, by doping of dicyanomethylene (DCM) and coumarin in an EML, the emission color was finely controlled with more than two times increase of η_{ext} [8]. At the same time, the concept of ETL showing the confinement of charge carriers and molecular excitons was confirmed by using novel electron transport materials.

In addition, not only small-molecular materials but also polymer materials are being applied to thin-film OLEDs. In the early study using poly(vinylcarbazole) (PVCz) as a host material, the basic operation of EL was confirmed [9]. In the 1990s,

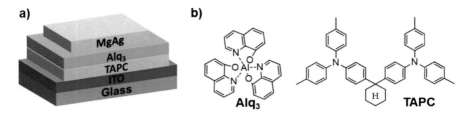

Fig. 2.3 (a) Structure of double-layer OLED reported by C. W. Tang et al. (b) molecular structures of tris(8-quinolinolato)aluminum(III) (Alq$_3$) as emitter and 4,4'-cyclohexylidenebis[N,N-bis(4-methylphenyl)benzenamine] (TAPC) as hole-transport layer

polymer OLEDs based on poly(p-phenylenevinylene) (PPV) as an EML were reported by J. H. Brruoughes [10]. After that, various studies on luminescent polymer materials including polyphenylenes and polythiophenes were actively developed.

In 1997, the first commercial application of OLEDs for a car radio system was realized. Nowadays, OLEDs have been widely used in mobile phones, tablet computers, lighting, and television. Since the first reports of OLEDs, a wide variety of organic semiconducting and luminescence materials, probably over a hundred thousand, have been designed and synthesized.

2.2.2 Luminescence Mechanisms of Organic Molecules and Solid Films

The luminescence phenomena in organic materials can be explained by the following photophysical interpretation when the materials are excited by a light source. The absorption of a photon in an organic material causes an electron to be excited from its ground state (S_0) to a singlet excited state (S_1). After photoexcitation, the excited electron losses its energy through photoluminescence (PL), intermolecular energy transfer, intramolecular energy transfer, isomerization, or dissociation (Fig. 2.4) [11].

Organic luminescent materials can be characterized by absolute PL quantum efficiency (Φ_{PL}) which is defined as photons emitted per photons absorbed by molecules, transient lifetime (τ), and emission spectrum. The presence of the nonradiative decay process decreases both Φ_{PL} and τ. The luminescence can be more specifically described as fluorescence or phosphorescence. These two emission processes can be distinguished by τ. While τ for fluorescence ranges from 10^{-9} s to 10^{-8} s, τ for phosphorescence is much longer and ranges from 10^{-6} s to 10^2 s (Table 2.1). In the case of fluorescence, excited electrons relax directly from an S_1 to an S_0 to produce light emission Eq. (2.1). On the other hand, phosphorescence occurs from a triplet excited state (T_1), which is created by intersystem crossing (ISC) from

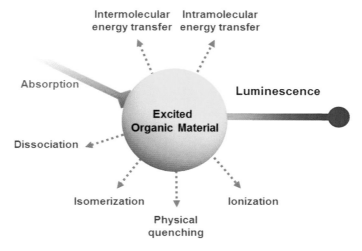

Fig. 2.4 Relaxation routes of optically excited organic molecules

an S_1 (Eq. (2.2)). Since phosphorescence is a spin-forbidden process, the lifetime is usually longer than that of fluorescence.

$$\text{(Fluorescence)} \quad S_0 + h\nu_{ex} \rightarrow S_1 \rightarrow S_0 + h\nu_{em} \quad (2.1)$$

$$\text{(Phosphorescence)} \quad S_0 + h\nu_{ex} \rightarrow S_1 \rightarrow T_1 \rightarrow S_0 + h\nu_{em} \quad (2.2)$$

($h\nu_{ex}$: excitation wavelength, $h\nu_{em}$: emission wavelength)

On the other hand, when luminescence is produced by electrical excitation, i.e., EL, the emission can again result from both fluorescence and phosphorescence (Fig. 2.5). Fluorescent OLEDs utilize singlet excitons for emission, while phosphorescent OLEDs utilize triplet excitons for emission. Due to the branching ratio of singlet and triplet excitons under electrical excitation [2], the production efficiency of singlet excitons is limited to the low value of 25%. In contrast, phosphorescent OLEDs can utilize both singlet and triplet excitons for phosphorescence emission by taking advantage of nearly 100% ISC in metal complexes. Therefore, nearly 100% of electro-generated excitons can be harvested for EL. Here, η_{ext} of OLEDs is given by the following equation:

$$\eta_{ext} = \eta_{int} \times \eta_{out} = \gamma \times \eta_{ST} \times \Phi_{PL} \times \eta_{out}, \quad (2.3)$$

where γ is the charge-balance factor, η_{ST} is exciton production efficiency, η_{int} is internal quantum efficiency, and η_{out} is light out-coupling efficiency (Fig. 2.6). To maximize η_{ext}, all of these factors should be nearly 100%. A high γ can be attained by the construction of appropriate multilayer structures [8, 12]. High Φ_{PL} can be achieved by using emissive materials that have been developed by suppressing nonradiative recombination, and these materials are often doped into a wide energy

Table 2.1 Fundamental photoluminescence properties of TPA-PZTCN

Sample type	Absorption (nm)	PL (nm)	PLQY prompt/delay (%)	τ_p (ns)[a]	τ_d (μs)[b]	k_r (10^7 s^{-1})	k_{ISC} (10^8 s^{-1})	k_{RISC} (10^5 s^{-1})	k_{nr}^T (10^4 s^{-1})
10^{-5} M in toluene	550	674	53.7/24.0	–	–	–	–	–	–
1 wt%-doped in mCBP	–	672	32.5/45.6	9.8	19.2	3.3	0.7	1.1	1.7
10 wt%-doped in mCBP	–	729	19.1/21.7	7.3	18.6	2.6	1.1	0.8	3.9

[a] Prompt fluorescence lifetime
[b] Delayed fluorescence lifetime

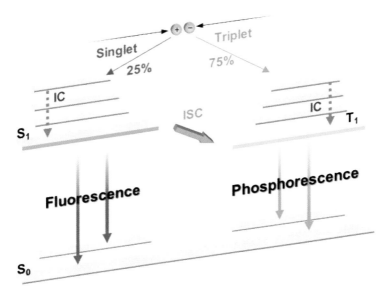

Fig. 2.5 Jablonski diagram of electronic transitions after recombination of hole and electron for fluorescence ($S_1 - S_0$), and phosphorescence ($T_1 - S_0$)

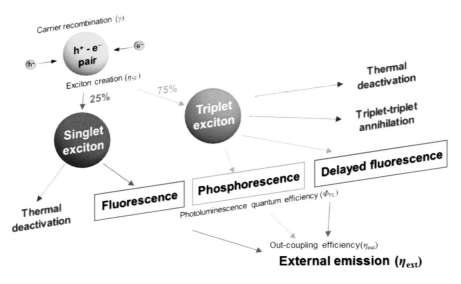

Fig. 2.6 Schematic representation of OLEDs process: charge carrier recombination, production of molecular excitons, internal emissions of fluorescence, phosphorescence, and delayed fluorescence, and external emission

gap host layer to minimize the concentration quenching of excitons that usually occurs at high concentrations. In the case of fluorescence-based OLEDs, the maximum η_{ext} is limited to 5% by assuming an η_{out} of 20%, which results from optical reflections and losses in organic layers [13]. Meanwhile, phosphorescent OLEDs can utilize both singlet and triplet excitons for emission, allowing η_{ST} and η_{ext} both to reach 20% [14]. However, we note that rather long τ of the phosphorescence process leads to strong nonradiative exciton annihilations, leading to roll-off in η_{ext} at high current density [15].

In OLEDs, it has been well established that the use of a guest-host (doping) system significantly contributes to enhancing η_{ext}. Since a neat film of emitter molecules most cases show rather strong concentration quenching, most emissive materials are doped in a host matrix at very low diluted concentrations, i.e., typically 1–2%. In a guest-host system, there are two kinds of energy transfer mechanisms from a host (exciton donor) to a guest (exciton acceptor): Förster and Dexter energy transfers (Fig. 2.7) [16, 17]. In the case of OLEDs, both Förster and Dexter processes contribute to energy transfer between the host and guest molecules because both singlet and triplet excitons are formed in host or guest molecules under electrical excitation.

Förster energy transfer is a long-range (up to ~10 nm) dipole-dipole coupling interaction between host and guest molecules [18]. Förster energy transfer is only allowed between a host singlet to a guest singlet state because the transitions between them having the same spin multiplicity are allowed, while the transitions between a singlet and a triplet states having different spin multiplicity are forbidden. The rate constant of Förster energy transfer, k_{FET}, for a guest-host system is given by the following equation:

$$k_{\text{FET}} = k_H \left(\frac{R_0}{R_{\text{HG}}}\right)^6, \qquad (2.4)$$

where k_H is the radiative decay rate of the host, R_{HG} is the separation between the host and guest molecules, and R_0 is the Förster transfer radius. R_0 can be calculated from the following equation:

$$R_0^6 = \frac{9000\kappa^2 \ln 10 \varphi_H}{128\pi^5 n^4 N_A} \int_0^\infty F_H(\lambda)\varepsilon_G(\lambda)d\lambda, \qquad (2.5)$$

where κ^2 is an orientation factor, φ_H is Φ_{PL} of the host, n is refractive intensity of the medium, N_A is the Avogadro constant, $F_H(\lambda)$ is normalized host PL intensity, $\varepsilon_G(\lambda)$ is molar absorption coefficient of the guest, and λ is wavelength. From these equations, the rate constant of Förster energy transfer can be determined by the fluorescence spectrum of the host and the absorption spectrum of the guest. To increase the Förster rate constant, a large overlap of host emission and guest absorption spectra, a large $\varepsilon_G(\lambda)$, and a small distance between host and guest molecules are required.

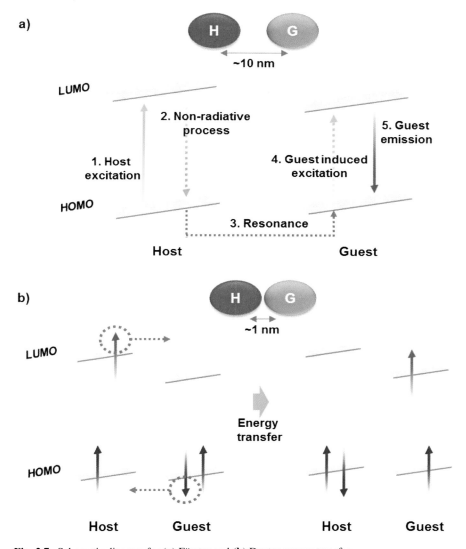

Fig. 2.7 Schematic diagram for (**a**) Förster and (**b**) Dexter energy transfer

Dexter energy transfer is a short-range (up to ~1 nm) intermolecular electron exchange process from host to guest. Dexter energy transfer is allowed between a host singlet and a guest singlet states, and a host triplet and a guest triplet states. The rate constant of Dexter energy transfer, k_{DEX}, is given by the following equation:

$$k_{\text{DEX}} = \left(\frac{2\pi}{h}\right)\kappa^2 \exp\left(\frac{-2R_{\text{HG}}}{L}\right) \int_0^\infty F_H(\lambda)\varepsilon_G(\lambda)d\lambda, \qquad (2.6)$$

where L is the sum of the Van der Waals radius for the two molecules. The rate constant of Dexter energy transfer strongly depends on the distance between the host and guest molecules and is also affected by the overlap between $F_H(\lambda)$ and $\varepsilon_G(\lambda)$. Therefore, not only PL decay processes in a single molecule but also the energy transfer processes in emitter layers should be considering to maximize OLED performance.

2.2.3 Efficiency Roll-Off in OLEDs

Roll-off of η_{ext}, an efficiency decrease with an increase of current density, has been widely observed in OLEDs. To realize high-performance OLEDs in displays and lighting sources, high brightness with low roll-off must be achieved. Roll-off characteristics can be characterized by using the critical current density ($J_{50\%}$) at which the η_{ext} is dropped to 50% of its maximum value [15].

Figure 2.8 shows exciton annihilation processes leading to roll-off. For instance, polaron/exciton and exciton/excitons annihilation, field-induced quenching, charge carrier imbalance, Joule heating, and degradation can affect η_{ext} [19]. The most relevant roll-off process in phosphorescent OLEDs is triplet-triplet annihilation (TTA) because of the long τ of phosphorescence emission and the high density of triplet excitons [20]. Reducing the molecular distance between dopant molecules in a guest-host system can further decrease TTA [21]. The process of TTA can be characterized by the following equations:

$$4\left(^3M^* + {}^3M^*\right) \to {}^1M^* + {}^3M^* + 4M, \qquad (2.7)$$

$$\frac{d\left[^3M^*\right]}{dt} = -\frac{[^3M^*]}{\tau} - \frac{k_q}{2}\left[^3M^*\right]^2 + \frac{J}{qd}, \qquad (2.8)$$

$$\frac{\eta_{50\%}}{\eta_0} = \frac{J_{50\%}}{4J}\left[\sqrt{1 + 8\frac{J}{J_{50\%}}} - 1\right], \qquad (2.9)$$

where $^3M^*$ and M refer to molecules in the triplet excited and ground states, respectively, $[^3M^*]$ is triplet exciton density, d is the width of the recombination zone, k_q is TTA rate constant, q is fundamental electric charge, and J is current density and η_0 is maximum η_{ext}.

Among these processes, singlet–singlet annihilation (SSA) has been observed in anthracene crystal and affects the singlet excitons density and lifetime [22]. Singlet–

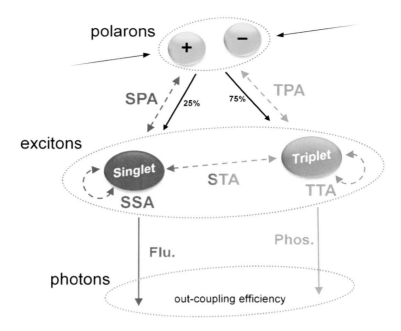

Fig. 2.8 Schematic illustration of roll-off process in OLEDs SPA singlet–polaron annihilation, SSA singlet–singlet annihilation, STA singlet–triplet annihilation, TPA triplet–polaron annihilation, TTA triplet–triplet annihilation

triplet annihilation (STA) usually occurs at high concentrations of guest doping and high current densities over 100 mA cm^{-2} [23]. Polarons can also interact with both singlet and triplet excitons, leading to exciton–polaron annihilation. Triplet–polaron annihilation occurs in both fluorescent and phosphorescent OLEDs because of the long excited state lifetime of triplets. Holes and electrons generally lead to stronger triplet–polaron annihilation. Further, singlet–polaron annihilation is of great concern in organic lasers, since very high current densities are applied [24]. Further, electrical filed-induced quenching, charge carrier imbalance, Joule heating, and degradation can affect OLED efficiency [19].

To suppress the roll-off and achieve high brightness in OLEDs, decreasing the local exciton density is crucial. Decreasing the luminescence exciton lifetime [24], reducing molecular aggregation [25], and broadening the recombination zone [26] are all routes for reducing the exciton density. Moreover, reducing the spectral overlap between emission and polaron absorption can decrease polaron annihilation [24].

2.2.4 White OLEDs

White OLEDs (WOLEDs) are receiving significant attention and are being developed as the next generation of solid-state light sources. However, their low brightness, device stability, and expensive manufacturing cost currently limit their application. To overcome these disadvantages, many researchers have been studying the development of effective emitter molecules and optimized device structures.

For white emission, the emission spectrum should cover large visible regions from 400 nm to 800 nm. Typically, emissive molecules show a narrow emission with a width of 50–100 nm that can be recognized as colors such as R, G, and B. To realize white emission, different colors of emission should be mixed to cover a broad emission spectrum, for example, by combining red, green, and blue or blue and orange emissions. The combination of each emission can be controlled by various device structures for high efficiency in white emission (Fig. 2.9) [27, 28].

The simplest method to obtain white emission is by blending different color luminophores in a single layer. The first single-layer WOLEDs were fabricated by blending of PVCz with orange, green, and blue dyes using solution-processing

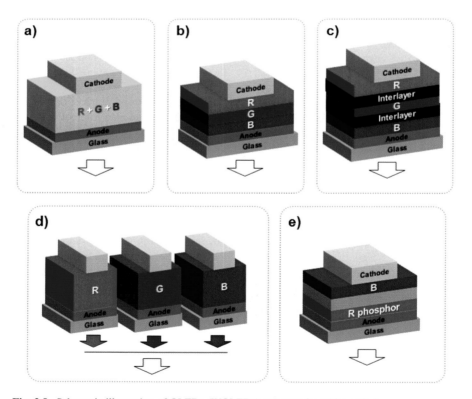

Fig. 2.9 Schematic illustration of OLEDs (WOLEDs) structure for white emission

[29]. After that, WOLEDs by vacuum depositing a blend of three kinds of phosphorescent luminophores have been reported [30]. Another method for fabricating WOLEDs is using a multilayer device structure [31, 32]. In this device structure, excitons can be created in an EML by direct recombination of holes and electrons or by energy transfer from a neighboring EML. This exhibits the best efficiency, spectral coverage, and device lifetime compared to other WOLED architectures. While the emission spectrum often changes depending on the applied driving voltage in the multilayer devices, this can be solved by controlling the recombination zone and width of the EMLs [33].

The EMLs can be further isolated to reduce their interaction and suppress brightness-dependent color shifts even at very high operating voltages [34]. In the extreme case of isolating the EMLs, white emission can be generated from monochrome pixels consisting of OLEDs (or striped WOLEDs). Striped WOLEDs generate white emission using independent red-, green-, and blue-emitting devices. However, the manufacturing process of these types of devices is rather complicated, resulting in high costs.

2.2.5 Solution-Processed OLEDs

The main methods for fabricating multilayer OLEDs are thermal evaporation under high vacuum [4] and solution processing [7]. Thermal evaporation is the most widely used method for the fabrication of efficient and stable OLEDs, while it has disadvantages of complexity and high production costs [29]. Solution-processing methods such as spin-coating [35, 36], ink-jet printing [37], and spray process [38] have attracted great attention for the fabrication of flexible and large-area OLEDs because of the simplicity and low cost.

Many phosphorescent polymers and small molecules have been studied as emitters for highly efficient solution-processed OLEDs. After the first reports of solution-processed OLEDs using the polymer PPV [7], phosphorescent conjugated polymers having high solubility and good morphology were developed, aimed at highly efficient polymer-based OLEDs [39–41]. However, the long conjugation length of the polymers led to a low T_1 energy level and decreased device efficiency [42, 43]. To solve these problems, phosphorescent emitters have been dispersed into nonconjugated polymers, resulting in improved efficiency [44, 45].

Small molecules are mainly used in vacuum thermal evaporation and have the advantages of having high Φ_{PL} and being easily purified. For the application of small molecules in solution-processed OLEDs, the problems of low solubility, poor morphology, and strong crystallization should be solved. Newly designed small molecules incorporating alkyl or alkoxy and flexible groups have been reported for use with solution processes [46, 47]. Rigid molecules such as spirofluorene have been introduced in the core unit to increase the glass transition temperature (T_g), which suppresses crystallization [48]. Using small molecules, highly efficient solution-processed OLEDs based on Ir(III) and Pt(II) complexes have been reported in many studies [49, 50].

2.2.6 Some Critical Issues in OLEDs

OLEDs have many excellent basic characteristics, such as self-luminescence, surface luminescence, high flexibility, high resolution, and high EL efficiency. Through R&D over the past 25 years, device characteristics superior to those of liquid crystals have been obtained at some points. However, OLEDs still have several problems to be solved to improve their performance: high cost due to the use of noble metals such as Ir and Pt as emitting materials, difficulty in achieving stable blue light emission, and low device stability and high-cost productivity because of the use of organic ultrathin films with a thickness of approximately 100 nm. Thus, the potential excellent device characteristics of organic semiconductors have not yet been fully obtained. In addition, improving the efficiency of extracting light from thin films by introducing photonic crystals and light scattering techniques has been widely discussed; however, no decisive solutions have been obtained [51]. Moreover, the high definition required for medium- and small-sized OLED displays is the greatest challenge. To realize this, RGB coloring at a 10 μm order of accuracy is required, and an innovative process must be developed to realize low-cost mass production. Currently, the following four points are necessary to realize next-generation OLEDs: (1) realization of highly efficient luminescence without using phosphorescent materials, (2) utilization of the intrinsic optical and electronic anisotropies of molecules, (3) development of an RGB coloring process for achieving high definition, (4) development of thick film devices having the organic layers thicker than 10 μm, and (5) creation of a low-cost fabrication process.

2.3 Thermally Activated Delayed Fluorescence for OLEDs

2.3.1 Principle of TADF in OLEDs

At present, OLEDs are expected to be practically applied to flat panel displays and illumination sources because they have unique characteristics, such as high EL efficiency and flexibility, and can be processed at low temperatures. Thus far, various fluorescent and phosphorescent materials have been developed to improve the EL efficiency of OLEDs [20, 52, 53]. As a result, highly durable and practically applicable OLEDs using fluorescent materials have been realized. However, the internal quantum efficiency (η_{int}), which is defined as the ratio of the number of photons that can be extracted for EL to the injected current (i.e., the number of injected carriers), was only 25% because of the limit imposed by the statistics of the electron spin state under electrical excitation [54, 55]. In contrast, OLEDs using phosphorescent materials based on luminescence from the triplet state can achieve $\eta_{int} = 100\%$ [14]. However, the design of molecules used in such OLEDs is greatly limited because the heavy atom effect (spin-orbital interaction) must be induced, for example, by using rare metals, to realize a highly efficient radiative transition from the triplet excited state to the ground state.

Previously, researchers reported several methods of achieving an η_{int} higher than the theoretical limit (25%) for OLEDs without using Ir complexes. Specifically, a method of developing phosphorescent materials that are free of rare metals [56, 57] and a method of generating a TTA [58] have been reported. For OLEDs using phosphorescent materials, TTA leads to the deactivation of triplet excitons and hence decreases the EL quantum efficiency; however, for OLEDs using fluorescent materials, the concentration of singlet excitons can be increased by TTA. Delayed fluorescence attributed to TTA was already confirmed from the EL phenomenon observed in an anthracene single crystal in the 1960s, indicating that triplet excitons actively move around and interact in an organic solid thin film. Hence, scientists have examined the application of the generation of single excitons attributed to TTA for improving EL efficiency [58]. In actuality, the η_{ext} has been reported to exceed its theoretical limit (5%) in some OLEDs using fluorescent materials, demonstrating the applicability of TTA as seen in the transient EL characteristics. However, the efficiency of generating single excitons by TTA is only 37.5% at maximum, and a novel luminescence mechanism is required.

Recently, our research group has proposed a method of achieving an η_{int} of 100% through up-conversion from the triplet excited state, which is generated at a probability of 75% upon electrical excitation, to the singlet excited state (Fig. 2.10) [59]. This method can be used to cause the triplet excited state to contribute to

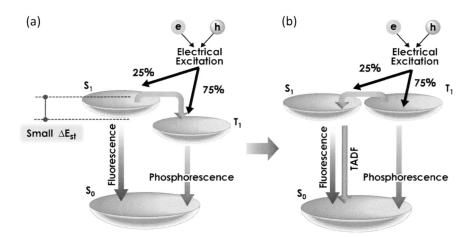

Fig. 2.10 Generation process of singlet and triplet excitons upon electrical excitation. (**a**) Conventional fluorescent and phosphorescent materials. (**b**) For TADF materials, the S_1 and T_1 levels are close to each other, and reverse energy transfer occurs with a high efficiency, enabling highly efficient EL from the S_1 level

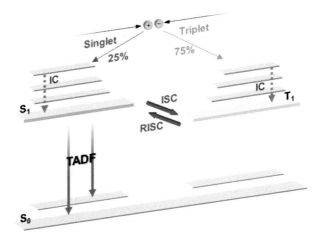

Fig. 2.11 Jablonski diagram of electronic transitions after recombination of holes and electrons for TADF

luminescence without using rare metals. The method involves the up-conversion of triplet excitons to singlet excitons using thermal energy and has long been known as E-type delayed fluorescence in the field of photochemistry. Well-known materials that exhibit thermally activated delayed fluorescence (TADF) are eosin [60], fullerene [61], and porphyrin [62] derivatives. Although the TADF process was considered to show a low power-conversion efficiency because it is generally an endothermic reaction, recent studies have revealed that highly efficient delayed fluorescence can be achieved by optimizing the molecular design.

TADF characteristics depend on the probability of reverse intersystem crossing (RISC) from the triplet to singlet excited states (Fig. 2.11). The EL efficiency increases as the energy difference between the singlet and triplet excited states (ΔE_{ST}) decreases. For example, condensed polycyclic aromatic compounds, such as anthracene, exhibit very intense fluorescence but are not expected to show efficient TADF because their ΔE_{ST} exceeds 1 eV. In contrast, ketone-based materials, such as benzophenone derivatives, have a relatively small ΔE_{ST} (= 0.1–0.2 eV) but do not exhibit intense luminescence at room temperature and only exhibit phosphorescence at low temperatures [63].

Here, the dominant factor of ΔE_{ST} is examined from the viewpoint of quantum chemistry. To obtain TADF with a high conversion efficiency, a small energy gap (ΔE_{ST}) between S_1 and T_1 levels is necessary for the luminescent materials, which can be realized by having a small orbital overlap between the highest occupied molecular orbital (HOMO) and the lowest unoccupied molecular orbital (LUMO) of emitter molecules [64]. The ΔE_{ST} of luminescent molecules can be described by the following equations:

$$E_S = (E_U - E_L) + K_{LU} \tag{2.10}$$

$$E_T = (E_U - E_L) - K_{LU} \qquad (2.11)$$

$$\Delta E_{ST} = E_S - E_T = 2K_{LU} \qquad (2.12)$$

where E_U is a ground level (the highest occupied molecular level, U-level), E_L is an excited level (the lowest unoccupied molecular level, L-level), and K_{LU} is an exchange energy. Furthermore, K_{LU} is given by the following equation [64]:

$$K_{LU} = \iint \varphi_U(1)\, \varphi_L(2) \frac{1}{r_{12}} \varphi_U(2) \varphi_L(1) d\tau_1 \tau_2 \qquad (2.13)$$

where φ_L and φ_U are the wave functions of the U- and L-level orbitals, respectively, and r_{12} is the distance between the electron (1) and is the gas constant. Then, the Berberan-Santos (2). As shown in Eq. (2.8), K_{LU} is determined by φ_L and φ_U and should generally decrease as their overlap decreases.

Evidence of this relationship can be found in several known materials. A small ΔE_{ST} can be obtained as a consequence of the orthogonal overlap between n and π^* orbitals (Fig. 2.12a). In contrast, π-π^* transitions normally possess large ΔE_{ST} because of the parallel overlap between π and π^* orbitals (Fig. 2.12b). Thus, small orbital overlapping between the HOMO and the LUMO should be necessary for small ΔE_{ST}. However, a high luminescence efficiency could not be attained in conventional materials with small ΔE_{ST} because n-π^* characteristics usually cause a small radiative decay rate (k_r). Therefore, the development of new luminescent materials having both small ΔE_{ST} and large k_r is required for realizing highly efficient TADF.

Here, a relatively large radiative rate constant (k_r) is required to obtain a high EL efficiency. In the case of the abovementioned aromatic compounds, however, a large overlap between the wave functions of the ground and excited states is required. Therefore, it is necessary to design molecules that exhibit highly efficient EL while maintaining a small ΔE_{ST}. Because large k_r and small ΔE_{ST} are conflicting, fine molecular design is required to realize them simultaneously. We carefully designed and created a novel molecule that has a small ΔE_{ST} and can exhibit highly efficient EL. Specifically, we designed and synthesized novel compounds that contain both electron-donating and electron-accepting substituents and successfully created a luminescent molecule showing an EL efficiency of nearly 100% while maintaining a very small ΔE_{ST} (<0.2 eV). This molecule allows the up-conversion from the triplet excited state generated upon electrical excitation to the singlet excited state, enabling the realization of highly efficient EL equivalent to that of phosphorescent devices from the singlet excited state. In particular, an ultimate level of external efficiency close to 20% was achieved in the green range (wavelength, 530 nm) for an OLED employing carbazolyl dicyanobenzene (CDCB) as an emitting layer [64]. In addition, we found that an intrinsically small ΔE_{ST} can be achieved in exciplexes, which are intermolecular complexes, and successfully obtained an η_{ext} higher than 10% by selecting the optimal materials.

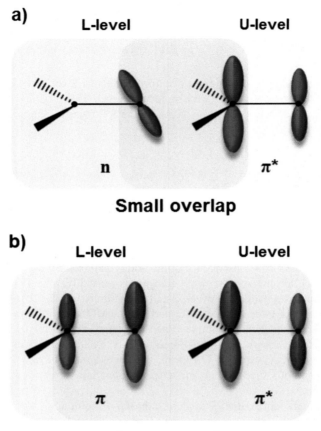

Fig. 2.12 n-π* orbital overlap (**a**) and π-π* orbital overlap (**b**) in benzophenone

In the following sections, we will introduce a method of up-conversion based on intramolecular charge transfer (ICT) that achieves a high η_{int} that exceeds its theoretical limit of 25% for traditional fluorescent OLEDs, new organic light-emitting materials suitable for this approach, and the organic EL characteristics of devices employing these new materials. I will also introduce an exciplex system, which is the excited state formed between donor and acceptor molecules, as an alternative method to improving η_{int} through up-conversion.

2.3.2 TADF Characteristics of Triazine Derivatives

In the development of high-efficiency TADF materials, it is necessary to design molecules with excellent EL characteristics while maintaining a small ΔE_{ST}. Namely, the key is the design of molecules that can simultaneously achieve both large k_r and small ΔE_{ST}. Here, the design of molecules that achieve small ΔE_{ST} is examined. In general, HOMO is distributed in the electron-donating units, whereas LUMO is distributed in the electron-accepting units. Therefore, a small ΔE_{ST} can be obtained by introducing electron-donating and electron-accepting groups inside the molecules to decrease the spatial overlap between the HOMO and LUMO. On the basis of this direction of molecular design, we designed and synthesized a novel molecule, 2,4-bis{3-(9H-carbazol-9-yl) -9H-carbazol-9-yl}-6-phenyl-1,3,5-triazine (CC2TA), in which the acceptor phenyltriazine unit is used as the central skeleton and donor bicarbazole units are bonded to both ends of the skeleton (Fig. 2.13) [65]. Molecular orbital calculations reveal that the HOMO and LUMO are locally distributed mainly in the bicarbazole and phenyltriazine units, respectively. As a result, the molecule successfully showed a very small ΔE_{ST} of 0.06 eV. In CC2TA, the introduction of bicarbazole with a Wurster structure (>N-aryl-N<) as a donor was the key to obtaining a small ΔE_{ST}. If a simple carbazole was bonded as a donor,

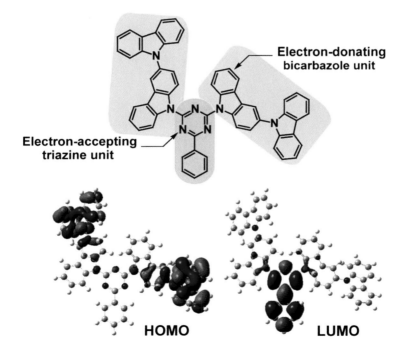

Fig. 2.13 Molecular structure and orbital of TADF material with triazine skeleton (CC2TA)

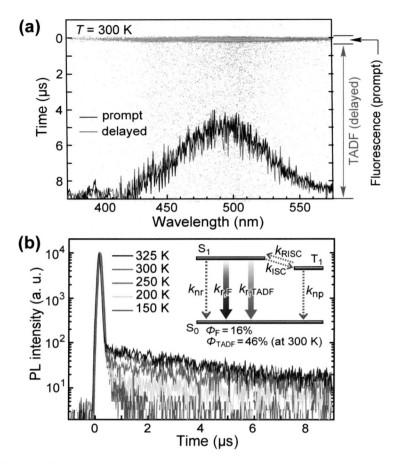

Fig. 2.14 (a) Time-resolved PL spectra of 6 wt%-CC2TA:DPEPO co-evaporated thin film and (b) temperature dependence of transient PL waveform

the HOMO and LUMO would overlap to some extent, resulting in an increased ΔE_{ST} of 0.35 eV.

To evaluate the TADF characteristics of CC2TA, the PL characteristics of a co-evaporated thin film, obtained by dispersing 6 wt% CC2TA into a bis [2-(diphenylphosphino)phenyl]ether oxide (DPEPO) host, are shown in Fig. 2.14. The triplet energy (E_T) of the DPEPO host is 3.1 eV, so triplets in CC2TA ($E_T = 2.85$ eV) can contribute to efficient luminescence without being transferred to the host. Figure 2.14a shows time-resolved PL spectra (streak images) of the CC2TA:DPEPO co-evaporated thin film. At room temperature, both prompt (lifetime, $\tau = 27$ ns) and delayed ($\tau = 22$ μs) PL components are observed in the same wavelength range. Figure 2.14b shows the temperature dependence of the transient PL waveform. The intensity of the delayed PL component markedly increases with increasing temperature. This result suggests that RISC from triplet to singlet excited

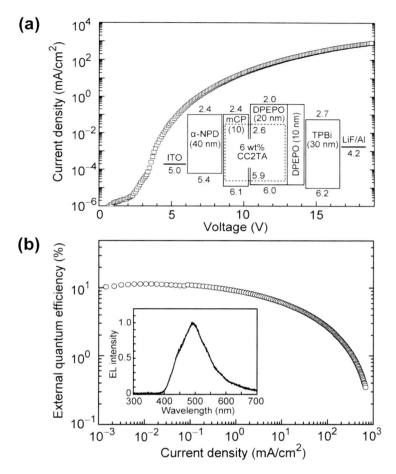

Fig. 2.15 (a) Device structure and relationship between current density and voltage and (b) dependence of external quantum efficiency (η_{ext}) on current density for TADF-OLED employing CC2TA as an emitting layer

states actively occurs in the temperature range around room temperature owing to thermal energy, resulting in efficient TADF. In contrast, the triplet excitons cannot overcome the energy barrier to the singlet excited state at a low temperature (150 K). Therefore, RISC scarcely occurs, and intense delayed fluorescence is not observed.

Next, the characteristics of an OLED employing the CC2TA:DPEPO co-evaporated thin film as an emitting layer are shown (Fig. 2.15). The maximum external quantum efficiency reaches $\Phi_{EL(ext)} = 11\%$. Here, the efficiency of EL involving TADF can be expressed by Eq. (2.3). Here, $\eta_{r,S}$ and $\eta_{r,T}$ are the probabilities of generation of the singlet and triplet excitons ($\eta_{r,S} = 25\%$, $\eta_{r,T} = 75\%$), Φ_F is the fluorescence quantum yield ($\Phi_F = 16\%$), Φ_{TADF} is the TADF quantum yield ($\Phi_{TADF} = 46\%$), and Φ_{ISC} is the probability of intersystem crossing ($\Phi_{ISC} = 84\%$).

From Eq. (2.1), the theoretical η_{int} is calculated to be 56%. Assuming the light-extracting efficiency to be 20%, the theoretical external quantum efficiency (η_{ext}) is 11%, which is in good agreement with the value obtained in our experimental device.

2.3.3 TADF Characteristics of Spiro Derivatives.

Spiro compounds are expected to exhibit TADF because donor and acceptor units can be introduced into their orthogonal π-conjugated system and the HOMO and LUMO can be spatially separated [66–68]. Figure 2.16 shows the molecular structure of a spirobifluorene derivative (Spiro-CN) that exhibits TADF. This molecule has two donor triarylamino groups and two acceptor cyano groups. The molecules between these groups are distorted by the spirobifluorene skeleton to form a twisted steric structure. As shown in the result of molecular orbital calculations in Fig. 2.16, the HOMO and LUMO are locally distributed in the triarylamino-based fluorene and cyano-based fluorene, respectively.

Figure 2.17a shows the PL spectrum of a Spiro-CN:m-CP co-evaporated thin film, in which 6 wt% Spiro-CN guest is dispersed in an m-CP host. The transient waveform consists of a prompt PL component with $\tau \approx 24$ ns and a delayed PL component with $\tau \approx 14$ μs. Figure 2.17b shows the temperature dependence of PL characteristics for the 6 wt%-Spiro-CN:m-CP co-evaporated thin film. The prompt fluorescence component shows no temperature dependence, whereas the PL intensity of the delayed fluorescence component clearly increases with temperature. This indicates that RISC from triplet to singlet excited states actively occurs with increasing temperature. The results of evaluating ΔE_{ST} using the Berberan-Santos equation on the basis of the abovementioned temperature dependence of PL characteristics are shown below. Here, it is assumed that Φ_{prompt} is the prompt fluorescence quantum yield, $\Phi_{delayed}$ is the delayed fluorescence quantum yield, Φ_T is the efficiency of generating triplet excitons, k_p is the radiative rate constant from the triplet excited state, k_{nr} is the nonradiative rate constant from the triplet excited state, k_{RISC} is the

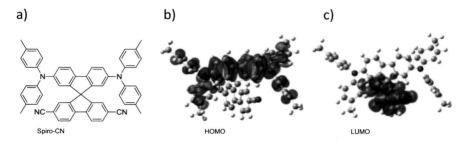

Fig. 2.16 (a) Molecular structure, (b) HOMO, and (c) LUMO of Spiro-CN

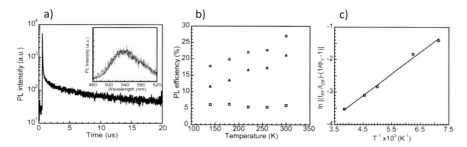

Fig. 2.17 (**a**) Transient PL spectrum of 6 wt%-Spiro-CN:*m*-CP co-evaporated thin film. The inset shows time-resolved PL spectra (red line, PL spectrum attributed to prompt fluorescence component; black line, PL spectrum attributed to delayed fluorescence component). (**b**) Temperature dependence of PL quantum yield (open circle, total; open triangle, delayed fluorescence component; open square, prompt fluorescence component). (**c**) Berberan-Santos plots of temperature dependence of PL intensity

rate constant for RISC, and R is the gas constant. Then, the Berberan-Santos equation is given by

$$\ln\left[\frac{\Phi_{\text{prompt}}}{\Phi_{\text{delayed}}}-\left(\frac{1}{\Phi_\tau}-1\right)\right] = \ln\left(\frac{k_p+k_{nr}}{k_{RISC}}\right) + \frac{\Delta E_{ST}}{RT} \qquad (2.14)$$

Figure 2.17c shows the Berberan-Santos plots of PL characteristics of the 6 wt%-Spiro-CN:*m*-CP co-evaporated thin film. ΔE_{ST} of Spiro-CN is estimated to be 0.057 eV from the slope of the straight line in the figure. This demonstrates that ΔE_{ST} of Spiro-CN is much smaller than that of conventional delayed fluorescent materials, such as C_{70} (0.26 eV) and tin(IV)fluoride-porphyrin (0.24 eV).

Next, the PL characteristics of an OLED employing the 6 wt%:Spiro-CN:*m*-CP co-evaporated thin film as an emitting layer were measured. A maximum η_{ext} of 4.4% was obtained for an OLED with a structure of indium tin oxide (ITO)/α-N, N′-Di(1-naphthyl)-N,N′-diphenylbenzidine (NPD)/6 wt%-Spiro-CN:*m*-CP/Bphen/MgAg/Ag. This value greatly exceeds the theoretical value of η_{ext} when a conventional fluorescent material with a fluorescence quantum yield of 27% is used as an emitting layer (i.e., $\eta_{\text{ext}} = 1.4\%$), meaning that the efficiency of generating excitons is increased by TADF.

2.3.4 TADF Characteristics of CDCB Derivatives

Although a small ΔE_{ST} can be achieved by spatially separating the HOMO and LUMO as mentioned above, this spatial separation generally decreases the transition dipole moment μ and hence decreases the EL quantum yield. To realize highly efficient TADF, a small ΔE_{ST} and a large μ must be simultaneously achieved while maintaining the appropriate level of overlap between the HOMO and LUMO.

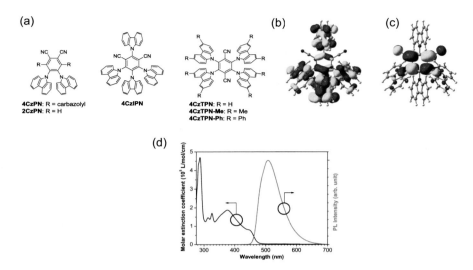

Fig. 2.18 (a) Molecular structure of CDCB derivatives and (b) HOMO, (c) LUMO, and (d) absorption and PL spectra of 4CzIPN

Figures 2.18a–c show the molecular structure, HOMO, and LUMO, respectively, of a CDCB derivative designed following the above-mentioned theory [64]. The HOMO and LUMO are locally distributed in the donor carbazolyl groups and the acceptor dicyanobenzene units, respectively. It is apparent that the HOMO and LUMO moderately overlap on the central benzene ring. This indicates that the CDCB derivative achieves both small ΔE_{ST} and large μ. Quantum chemical calculations also reveal that the change in the molecular structure among the S_0, S_1, and T_1 states is small in 4CzIPN. From this result, 4CzIPN is expected to achieve a high EL quantum yield as a result of the suppressed nonradiative deactivation process.

Figure 2.18d shows the absorption and PL spectra of 4CzIPN in a toluene solution. 4CzIPN has a wide PL band of approximately 507 nm attributed to ICT. The Stokes shift of 4CzIPN is smaller than that in conventional ICT-based luminescence. This means that the change in the molecular structure of 4CzIPN is small in electron excitation from the S_0 to S_1 states.

Next, the TADF characteristics of a 6 wt%-4CzIPN:CBP co-evaporated film will be discussed. Figure 2.19a shows the PL transient decay characteristics of this film at 100 K, 200 K, and 300 K. PL transient decay characteristics comprising two types of component, a nanosecond-order short-lifetime component and a microsecond-order long-lifetime component, are observed at any temperature. Figure 2.19b shows time-resolved PL spectra of the short- and long-lifetime components at 300 K. Luminescence attributed to the long-lifetime component is delayed fluorescence because the PL spectrum from the long-lifetime component is in agreement with the fluorescence spectrum from the short-lifetime component. Figure 2.19c shows the temperature dependencies of the PL quantum yield of the prompt and delayed fluorescence components. The prompt fluorescence quantum yield slightly increases with

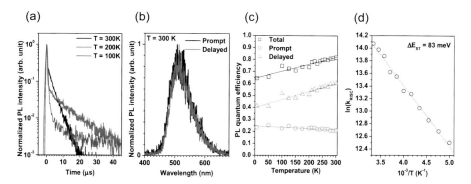

Fig. 2.19 (**a**) PL transient decay, (**b**) time-resolved PL spectra, (**c**) temperature dependence of PL quantum yield, and (**d**) Arrhenius plots of the rate constant for RISC for 6 wt% 4CzIPN:CBP co-evaporated film

decreasing temperature. This is due to the suppressed nonradiative deactivation process. In contrast, the delayed fluorescence quantum yield markedly decreases with decreasing temperature. This is due to the suppression of RISC from the T_1 to S_1 states with decreasing temperature.

The ΔE_{ST} of 4CzIPN can be estimated from the temperature dependence of the PL quantum yield observed above. Here, the rate constant for RISC from the T_1 to S_1 states (k_{RISC}) is expressed using the rate constant for prompt fluorescence (k_p), the rate constant for delayed fluorescence (k_d), the prompt fluorescence quantum yield (Φ_p), the delayed fluorescence quantum yield (Φ_d), and the rate constant for intersystem crossing (k_{ISC}) from the S_1 to T_1 states as follows.

$$k_{RISC} = \frac{k_p k_d}{k_{ISC}} \frac{\Phi_d}{\Phi_p}. \tag{2.15}$$

Since k_p and k_d can be determined from the transient PL curve and Φ_p and Φ_d can be measured, k_{RISC} can be evaluated using Eq. (2.3). k_{RISC} and ΔE_{ST} have a relationship that can be expressed by $k_{RISC} \sim A\exp(-\Delta E_{ST}/k_B T)$, where A is a constant, k_B is the Boltzmann constant, and T is the temperature. Therefore, ΔE_{ST} can be calculated from the temperature dependence of k_{RISC}. Figure 2.19d shows the Arrhenius plots of k_{RISC} between 200 K and 300 K. ΔE_{ST} is calculated to be 83 meV from the slope of the straight line in the figure. This demonstrates that the energy difference between the S_1 and T_1 states is small for 4CzIPN.

Figure 2.20 shows the characteristics of OLEDs employing CDCB derivatives as an emitting layer with the structure of ITO/α-NPD (35 nm)/6 wt%-CDCB:CBP (15 nm)/TPBi (65 nm)/LiF (0.8 nm)/Al (80 nm). The external quantum efficiencies of 4CzPN, 4CzIPN, and 4CzTPN are all very high, 17.8%, 19.3%, and 17.1%, respectively. This means that 4CzIPN, which shows a particularly high external quantum efficiency, has an internal quantum efficiency of nearly 100%.

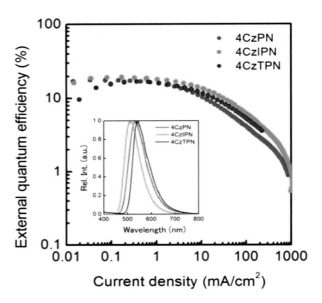

Fig. 2.20 η_{ext}-J (current density) characteristics of OLED employing CDCB derivative as an emitting layer. The inset shows EL spectra

2.3.5 TADF Characteristics of Exciplexes.

In the radiation transition of organic compounds, electrons generally transit from the LUMO to the HOMO within a single molecule. In the excited state formed within a single molecule, therefore, the HOMO and LUMO are confined in the molecule, causing the electron exchange integral to be large. This results in an increased ΔE_{ST}. We focused on the exciplex state, which is the excited state formed between electron-donating and electron-accepting molecules. In the radiation process of exciplexes, charges transit from the LUMO of the electron-accepting molecule to the HOMO of the electron-donating molecule. Therefore, the electron exchange integral of the exciplex is small because the HOMO and LUMO are spatially separated, resulting in a very small ΔE_{ST}. This increases the probability of up-conversion from the triplet to singlet excited states.

Here, the up-conversion from the triplet to singlet excited states is explained using the exciplex formed between electron-donating 4,4′,4″-tris[3-methylphenyl (phenyl)amino] triphenylamine (m-MTDATA) and electron-accepting 2-(biphenyl-4-yl)-5-(4-tert-butylphenyl)-1,3,4-oxadiazole (t-Bu-PBD) [69]. Figure 2.21a shows the PL spectrum of a 50 mol%-m- MTDATA:t-Bu-PBD co-evaporated film as well as the fluorescence and phosphorescence spectra of m-MTDATA and t-Bu-PBD single thin films. The PL intensity peak for the co-evaporated film appears at a wavelength of 540 nm, which is longer than that of the peaks for fluorescence from the m-MTDATA and t-Bu-PBD thin films. This is because an exciplex is formed between the molecules of m-MTDATA and t-Bu-PBD.

To achieve a high up-conversion probability, attention should be paid to the confinement of the triplet excited state of the exciplex resulting from the triplet

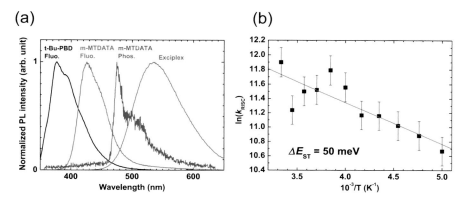

Fig. 2.21 (a) PL spectrum of 50 mol%-*m*-MTDATA:*t*-Bu-PBD co-evaporated film and fluorescence and phosphorescence spectra of *m*-MTDATA and *t*-Bu-PBD single thin films. (b) Temperature dependence of k_{RISC}

excited states of electron-donating and electron-accepting materials because energy can be transferred from the triplet excited state of the exciplex to the triplet excited states of electron-accepting and electron-donating materials to cause nonradiative deactivation at each triplet excited state, resulting in a markedly decreased up-conversion probability. The phosphorescence intensity peak for the *m*-MTDATA thin film appears at 475 nm, which means that the triplet excited state of the exciplex is sufficiently confined. On the other hand, the phosphorescence intensity peak for the *t*-Bu-PBD thin film appears at approximately 510 nm [70], again indicating that the triplet excited state of the exciplex is sufficiently confined by that of *t*-Bu-PBD and that a high up-conversion probability can be expected. In practice, delayed fluorescence resulting from the fluorescence component of the exciplex and the up-conversion from the triplet to singlet excited states has been confirmed in the transient PL characteristics of the 50 mol%-*m*-MTDATA:*t*-Bu-PBD co-evaporated film at room temperature.

ΔE_{ST} of the exciplex formed between *t*-Bu-PBD and *m*-MTDATA was estimated by measuring the temperature dependence of the rate constant for RISC, k_{RISC}, which is given by
Eq. (2.16)

$$k_{RISC} = A \, \exp\left(\frac{\Delta E_{ST}}{k_B T}\right). \qquad (2.16)$$

Here, A is a constant, k_B is the Boltzmann constant, and T is temperature. Using Eq. (2.4), ΔE_{ST} can be estimated from the temperature dependence of k_{RISC}. k_{RISC} can be estimated from the rate constants and the PL quantum yields of the prompt and delayed fluorescence components using Eq. (2.3) [69]. k_p, k_d, Φ_p, and Φ_d can be experimentally determined from the PL decay curve and the temperature dependence of the PL intensity. The Arrhenius plots of k_{RISC} calculated using Eq. (2.3) are shown

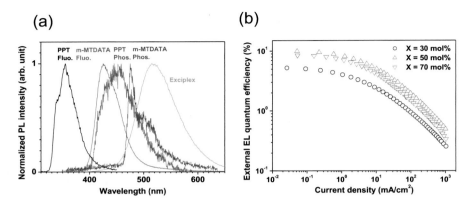

Fig. 2.22 (a) PL spectrum of 50 mol%-m-MTDATA:PPT co-evaporated film and fluorescence and phosphorescence spectra of m-MTDATA and PPT single thin films. (b) Dependence of η_{ext} on current density for OLED with the structure of ITO/m-MTDATA (35 nm)/X mol%-m-MTDATA: PPT (30 nm)/PPT (35 nm)/LiF/Al

in Fig. 2.21b. Here, k_{RISC} is assumed to be independent of temperature. The activation energy was 50 meV, demonstrating that the singlet and triplet excited states of the exciplex are very close.

Although ΔE_{ST} of the exciplex formed between m-MTDATA and t-Bu-PBD is found to be small, the EL efficiency of OLEDs with the m-MTDATA:t-Bu-PBD co-evaporated film as an emitting layer is still low (~2%). We examined delayed fluorescence from the exciplex formed between various donor and acceptor materials. As a result, intense delayed fluorescence was observed in the exciplex formed between m-MTDATA as the donor material and 2,8-bis(diphenyl phosphoryl) dibenzo[b,d]thiophene (PPT) as the acceptor material [29]. Figure 2.22a shows the PL spectrum of a 50 mol%-m-MTDATA:PPT co-evaporated film as well as the fluorescence and phosphorescence spectra of m-MTDATA and PPT single thin films. The PL intensity peak of the co-evaporated film appears at a wavelength of 520 nm, which is longer than the fluorescence intensity peaks from the m-MTDATA and PPT single thin films, meaning that exciplexes are formed in the co-evaporated film. The PL quantum yield of the exciplex formed between m-MTDATA and PPT is 28.5% and is higher than that of the exciplex formed between m-MTDATA and t-Bu-PBD (19.6%). In addition, the PL quantum yield of the delayed fluorescence component is 25.4% and is higher than that for the exciplex formed between m-MTDATA and t-Bu-PBD (8.2%). To evaluate the effect of the increased PL quantum efficiency of delayed fluorescence, the characteristics of OLEDs with the structure of ITO/m-MTDATA (35 nm)/X mol%-m-MTDATA: PPT (30 nm)/PPT (35 nm)/LiF/Al are shown in Fig. 2.22b. η_{ext} of the OLED is 10.0%, which exceeds the theoretical limit of 5% for devices using fluorescent materials. Thus, delayed fluorescence from the exciplex state as an intermolecular charge-transfer excited state is enhanced by selecting appropriate donor and acceptor materials (Figs. 2.23, 2.24, 2.25 and 2.26).

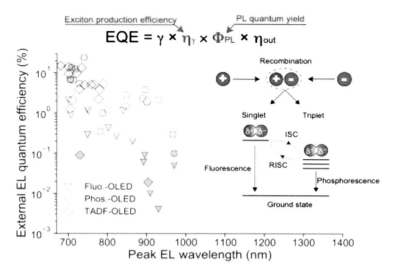

Fig. 2.23 η_{ext} against EL peak wavelength in reported NIR-OLEDs and schematic view of exciton formation process in OLEDs

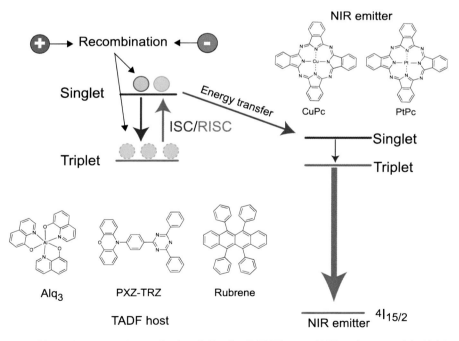

Fig. 2.24 Exciton harvesting mechanism (left) using TADF host, and NIR emitter materials (right)

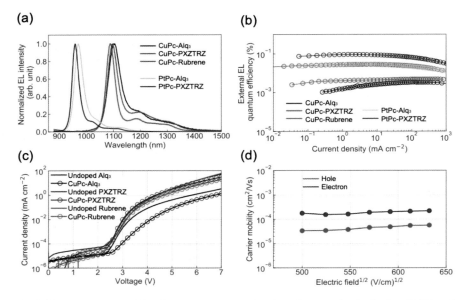

Fig. 2.25 (**a**) EL spectra of five OLEDs. (**b**) η_{ext} as a function of current density for five OLEDs. (**c**) Current density against as a function of applied voltage for the OLEDs with and without CuPc dopant. (**d**) Carrier mobilities of a PXZ-TRZ neat film for hole and electron obtained by TOF

2.3.6 Outlook of TADF

TADF technologies have been developed as technologies that can greatly change the performance of future OLEDs to enable their practical application. Also, from the viewpoint of photochemistry and material chemistry, TADF technologies have given rise to a new category of luminescent materials, contributing to the academic progress in this field. The attractiveness of organic compounds lies in the diversity of molecular structures, and TADF materials are considered as novel materials created by making full use of that diversity. Namely, new luminescent materials other than the previously known fluorescent and phosphorescent materials have become available. The TADF phenomenon was named hyperfluorescence. If the luminescent materials used for OLEDs greatly shift to third-generation TADF materials in the future, the problem of high cost due to the use of phosphorescent materials can be solved, and the resource-depletion risk can be avoided by the rare-metal-free element strategies. Moreover, TADF materials are expected to be applied to the creation of luminescent materials that exhibit highly efficient blue emission and to contribute to highly efficient cost-effective organic EL illumination, which will significantly contribute to the vitalization of the market of OLEDs in the future. Molecular materials can be designed in an infinite number of patterns. I hope that organic optoelectronics will be driven by the creation of new molecular materials.

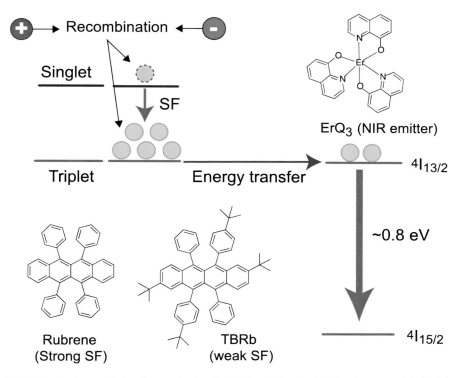

Fig. 2.26 Exciton multiplication mechanism (left) using SF host and NIR emitter materials (right)

2.4 NIR-OLED

2.4.1 Recent Progress of NIR-OLED

Light in the near-infrared (NIR) region (700–2000 nm) has unique characteristics such as permeability and noninvasiveness in living organisms, low light scattering properties in the atmosphere, and low optical waveguide loss in Si fibers. Therefore, by using inorganic light-emitting diodes (LEDs), inorganic laser diodes (laser diodes: LDs), and inorganic semiconductor light-receiving devices, there is a wide variety of applications in sensing and imaging of biological information, distance measurement, and optical fiber optical communication represented by LiDAR. In particular, photoelectric pulse wave measurement and blood oxygen saturation measurement using NIR light have become essential medical technologies for patient health management and disease diagnosis, and these attract intense attention for monitoring biological information in daily life.

Since OLEDs can employ a wide variety of luminescent materials, it is possible to develop OLEDs that exhibit EL not only in visible light but also in ultraviolet to NIR regions. In particular, the realization of OLEDs that exhibit NIR EL is expected not only to increase the value of OLEDs as display devices but also to open new possibilities as sensing light sources. However, reports of OLEDs that emit light in the NIR region are still limited (Fig. 2.23). Thus, there is a strong need to establish comprehensive design guidelines for NIR organic light-emitting materials and NIR OLEDs.

As mentioned in Eq. (2.3), η_{ext} of OLEDs is expressed as the product of four factors: (1) charge injection/transport balance, (2) photoluminescence (PL) quantum yield, (3) exciton generation efficiency, and (4) light extraction efficiency. The major problems that hinder the improvement of η_{ext} of NIR OLEDs are their low PL quantum yield of NIR organic light-emitting materials and the low exciton generation efficiency under current excitation (Fig. 2.23). For example, η_{ext} of OLEDs using copper phthalocyanine (CuPc) that emits light in the NIR region of 1.1 μm, which has been reported since the 2000 s, is as low as $6 \times 10^{-3}\%$ [71]. This is mainly due to the low PL quantum yield of CuPc: < 0.1% [72]. In general, the dramatic drop in PL quantum yield in the NIR region is explained by the energy gap law [73]. In the NIR emission process, as the energy gap between the ground state and the excited state decreases, the wave functions of the two states overlap significantly, causing an exponential increase in the nonradiative deactivation rate constant.

In OLED, which is a current-driven device, the lowest singlet excited state, S_1, and the lowest triplet excited state, T_1, are generated simultaneously with a branching ratio of 25% and 75%, respectively [74]. Therefore, in OLEDs using fluorescent materials as emitters, only fluorescence based on the transition from S_1 to S_0, which is a spin-allowing process, is observed, and the theoretical maximum exciton generation efficiency remains at 25%. Up to now, the development of materials aiming at the utilization of triplet excitons has been vigorously pursued to improve exciton generation efficiency. One of them is the use of room-temperature phosphorescent materials. Phosphorescent materials exhibit strong spin-orbital interactions due to the internal heavy atom effect such as iridium (Ir) and platinum (Pt). As a result, the spin transition of the intersystem crossing (ISC) from S_1 to the higher triplet excited state (T_n) and the phosphorescence process from T_1 to S_0 restrictions are relaxed. In fact, phosphorescent OLEDs have achieved high η_{ext} reaching the theoretical upper limit from the blue to the NIR region [75–77]. In recent years, TADF materials have been rapidly developed to utilize the triplet excited state for light emission without using rare and precious metals. Generally, in the material design of TADF materials, HOMO and LUMO are spatially separated by introducing a steric hindrance between the electron-donating and electron-accepting units. This separation reduces the energy difference (ΔE_{ST}) between S_1 and T_1, which is twice the exchange integral, and enables the expression of an efficient reverse intersystem crossing (RISC) process of T_1 to S_1. In OLEDs using TADF materials, high η_{ext} approaching the theoretical upper limit is obtained from blue to deep red, demonstrating its high potential [64, 78, 79].

However, there are limited reports on TADF-OLEDs in the NIR region with wavelengths longer than 700 nm. Particularly in NIR OLEDs, an increase in driving voltage due to charge trapping in light-emitting materials with a narrow bandgap is a problem. Thus, an increase in η_{ext} is essential to reduce power consumption.

Therefore, it is strongly desired to achieve an exciton generation efficiency of ~100% by utilizing the RISC process of TADF materials. In addition, to further increase the output of NIR OLEDs, it is essential to develop exciton-sensitized materials beyond TADF materials, i.e., hyperfluorescence, to realize exciton generation efficiency exceeding 100%.

In this section, we introduce the use of an exciton sensitization process in NIR OLEDs and the development of NIR TADF molecules with high PL quantum yields to increase the η_{ext} of NIR OLEDs.

2.4.2 Utilization of Exciton Sensitization Process by TADF Molecules

First, a molecule exhibiting TADF properties is used as a host material, and NIR light is obtained using exciton sensitization by transferring the energy of a singlet exciton generated by the RISC process in the TADF molecule to a guest molecule (Fig. 2.24) [80]. Copper (II) phthalocyanine (CuPc) and platinum (II) phthalocyanine (PtPc), which are NIR phosphorescent dyes in the 1.1 μm band, were used as luminescent molecules. The TADF material of 2-phenoxazine-4,6-diphenyl-1,3,5-triazine (PXZ-TRZ) [68] was used as the host material for the emission layer, and a fluorescent molecule was used for comparison. As a reference, tris-(8-hydroxyquinolinato)aluminum (Alq_3) and 5,6,11,12-tetraphenylnaphthacene (rubrene) are also used as hosts. In these combinations of host and guest materials, highly efficient energy transfer (~90%) from the host molecule to the guest molecule is observed under photoexcitation. In addition, from the temperature dependence measurement of NIR emission intensity in the TADF host under photoexcitation, thermally activated luminescence behavior was observed at 5–100 K, in contrast to that in the fluorescent host. This suggests that the RISC process in the TADF molecule contributes to the NIR emission from the guest molecule.

Figure 2.25 shows (a) EL spectra and (b) η_{ext} vs. current density characteristics for each OLED. EL emission derived from CuPc or PtPc was observed from each device. In a device using CuPc as a NIR light-emitting guest molecule and a TADF material of PXZ-TRZ as a host molecule, $\eta_{ext} = 0.03\%$, which is about nine times that of the fluorescent host of Alq_3 and about six times that of the fluorescent host of rubrene. Also, in an OLED using PtPc as a guest molecule, $\eta_{ext} = 0.1\%$ was obtained by using a TADF host, which is significantly higher than that with the fluorescent host.

Also, in the analysis of the current density-voltage characteristics of each CuPc device shown in Fig. 2.25c, a significant increase in the driving voltage was observed only when CuPc molecules were doped into the fluorescent host of Alq_3. This is because the direct charge trapping process to the guest molecule is dominant in the Alq_3-hosted device, and the energy transfer process from the host molecule to the guest molecule is dominant in the rubrene host device and the PXZ-TRZ host device. Charge mobility evaluation using time-of-flight (TOF) measurements shown in Fig. 2.25d confirms bipolar transport originating from electron-donating PXZ and electron-accepting TRZ sites in the PXZ-TRZ host medium. This contributes to the suppression of charge trapping to the NIR organic dye.

The difference in these emission mechanisms is due to (1) hole trapping due to the difference in HOMO levels between the host and guest molecules and (2) the difference in charge transport properties of the host. The improvement of η_{ext} by using the TADF host PXZ-TRZ can be attributed to the improvement of the charge balance for the Alq$_3$ host and the improvement of the exciton generation efficiency for the rubrene host.

2.4.3 Utilization of Exciton Sensitization Process by Singlet Fragmentation

By using TADF molecules as a host material, ideal exciton generation efficiency can be obtained even in NIR OLEDs. However, the theoretical upper limit of the exciton generation efficiency of conventional OLEDs is 100%, and it is necessary to develop a new exciton sensitization mechanism to aim for further improvement in the efficiency of OLEDs. As a promising mechanism to realize this, we introduce a type of multiple exciton generation processes in organic molecules called singlet exciton fission (SF), $S_1 + S_0 \rightarrow T_1 + T_1$, leading to a photophysical process in which two triplet excitons are generated from one singlet exciton and the theoretical upper limit of (25% × 2) + 75% = 125% [81]. In fact, research on this process has been actively conducted with the aim of improving the quantum efficiency of organic thin-film solar cells [82]. However, there have been no reports of the use of the SF process in OLEDs, while the research has been carried out on utilizing TTA processes. We introduce a NIR OLED that exhibits more than 100% exciton generation efficiency by using the material as a guest molecule (Fig. 2.26) [83].

Rubrene has been reported to exhibit efficient SF processes in amorphous thin films and is expected to be used as a host material for light-emitting layers [84]. In addition, since the T_1 energy level of rubrene is about 1.14 eV, tris-(8-hydroxyquinolinato)erbium(III) (ErQ$_3$), which has an emission level of 0.80 eV, can be used as a NIR luminescent material that accepts the T_1 energy of rubrene [85]. As a host material, 2,8-di-tert-butyl-5,11-bis(4-tert-butylphenyl)-6,12-diphenyltetracene (TBRb) [86], which shows almost no SF process, was used for comparison.

Figure 2.27a shows the emission spectrum of a single rubrene film and the absorption spectrum of ErQ$_3$ in a 5 × 10^{-6} M/DMSO-d$_6$ solution. The emission peak wavelength of rubrene single film is 562 nm, and there is no significant overlap with the absorption spectrum of ErQ$_3$. This means that the rate constant of dipole-dipole energy transfer (FRET) from the S_1 state of rubrene, the SF host molecule, to the S_1 state of the ErQ$_3$ molecule is extremely small. In the ErQ$_3$-rubrene

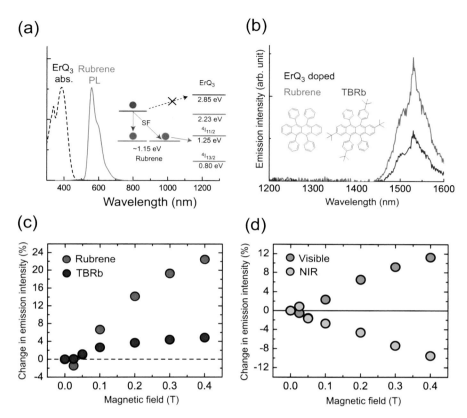

Fig. 2.27 (a) Absorption spectrum of ErQ$_3$ in solution and PL spectrum of a rubrene neat film. inset) Jablonski diagram for harvesting of triplets produced by singlet fission in rubrene. (b) The NIR-PL spectra of ErQ$_3$ in rubrene and TBRb host matrices. (c) The steady-state change in fluorescence intensity from a rubrene neat film and a TBRb neat film as a function of external magnetic field. (d) The steady-state change in fluorescence and NIR emission intensity from a 2 mol %-ErQ$_3$-rubrene film as a function of external magnetic field

co-evaporated thin film, the SF process between rubrene molecules appears predominantly over the FRET process from rubrene to ErQ$_3$.

Figure 2.27b shows NIR PL spectra of co-evaporated thin films in which ErQ$_3$ was dispersed in each host material at a concentration of 2 mol%. In this measurement, although a CW laser light source with an excitation wavelength of 514 nm that selectively excites only the SF host molecule is used, a clear NIR emission with the emission peak at 1530 nm derived from the $^4I1_{3/2} \rightarrow {}^4I1_{5/2}$ transition of the ErQ$_3$ molecule was observed. This indicates that in the ErQ$_3$-rubrene and ErQ$_3$-TBRb co-evaporated thin films, the luminescence from ErQ$_3$ molecules was observed through an energy transfer process which is different from the FRET process. Here, considering the energy levels of each material (Fig. 2.27a inset), it is expected that energy transfer occurs from the T_1 state of rubrene and TBRb to the central metal Er^{3+} of ErQ_3. The NIR PL quantum yield of the ErQ_3-rubrene thin film is $7 \times 10^{-3}\%$, which is comparable to the PL quantum yield of the ErQ_3-TBRb thin film: $4 \times 10^{-3}\%$.

Figure 2.27c shows the results of magnetic field dependence measurements of the visible luminescence intensity for rubrene and TBRb monolayers. With the application of a magnetic field, a remarkable increase in visible fluorescence intensity derived from the S_1 state is observed, especially in rubrene monolayers. Regarding the SF process in rubrene molecules, it is known that the generation of the singlet-triplet exciton pair $^1(TT)$ that mediates the SF process is inhibited by the application of a magnetic field [82]. Therefore, the results in Fig. 2.27c indicate that the SF process occurs efficiently in the rubrene single film, while in the TBRb thin film, the presence of bulky tert-butyl groups suppresses the intermolecular interaction and inhibits the SF process.

Figure 2.27d shows magnetic field dependence measurements of visible and NIR luminescence intensities in ErQ_3-rubrene thin films. An increase in visible fluorescence intensity due to the S_1 state of the rubrene molecule and a decrease in NIR emission intensity due to the $^4I_{13/2} \rightarrow {}^4I_{15/2}$ transition of the ErQ_3 molecule are observed with the application of a magnetic field. The observed increase in rubrene's visible fluorescence intensity upon application of a magnetic field is a result of efficient SF processes in the rubrene host medium as in the case of rubrene monolayers. In addition, the decrease in NIR emission intensity of ErQ_3 with the application of a magnetic field is attributed to the decrease in rubrene triplet excitons due to the inhibition of the SF process, which was generated by the SF process in rubrene molecules in the ErQ_3-rubrene thin film. Also, the SF efficiency calculated from the rate of change in visible/NIR emission intensity is 54.3%. Combining the above results, we obtained NIR luminescence derived from the SF process in the two types of ErQ_3-doped films.

Next, the characteristics of a NIR SF-OLED using an ErQ_3-rubrene co-evaporated thin film, which has been confirmed to be useful for photoexcitation. Figure 2.28a shows the device structure of the fabricated NIR SF-OLED, and Fig. 2.28b shows the current density-voltage characteristics. Comparing the current density-voltage characteristics in the rubrene device and the ErQ_3-rubrene device, we found that the rising voltage and the drive voltage match very well, indicating that the exciton generation process is taking place at the SF host: rubrene.

Figure 2.28c shows the NIR EL spectra at 50 mA cm^{-2} for the ErQ_3-rubrene and ErQ_3-TBRb devices. In both devices, clear NIR luminescence with an emission peak at 1530 nm derived from the $^4I_{13/2} \rightarrow {}^4I_{15/2}$ transition of the ErQ_3 molecule was observed. In addition, the ErQ_3-rubrene device has a NIR EL intensity about 1.4 times higher than that of the ErQ_3-TBRb device. Figure 2.28d shows magnetic field dependence measurements of the visible and NIR luminescence intensity of the ErQ_3-rubrene device under a constant current drive. While almost no change in driving voltage due to magnetic field application is observed, an increase in visible fluorescence intensity derived from rubrene molecules and a decrease in NIR emission intensity derived from ErQ_3 molecules are observed. This indicates that in the ErQ_3-rubrene device, triplet excitons generated by the SF process in the

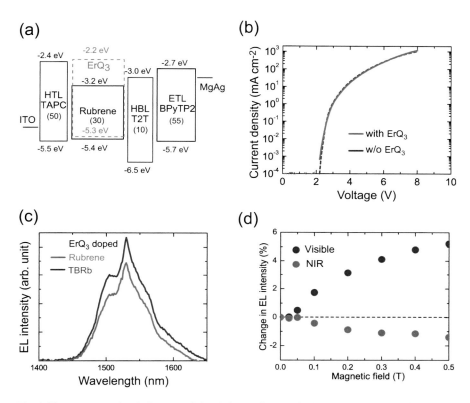

Fig. 2.28 (a) Energy level diagram of the designed OLED. (b) Current density as a function of applied voltage for the OLEDs with and without ErQ$_3$ dopant. (c) EL spectra of tested OLEDs. (d) The steady-state change in fluorescence and NIR emission intensity from a 2 mol%-ErQ$_3$-rubrene-based OLED as a function of external magnetic field

rubrene molecule transfer energy to the ErQ$_3$ molecule, leading to the NIR luminescence, as in the case of photoexcitation. In addition, the exciton generation efficiency calculated from the rate of change in visible and NIR emission intensity is 100.8%, and the exciton generation efficiency exceeding the conventional theoretical limit of 100% has been achieved by using the SF process. In the future, by using TADF molecules as hosts and SF molecules and phosphorescent molecules as guest molecules, it is expected to generate 100% singlet excitons on TADF molecules and double the singlet excitons using the SF process. Thus, 200% exciton generation efficiency is expected by splitting into triplet excitons and finally energy transfer to phosphorescent molecules (Fig. 2.29).

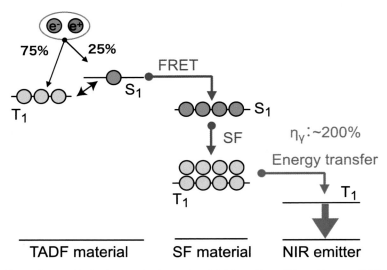

Fig. 2.29 Exciton multiplication mechanism using TADF sensitizer, SF sensitizer, and NIR emitter materials

2.4.4 Development of TADF Molecules Exhibiting High-Efficiency Near-Infrared Luminescence

So far, we have mentioned that the use of exciton sensitization processes such as the phosphorescence process, the TADF process, and the SF process are promising. However, to improve the η_{ext} in NIR OLEDs, it is essential to improve not only the exciton generation efficiency but also the PL quantum yield of the NIR light-emitting material itself. In fact, the PL quantum yield of ErQ_3, the NIR light-emitting material used in Sect. 2.4.2, is extremely low at around 10^{-3}–10^{-2}%. On the other hand, phosphorescent OLEDs using Pt complexes with high PL quantum yield have achieved high η_{ext} of 24% and 3.8% at the EL wavelength of 740 nm [76] and 900 nm, respectively [87]. OLEDs using rare-metal-free TADF materials have also achieved η_{ext} of over 20% up to deep red [79]. However, there are very few reports of TADF materials that exhibit high-efficiency NIR emission in the wavelength region of 700 nm or longer, and their η_{ext} remains as low as 10% or less [88–90]. In addition, NIR TADF-OLEDs using a pyrazinoacenaphthene skeleton have been reported to have an η_{ext} of 14.1% at an EL wavelength of 700 nm [91], but many problems remain such as roll-off at practical driving current region. This roll-off originates from the small RISC rate constant $(k_{RISC}) = 10^2 \sim 10^3$ s^{-1} of the NIR TADF material. That is, exciton annihilation occurs remarkably because of a large accumulation of triplet excitons during device operation [92]. Therefore, it is essential to develop NIR TADF materials that exhibit high PL quantum yield and efficient RISC to achieve high η_{ext} at practical drive current region. In this section, we introduce the photophysical and OLED properties of the NIR TADF material:

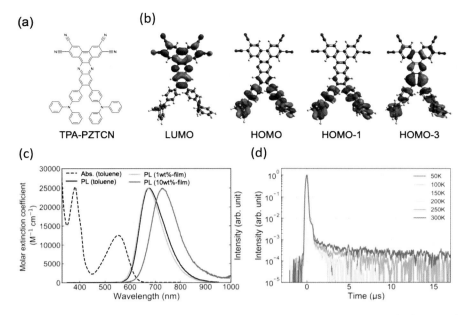

Fig. 2.30 (**a**) The chemical structure of TPA-PZTCN. (**b**) HOMO and LUMO distribution of TPA-PZTCN. (**c**) PL spectra of TPA-PZTCN in dilute toluene solution, 1 wt%-doped in mCBP film, and 10 wt%-doped in mCBP film. Black dashed line indicates the absorption spectrum of TPA-PZTCN solution. (**d**) Temperature-dependent transient PL decay curves of a 10 wt%-TPA-PZTCN-mCBP film

Dibenzo[a,c]phenazine-2,3,6,7-tetracarbonitrile (TPA-PZTCN) [93]. In addition, we describe the results of a study on the use of TPA-PZTCN as a TADF-assisted dopant [94] for fluorescent materials in the 900 nm band, and its application to photoplethysmography (PPG) devices using high-efficiency NIR OLEDs.

PZTCN is a promising electron-accepting scaffold with strong electron-withdrawing properties. Figure 2.30a shows the chemical structure of TPA-PZTCN combined with TPA, the donor scaffold, and Fig. 2.30b shows the distribution of HOMO and LUMO. TPA-PZTCN has a structure in which the 11,12-positions of PZTCN, a rigid electron-accepting skeleton, are substituted with the electron-donating group triphenylamine (TPA). From DFT calculations, the dihedral angle between the TPA unit and the PZTCN unit in the ground state is 48.5°. This steric twist effectively separates HOMO and LUMO. Therefore, a small energy difference between S_1 and T_1: $\Delta E_{ST} = 0.14$ eV is formed, allowing efficient RISC processes to occur. The cyclic voltammogram (CV) curves are reversible to both oxidation and reduction, indicating that TPA-PZTCN is electrochemically stable. The experimental values of HOMO and LUMO calculated from the oxidation/reduction peak potentials are -5.66 eV and -3.84 eV, respectively.

Figure 2.30c shows the absorption and PL spectra of the TPA-PZTCN co-deposited thin film and in toluene solution, and Table 2.1 summarizes the related photophysical properties. Emission was observed around 670 nm in the toluene solution and 1 wt% co-evaporated thin film, and a high PL quantum yield of about 80% was obtained. Further, even with the 10 wt% co-evaporated thin film, the PL quantum yield exceeded 40%, and high NIR emission characteristics were obtained, despite the PL peak being elongated to 729 nm.

Figure 2.30d shows the results of temperature dependence measurements of transient PL decay in 10 wt%-TPA-PZTCN-mCBP thin films. A delayed component was observed in the high-temperature range of 200–300 K, which clearly indicates the existence of the RISC process in the TPA-PZTCN molecule. Also, $k_{RISC} = 7.6 \times 10^4$ s^{-1} was calculated from the transient PL decay at room temperature. This is about two orders of magnitude higher than the k_{RISC} of the reported high-efficiency NIR TADF material. TD-DFT calculations suggest the contribution of the localized-charge-transfer mixed state: ^{3}HLCT$_2$ in the RISC process of ^{3}CT$_1 \rightarrow$ ^{1}CT$_1$, which is a charge-transfer (CT) excited state. It is considered a factor for large k_{RISC} [92, 95].

Further, TPA-PZTCN exhibits highly efficient NIR emission and RISC process in the 700 nm band, and it can also be used as a TADF-assisted dopant [94] for the NIR fluorescent material in the 900 nm band of BBT-TPA [96] in the 1 wt%-BBT-TPA-10 wt%-TPA-PZTCN-mCBP co-evaporated thin film. NIR EL with a peak at 874 nm derived from the BBT-TPA molecule was observed (Fig. 2.31a). A high quantum yield of 12.9% was also obtained.

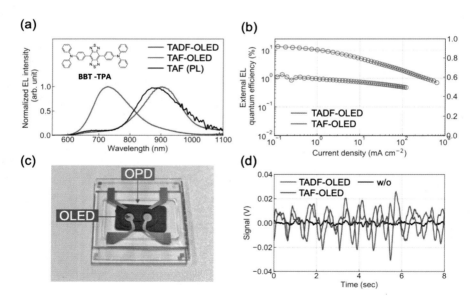

Fig. 2.31 (a) EL spectra of tested OLEDs and PL spectrum of a TAF-film. (b) η_{ext} as a function of current density for two OLEDs. (c) Photograph of an OLED/OPD integrated device. (d) PPG signals from an OLED/OPD integrated device

Table 2.2 EL properties of NIR-OLEDs

OLED type	EL (nm)[a]	EQE_{max} (%)	EQE (%)[a]	EQE (%)[b]	Voltage (V)[b]	Output power (mW cm^{-2})[b]	LT95 (h)[b]
TADF	734	13.4	10.2	4.9	7.6	0.9	168
TAF	901	1.1	1.0	0.8	10.4	0.1	>600

[a] Measured at current density of 1 mA cm^{-2}
[b] Measured at current density of 10 mA cm^{-2}

Next, the EL characteristics of two types of OLEDs (TADF-OLED: without BBT-TPA, TAF-OLED: with BBT-TPA) having a 10 wt%-TPA-PZTCN-mCBP co-evaporated thin film as a light-emitting layer are shown in Fig. 2.31. It summarizes (a) EL spectra and (b) η_{ext}-current density characteristics of each device, and Table 2.2 summarizes the relevant OLED characteristics. TADF-OLED and TAF-OLED showed EL peaking at 734 nm and 901 nm, respectively, with the maximum η_{ext} of 13.4% and 1.1%, respectively. These are the highest η_{ext} values for NIR TADF-OLED in each wavelength region. Further, the fabricated TADF-OLEDs maintain an η_{ext} of more than 10% even at a current density of 1 mA cm^{-2}, suggesting that the efficient RISC process in the TPA-PZTCN molecule contributes to the reduction of triplet exciton density and the suppression of roll-off has been achieved. In addition, the TAF-OLED has extremely good device durability of $LT_{95} > 600$ hours even when driven at a constant current of 10 mA cm^{-2}, and the development of practical devices is expected in the future.

Finally, we show the results of reflection-type pulse wave sensing of the finger joint using an on-chip device that integrates the developed NIR TADF-OLED and an organic photodiode (OPD) [97] (Fig. 2.31c). It has been confirmed that the OPD fabricated exhibits an external quantum efficiency of 8.3% or more in the wavelength region of 700–900 nm even when 0 V is applied and has a high specific detection capability of about 10^{11}–10^{12} Jones. As shown in Fig. 2.31d, when both TADF-OLED (EL: 734 nm) and TAF-OLED (EL: 901 nm) were used, NIR changes in scattered light intensity were observed, and pulse wave sensing at two wavelengths was successful. Until now, there have been no reports of pulse wave sensing with an all-organic device, especially in the 900 nm band.

Unlike conventional display applications, NIR OLEDs are expected to open up new industrial fields such as sensors. Fortunately, NIR OLEDs exhibit extremely durable properties and are of great interest as practical devices. In the future, further performance improvement can be expected by suppressing nonradiative deactivation based on the viewpoint of material chemistry and developing new exciton mechanisms such as SF.

Acknowledgment The authors deeply acknowledged Dr. Saeyoun Lee, Dr. Tetsuya Nakagawa, Dr. Katsuyuki Shizu, Dr. Takuma Yasuda, and Dr. William Potscavage for their assistance in the preparation of this manuscript. A part of the description was reproduced from Jpn. J. Appl. Phys., 53, 6, 060101, (2014).

References

1. H. Akamatsu, H. Inokuchi, J. Chem. Phys. **18**, 810 (1950)
2. M. Pope, C.E. Swenberg, *Electronic Processes in Organic Crystals and Polymers* (Oxford University Press, 1999)
3. C. Adachi, T. Tsutsui, S. Saito, Appl. Phys. Lett. **55**, 1489 (1989)
4. W. Helfrich, W.G. Schneider, *Phys. Rev. Lett.* **14**, 229 (1965)
5. W. Helfrich, W.G. Schneider, *J. Chem. Phys.* **44**, 2902 (1966)
6. P.S. Vincentt, W.A. Barlow, R.A. Hann, G.G. Roberts, *Thin Solid Films* **94**, 171 (1982)
7. C.W. Tang, S.A. Vanslyke, *Appl. Phys. Lett.* **51**, 913 (1987)
8. C.W. Tang, S.A. Vanslyke, C.H. Chen, *J. Appl. Phys.* **65**, 3610 (1989)
9. R.H. Partridge, Polymer **24**, 748 (1983)
10. J.H. Burroughes, D.D. Bradley, A.R. Brown, R.N. Marks, K. Mackay, R.H. Friend, P.L. Burns, A.B. Holmes, Nature **347**, 539 (1990)
11. C.E. Wayne, R.P. Wayne, *Photochemistry* (Oxford Chemistry Primers, 1995)
12. E. Aminaka, T. Tsutsui, S. Saito, *J. Appl. Phys.* **79**, 8808 (1996)
13. J.S. Kim, P.K.H. Ho, N.C. Greenham, R.H. Friend, *J. Appl. Phys.* **88**, 1073 (2000)
14. C. Adachi, M.A. Baldo, M.E. Thompson, S.R. Forrest, *J. Appl. Phys.* **90**, 5048 (2001)
15. C. Adachi, M.A. Baldo, S.R. Forrest, *Phys. Rev.B* **62**, 10967 (2000)
16. T. Förster, *Discuss. Faraday Soc.* **27**, 7 (1959)
17. D.L. Dexter, *J. Chem. Phys.* **21**, 836 (1953)
18. B.P. Lyons, A.P. Monkman, *Phys. Rev. B* **71**, 235201 (2005)
19. C. Murawski, K. Leo, M.C. Gather, *Adv. Mater.* **25**, 6801 (2013)
20. M.A. Baldo, D.F. O'Brien, Y. You, A. Shoustikov, S. Sibley, M.E. Thompson, S.R. Forrest, *Nature* **395**, 151 (1998)
21. J. Kalinowski, J. Mezyk, F. Meinardi, R. Tubino, M. Cocchi, D. Virgili, *J. Appl. Phys.* **98**, 063532 (2005)
22. S.D. Babenko, V.A. Benderskii, V.I. Gol'Danskii, A.G. Lavrushko, V.P. Tychinskii, *Chem. Phys. Lett.* **8**, 598 (1971)
23. D. Kasemann, R. Brückner, H. Fröb, K. Leo, *Phys. Rev. B* **84**, 115208 (2011)
24. M.A. Baldo, R.J. Holmes, S.R. Forrest, *Phys. Rev. B* **66**, 035321 (2002)
25. S. Reineke, K. Walzer, K. Leo, *Phys. Rev. B* **75**, 125328 (2007)
26. D. Song, S. Zhao, H. Aziz, *Adv. Funct. Mater.* **21**, 2311 (2011)
27. M.C. Gather, A. Kohnen, K. Meerholz, *Adv. Mater.* **23**, 233 (2011)
28. S. Reineke, M. Thomschke, B. Lussem, K. Leo, *Rev. Mod. Phys.* **85**, 1245 (2013)
29. J. Kido, K. Hongawa, K. Okuyama, K. Nagai, *Appl. Phys. Lett.* **64**, 815 (1994)
30. H.A. Al Attar, A.P. Monkman, M. Tavasli, S. Bettington, M.R. Bryce, *Appl. Phys. Lett.* **86**, 121101 (2005)
31. G. Schwartz, K. Fehse, M. Pfeiffer, K. Walzer, K. Leo, Appl. Phys. Lett. **89**, 083509 (2006)
32. Y. Tomita, C. May, M. Toerker, J. Amelung, M. Eritt, F. Loeffler, C. Luber, K. Leo, K. Walzer, K. Fehse, Q. Huang, *Appl. Phys. Lett.* **91**, 253501 (2007)
33. Y.R. Sun, N.C. Giebink, H. Kanno, B.W. Ma, M.E. Thompson, S.R. Forrest, Nature **440**, 908 (2006)
34. T.W. Lee, T. Noh, B.K. Choi, M.S. Kim, D.W. Shin, J. Kido, *Appl. Phys. Lett.* **92**, 043301 (2008)
35. H. Kim, Y. Byun, R.R. Das, B.K. Choi, P.S. Ahn, *Appl. Phys. Lett.* **91**, 093512 (2007)
36. J.P.J. Markham, S.-C. Lo, S.W. Magennis, P.L. Burn, I.D.W. Samuel, *Appl. Phys. Lett.* **80**, 2645 (2002)
37. F. Villani, P. Vacca, G. Nenna, O. Valentino, G. Burrasca, T. Fasolino, C. Minarini, D.D. Sala, *J. Phys. Chem. C* **113**, 13398 (2009)
38. J. Ju, Y. Yamagata, T. Higuchi, *Adv. Mater.* **21**, 4343 (2009)
39. H. Yersin, *Highly Efficient OLEDs with phosphorescent Materials* (Willey-VCH, 2008)
40. W. Zhu, W. Mo, M. Yaun, W. Yang, Y. Cao, *Appl. Phys. Lett.* **80**, 2045 (2002)

41. C. Jiang, W. Yang, J. Pengm, S. Xiao, Y. Cao, *Adv. Mater.* **16**, 537 (2004)
42. M. Sudhakar, P.I. Djurovich, T.E. Hogen-Esch, M.E. Thompson, *J. Am. Chem. Soc.* **125**, 7769 (2003)
43. A. van Dijken, J.J.A.M. Bastiaansen, N.M.M. Kiggen, B.M.W. Langeveld, C. Rothe, A. Monkman, I. Bach, P. Stossel, K. Brunner, *J. Am. Chem. Soc.* **126**, 7718 (2004)
44. S.A. Choulis, V.E. Choong, A. Patwardhan, M.K. Mathai, F. So, *Adv. Funct. Mater.* **16**, 1075 (2006)
45. B.C. Krummacher, M.K. Mathai, V.E. Choong, S.A. Choulis, F. So, A. Winnacker, *Org. Electron.* **7**, 313 (2006)
46. J.A. Cheng, C.H. Chen, C.H. Liao, *Chem. Mater.* **16**, 2862 (2004)
47. J. Qiao, L.D. Wang, J.F. Xie, G.T. Lei, G.S. Wu, Y. Qui, *Chem. Commun.* **4560** (2005)
48. Y. Shirota, *J. Mater. Chem.* **15**, 75 (2005)
49. N. Rehmann, D. Hertel, K. Meerholz, H. Beckers, S. Heun, *Appl. Phys. Lett.* **91**, 103507 (2007)
50. G.J. Zhou, W.Y. Wong, B. Yao, Z. Xie, L. Wang, *J. Mater. Chem.* **18**, 1799 (2008)
51. Y.R. Do, Y.C. Kim, Y.W. Song, C.O. Cho, H. Jeon, Y.J. Lee, S.H. Kim, Y.H. Lee, *Adv. Mater.* **15**(14), 1214 (2003)
52. S. Lamansky, P. Djurovich, D. Murphy, F. Abdel-Razzaq, H.-E. Lee, C. Adachi, P.E. Burrows, S.R. Forrest, M.E. Thompson, *J. Am. Chem. Soc.* **123**, 4304 (2001)
53. N. Tessler, G.J. Denton, R.H. Friend, *Nature* **382**, 695 (1996)
54. T. Tsutsui, S. Saito, *Organic Multilayer-Dye Electroluminescent Diodes: Is There Any Difference with Polymer LED?* (Kluwer Academic, Dordrecht, 1993)
55. L.J. Rothberg, A.J. Lovinger, *J. Mater. Res.* **11**, 3174 (1996)
56. J.C. Deaton, S.C. Switalski, D.Y. Kondakov, R.H. Young, T.D. Pawlik, S.B. Harkins, A.J.M. Miller, S.F. Mickenberg, J.C. Peters, *J. Am. Chem. Soc.* **132**, 9499 (2010)
57. O. Bolton, L. Kangwon, H.-J. Kim, K.Y. Lin, J. Kim, *Nat. Chem.* **3**, 205 (2011)
58. D.Y. Kondakov, T.D. Pawlik, T.K. Hatwar, J.P. Spindler, *J. Appl. Phys.* **106**, 124510 (2009)
59. A. Endo, M. Ogasawara, A. Takahashi, D. Yokoyama, Y. Kato, C. Adachi, Adv. Mater. **21**, 4802 (2009)
60. C.A. Parker, *Photoluminescence of Solutions* (Elsevier, Amsterdam, 1968)
61. M.N. Berberan-Santos, J.M.M. Garcia, *J. Am. Chem. Soc.* **118**, 9391 (1996)
62. M. Furukawa and S. Igarashi, Presented at the 81st Spring Mtg. of the Chem. Soc. Jpn. No. 2, F7–30 (2002).
63. B. Valeur (ed.), *Molecular Fluorescence: Principles and Applications*, 1st edn. (Wiley-VCH, Wheinheim, 2002)
64. H. Uoyama, K. Goushi, K. Shizu, H. Nomura, C. Adachi, *Nature* **492**, 234 (2012)
65. S.Y. Lee, T. Yasuda, H. Nomura, C. Adachi, *Appl. Phys. Lett.* **101**, 093306 (2012)
66. T. Nakagawa, S.-Y. Ku, K.-T. Wong, C. Adachi, *Chem. Commun.* **48**, 9580 (2012)
67. G. Mehes, H. Nomura, Q. Zhang, T. Nakagawa, C. Adachi, *Angew. Chem. Int. Ed.* **51**, 11311 (2012)
68. H. Tanaka, K. Shizu, H. Miyazaki, C. Adachi, *Chem. Commun.* **48**, 11392 (2012)
69. K. Goushi, K. Yoshida, K. Sato, C. Adachi, *Nat Photonics* **6**, 253 (2012)
70. M. Suzuki, S. Tokito, F. Sato, *Appl. Phys. Lett.* **86**, 103507 (2005)
71. T.C. Rosenow, K. Walzer, K. Leo, *J. Appl. Phys.* **103**, 043105 (2008)
72. P.S. Vincett, E.M. Voigt, K.E. Rieckhoff, *J. Chem. Phys.* **55**, 4131 (1971)
73. R. Englman, J. Jortner, *Mol. Phys.* **18**, 145 (1970)
74. M.A. Baldo, D.F. O'Brien, M.E. Thompson, S.R. Forrest, *Phys. Rev. B* **60**, 14422 (1999)
75. C. Adachi, M.A. Baldo, M.E. Thompson, S.R. Forrest, *Appl. Phys. Lett.* **90**, 5048 (2001)
76. K.T. Ly, R.W.C. Cheng, H.W. Lin, Y.J. Shiau, S.H. Liu, P.T. Chou, C.S. Tsao, Y.C. Huang, Y. Chi, *Nat. Photonics*. **11**, 63 (2017)
77. X. Li, J. Zhang, Z. Zhao, L. Wang, H. Yang, Q. Chang, N. Jiang, Z. Liu, Z. Bian, W. Liu, Z. Lu, C. Huang, *Adv. Mater.* **30**, 1705005 (2018)
78. S. Hirata, Y. Sakai, K. Masui, H. Tanaka, S.Y. Lee, H. Nomura, N. Nakamura, M. Yasumatsu, H. Nakanotani, Q. Zhang, K. Shizu, H. Miyazaki, C. Adachi, *Nat. Mater.* **14**, 330 (2015)

79. Y.L. Zhang, Q. Ran, Q. Wang, Y. Liu, C. Hänisch, S. Reineke, J. Fan, L.S. Liao, *Adv. Mater.* **31**, 1902368 (2019)
80. R. Nagata, H. Nakanotani, C. Adachi, *Adv. Mater.* **29**, 1604265 (2017)
81. C.E. Swenberg, W.T. Stacy, *Chem. Phys. Lett.* **2**, 327 (1968)
82. D.N. Congreve, J. Lee, N.J. Thompson, E. Hontz, S.R. Yost, P.D. Reusswig, M.E. Bahlke, S. Reineke, T.V. Voorhis, M.A. Baldo, *Science* **340**, 331 (2013)
83. R. Nagata, H. Nakanotani, W.J. Potscavage Jr., C. Adachi, *Adv. Mater.* **30**, 1801484 (2018)
84. G.B. Piland, J.J. Burdett, D. Kurunthu, C.J. Bardeen, *J. Phys. Chem. C* **117**, 1224 (2013)
85. W.P. Gillin, R.J. Curry, *Appl. Phys. Lett.* **74**, 798 (1999)
86. Y.S. Wu, T.H. Liu, H.H. Chen, C.H. Chen, *Thin Solid Films* **496**, 626 (2006)
87. K.R. Graham, Y. Yang, J.R. Sommer, A.H. Shelton, K.S. Schanze, J. Xue, J.R. Reynolds, *Chem. Mater.* **23**, 5305 (2011)
88. S. Wang, X. Yan, Z. Cheng, H. Zhang, Y. Liu, Y. Wang, *Angew. Chem. Int. Ed.* **127**, 13260 (2015)
89. C. Li, R. Duan, B. Liang, G. Han, S. Wang, K. Ye, Y. Liu, Y. Yi, Y. Wang, *Angew. Chem. Int. Ed.* **56**, 11525 (2017)
90. D.G. Congrave, B.H. Drummond, P.J. Conaghan, H. Francis, S.T.E. Jones, C.P. Grey, N.C. Greenham, D. Credgington, H. Bronstein, *J. Am. Chem. Soc.* **141**, 18390 (2019)
91. J. Xue, Q. Liang, R. Wang, J. Hou, W. Li, Q. Peng, Z. Shuai, J. Qiao, *Adv. Mater.* **31**, 1808242 (2019)
92. H. Noda, H. Nakanotani, C. Adachi, *Sci. Adv.* **4**, eaao6910 (2018)
93. U. Balijapalli, R. Nagata, N. Yamada, H. Nakanotani, M. Tanaka, A. D'Aléo, V. Placide, M. Mamada, Y. Tsuchiya, C. Adachi, *Angew. Chem. Int. Ed. Engl.*. https://doi.org/10.1002/anie.202016089
94. H. Nakanotani, T. Higuchi, T. Furukawa, K. Masui, K. Morimoto, M. Numata, H. Tanaka, Y. Sagara, T. Yasuda, C. Adachi, *Nat. Commun.* **5**, 4016 (2014)
95. H. Noda, X.K. Chen, H. Nakanotani, T. Hosokai, M. Miyajima, N. Notsuka, Y. Kashima, J.L. Brédas, C. Adachi, *Nat. Mater.* **18**, 1084 (2019)
96. G. Qian, B. Dai, M. Luo, D. Yu, J. Zhan, Z. Zhang, D. Ma, Z.Y. Wang, *Chem. Mater.* **20**, 6208 (2008)
97. X. Wang, H. Li, Z. Su, F. Fang, G. Zhang, J. Wang, B. Chu, X. Fang, Z. Wei, B. Li, W. Li, *Org. Electron.* **15**, 2367 (2014)

Chapter 3
Organic Solar Cells

Shuzi Hayase

Abstract Research trends on printable solar cells such as dye-sensitized solar cells, organic thin film solar cells, and perovskite solar cells are reviewed. The efficiency of the dye-sensitized solar cells and the organic thin film solar cells is now 15% and over 19%, respectively. The efficiency of the lead-based perovskite solar cells reached over 25% which is close to that of the Si solar cell, and the research target is shifting to the fabrication of the large modules and the module stability. Tin-based perovskite solar cells attract attention as the narrow bandgap perovskite solar cells with the bandgap of 1.0 eV. The efficiency is improved to 23–24%. To aim at further higher efficiency, perovskite tandem solar cells composed of perovskite/perovskite, perovskite/Si, and perovskite/CIGS are reported. The highest efficiency is over 33%. The mechanism of the charge separation and items to be solved is discussed.

Keywords Dye-sensitized solar cell · Organic thin-film solar cells · Perovskite solar cells · Printable · Efficiency · Tandem · Tin · Tin-lead · Lead · Silicon · CIGS

3.1 Introduction

Solar cells are composed of n-type layer or p-type layer/light-harvesting layer/p-type layer or n-type layer as shown in Fig. 3.1. Organic solar cells have organic molecules as the light-harvesting layer. Holes and electrons are prepared in the light-harvesting layer, and they are collected by p-type layers and n-type layers, respectively. For example, dye-sensitized solar cells, organic thin film solar cells, halide perovskite solar cells, quantum dot solar cells, and so on are classified in the organic solar cells. They are prepared by printing technologies and/or vacuum deposition technologies. Therefore, these organic solar cells are representative of the printable solar cells and flexible solar cells.

S. Hayase (✉)
The University of Electro-Communications, Info-Powered Energy System Research Center, Chofu, Tokyo, Japan
e-mail: hayase@uec.ac.jp

© The Author(s), under exclusive license to Springer Nature Japan KK 2024
S. Ogawa (ed.), *Organic Electronics Materials and Devices*,
https://doi.org/10.1007/978-4-431-56936-7_3

Fig. 3.1. Structure of solar cells

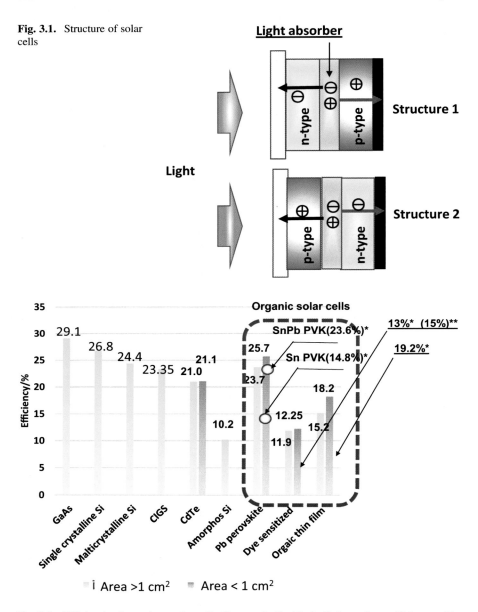

Fig. 3.2. Efficiencies for various solar cells. Bar graph: Certified efficiency from efficiency table 61 [1]. **Certified from NRWL efficiency chart [2], DSSC:15% [3], *From paper (not certified) SnPb PVK PV:23.6% [4], Sn PVK PV(14.8%) [5]

Figure 3.2 summarizes the certified efficiency and the efficiency of various solar cells including inorganic and organic solar cells reported in the papers [1]. The certified efficiency is measured in certified institutes, and the efficiency reported in the paper means that the efficiency was measured by their own solar simulators. In addition, the efficiency varied by the cell size. In Fig. 3.2, orange bar is the efficiency of the cell with larger than 1 cm^2, and these efficiencies are more reliable than those of the smaller cells. Among these organic solar cells, the halide perovskite solar cell gives the highest efficiency of 23.7% (1 cm^2) and 25.7% (0.096 cm^2). When the efficiency is compared with almost the same cell size, the efficiency of the perovskite solar cells is a little higher than that of the CIGS solar cell. 25.7% of the smaller size-perovskite solar cells implies that the perovskite solar cell has potentials to catch up the single silicon solar cells of 26.8% instead of the printed solar cells. Organic thin-film solar cells and dye-sensitized solar cells follow the perovskite solar cell with the certified efficiency of 18.2% [1] and 19.2% [2] and 12.25% [1] (15% in paper [3]).

In this report, the recent trends on the dye-sensitized solar cell and the organic thin-film solar cell are firstly described briefly. After that, the present and the future of the halide perovskite solar cells are discussed.

3.2 Mechanism of Charge Separation

These organic solar cells are at least composed of three layers, electron transfer layer (ETL), light harvesting layer(LHL), and hole transfer layer(HTL), as described in Fig. 3.1. These precise structures are different to each other; however, roughly speaking, the charge separation occurs as shown in Fig. 3.3. The LHL is excited by absorbing the sunlight, and excitons are formed in the LHL. By adding exciton

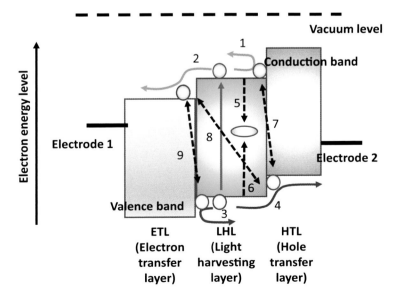

Fig. 3.3. Charge separation mechanism

Table 3.1 Solar cell composition

	Electrode 1	ETL (hole blocking)	LHL	HTL (electron blocking)	Electrode 2
Dye-sensitized solar cells	TFO ITO	TiO_2 SnO_2 ZnO	Organic dyes (Ru—dyes)	Redox species I^-/I_3^- Co^{2+}/Co^{3+} Cu^+/Cu^{2+}	Pt Carbon
Organic thin-film solar cells (Normal structure)	ITO	ZnO SnO_2	Organic n- and p-type compounds Polythiophene derivatives, P3HT, PM6, Y6, L8-BO	p-type organic or inorganic compound MoO_3	Ag, etc.
Perovskite solar cells	FTO ITO	Fullerenes (C60, PCBM, ICBA) TiO_2 SnO_2 ZnO	Halide perovskite	Organic p-type compounds Polythiophene compounds, P3HT, polyamine compound Spiro-MeOTAD PTTA MeO2PACz 2PACz	Au Ag Carbon Cu Al

dissociation energy, free carriers (electrons and holes) are formed. The electron is collected in the conduction band of the ETL, where the conduction band is deeper than that of the LHL (route 2). Since the conduction band of the HTL is shallower than that of the LHL, these electrons are kicked out to the ETL (route 1). In the same way, these holes are collected in the valence band of the HTL (route 3 and 4). Therefore, hole collection and electron collection occur selectively. Photovoltaic performance is degraded by several charge recombinations at the carrier trap sites in LHL (route 5 and 6), at the interface between LHL and HTL (route 7), and at the interface between LHL and ETL (route 9). If the thickness of the LHL is thin, the charge recombination at the interface between ETL and HTL must be considered. In summary, the following items are required for the devices to show photovoltaic performances.

1. Conduction band and valence band energy level is shallower in the following order: ETL, LHL, and HTL.
2. The interface between ETL/LHL and HTL/LHL must be optimized.
3. The structure defects creating the charge recombination sites should be avoided.
4. Fermi level of each layer is deeper in the following order, ETL, LHL, and HTL, which is not described in Fig. 3.3.

Table 3.1 and Fig. 3.4 summarize the materials used in these organic solar cells. The precise explanations are described in each session. Efficiency, open circuit voltage (Voc), and short circuit current (Jsc) can be estimated from the bandgap of LHL by using "Shockler-Quesser limit" [6] (https://github.com/marcus-cmc/

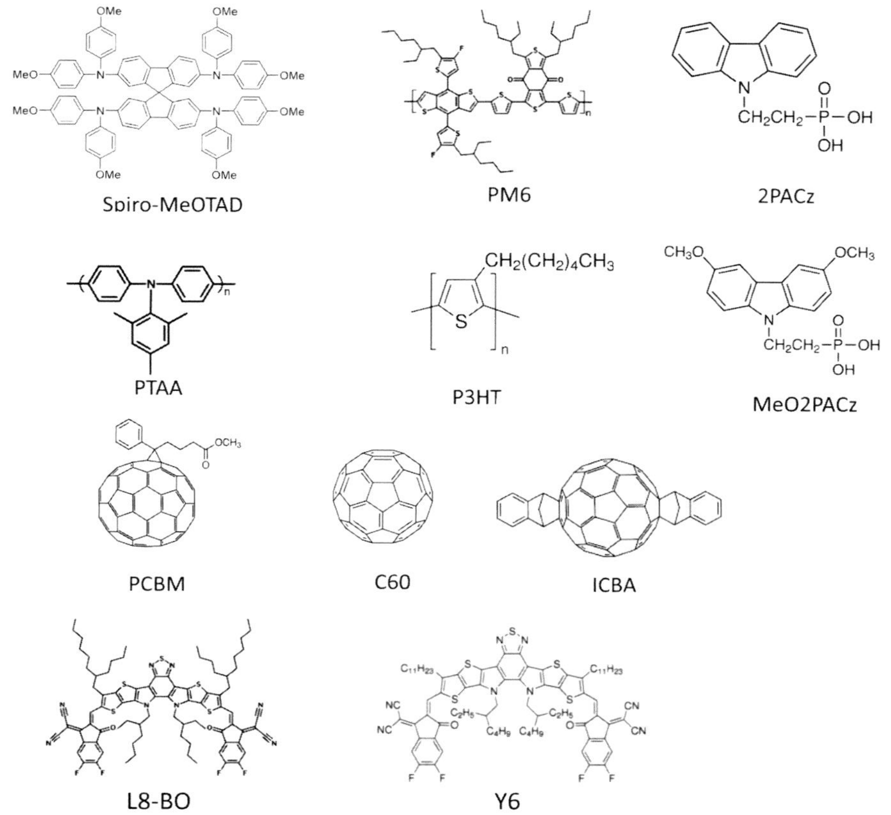

Fig. 3.4. Structures of abbreviation described in Table 3.1

Shockley-Queisser-limit/blob/master/SQ%20limit.csv). As the bandgap is shallower, the LHL can harvest a wide wavelength region of light. Therefore, the Jsc is larger; however, the Voc decreases. Namely, the theoretical efficiency has maximum in a certain bandgap. As summarized in Fig. 3.5, Jsc and fill factor (FF) increased as the bandgap is wider. The difference between the bandgap and Voc (Voc loss) is larger as the bandgap is wider. As shown in Fig. 3.4d, the maximum theoretical efficiency is obtained by LHL with the bandgap of about 1.2–1.4 eV. These theoretical limits are targets for the solar cell research.

3.3 Dye-Sensitized Solar Cells (DSSC)

Dye-sensitized solar cells are composed of n-type inorganic layer (TiO_2, SnO_2, ZnO)/organic dye (LHL)/redox shuttle I^-/I_3^- in solution (corresponding to p-type layer) as shown in Fig. 3.6 [7]. These are correspondents to ETL/LHL/HTL structure

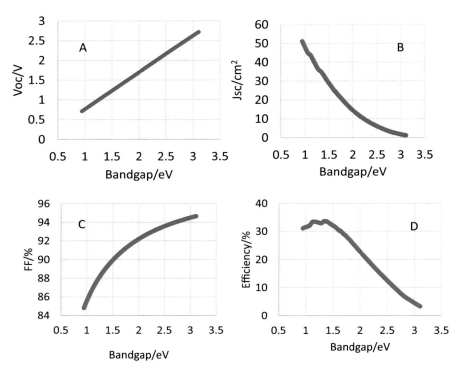

Fig. 3.5. Theoretical Jsc, Voc(A), Jsc(B), FF(C), and efficiency(D) estimated by Shockley-Queisser limit [6] (https://github.com/marcus-cmc/Shockley-Queisser-limit/blob/master/SQ%20limit.csv)

in Fig. 3.3. The TiO_2 layer is the aggregate of nano TiO_2 particles with 10–50 nm diameter. On the surface, dye molecules are bonded through carboxylates. Ru dyes are often used because of the broad visible light absorption. Because of the highly porous structure of the TiO_2 layer, a lot of the dye molecules are adsorbed on the surface of nano-porous TiO_2 layer, which makes the sufficient light harvesting possible, even if the dye adoption is monolayer. Light is harvested by the dye molecule, and the generated electrons are corrected by TiO_2 and diffused to the ITO electrode. The holes in the dye oxidize the I^- to I_3^-, and the resultant I_3^- diffuses to the counter electrode (Pt) and is reduced by the Pt electrode. The design of the dye molecules is significantly important because the dye monolayer determines the direction of the electron and hole injection. For example, the Ru dye in Fig. 3.7 is designed to push the electron toward the carboxylic moieties, and holes are pushed out from the Neocarzinostatin (NCS) moiety site at the excited state. The efficiency has been improved from 12.25% to 15.2% [3]. The recent progress and research directions have been shown by Graetzel and his coworkers [3]. In the case of the dye-sensitized solar cells, the valence band in Fig. 3.3 is correspondent to the redox potential of these redox shuttles. Since the maximum Voc of the DSSC is

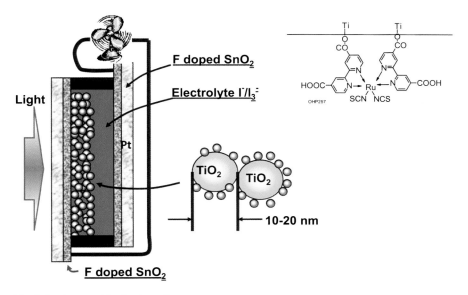

Fig. 3.6. Dye-sensitized solar cell structure

Fig. 3.7. Dye design

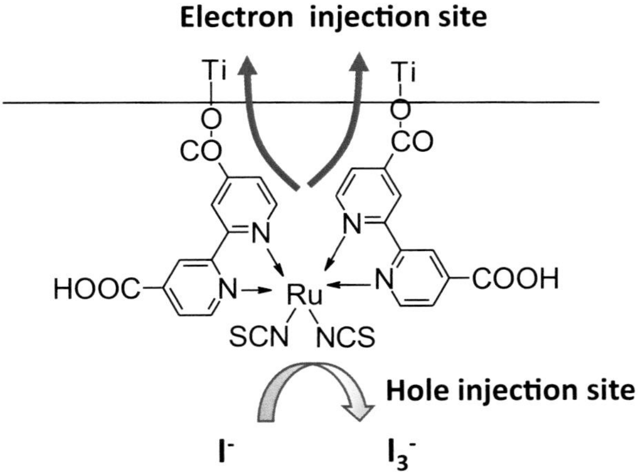

determined by the difference between the conduction band of the HTL and the redox potential, the combination of the ETL with shallow conduction band and the redox shuttle with deeper redox potential gives higher Voc. Previously, I_3^-/I^- redox shuttle has been employed for the research. Since Co^{3+}/Co^{2+}, and Cu^{2+}/Cu^+ redox species has deeper redox potential than I_3^-/I^-, the former is expected to give higher Voc. However, Co^{3+}/Co^{2+} and Cu^{2+}/Cu^+ redox species with the counter ions are large, which limits the diffusion of the electrolyte in the nano pores of the ETL such as nano-porous TiO_2. To solve the diffusion problem, the thickness of the nano-

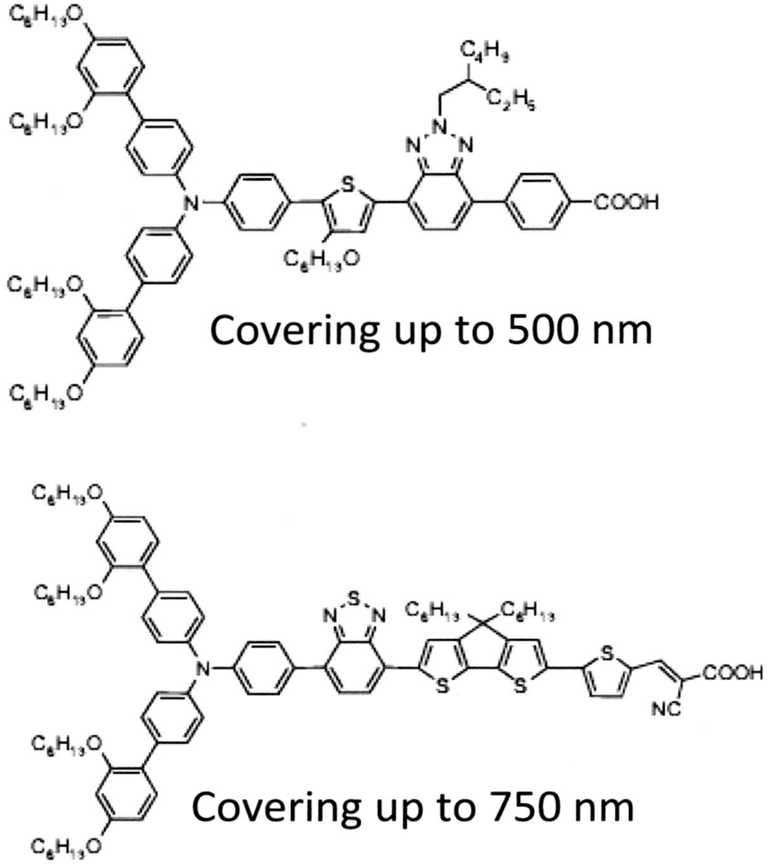

Fig. 3.8. Dyes for co-adsorption

porous ETL is reduced to around 5 microns from 50 microns conventionally employed. In addition, two types of organic dyes with high extinction coefficient (Fig. 3.8), instead of the conventional Ru dyes, are used to keep the light-harvesting properties. Since the two dyes cover different wavelength regions, respectively, these co-adsorbed dyes cover wide range of wavelength up to 750 nm. In DSSC, since the dye monolayer (ETL) separates the ETL and the HTL (redox), the charge recombination is apt to occur through route 8 in Fig. 3.3. Therefore, sufficient surface coverage of the ETL (TiO_2) with these dyes is necessary. They have reported a new dye-adsorption method using pre-adsorption and the substitution. The surface of the TiO_2 was covered by a hydroxamic acid group (pre-adsorption) followed by the dye substitution to give the compact dye-absorption layer on the TiO_2. Namely, the combination of the compact dye monolayer on the TiO_2 and Cu^{2+}/Cu^+ with the deep redox potential gives the high efficiency of 15.2% [3]. The Voc is close to

1.0 V. 14.2% efficiency has been reported by using Co^{3+}/Co^{2+} redox shuttle [8]. It has been reported that each 0.2 eV is needed for the electron and hole to inject efficiently from the dye to the ETL and the dye to the HTL, respectively. Totally, at least 0.4 eV Voc loss from the highest occupied molecular orbital (HOMO)—lowest unoccupied molecular orbital (LUMO) gap of the dye are needed. How to decrease the Voc loss, keeping the charge injection from the HTL and the ETL is the research item further. The efficiency has been improved step by step; however, the absolute efficiency is not as high as that of the halide perovskite solar cells. Therefore, they are aiming at applying these DSSC to indoor uses. They reported about 30% efficiency in ambient condition [3].

3.4 Organic Thin-Film Solar Cells

In this session, organic thin-film solar cells with bulk hetero junction are discussed. The solar cell efficiency has been remarkably improved. The light harvesting of the DSSC is carried out in the monolayer. When the LHL has a certain thickness, the mechanism of the charge separation is not the same as that of the DSSC. Figure 3.9 shows the carrier generation and the carrier diffusion in the LHL. After the LHL is excited, excitons (electron-hole pair) form. By adding some energy called the exciton dissociation energy, the exciton gives free carrier of electron and hole. Both the hole and the electron diffuse in the LHL and collected in each electrode. Since the exciton dissociation occurs at the interface of the p-type layer and n-type layer, the exciton must diffuse in the LHL to the interface. At least 100 nm thickness is needed for the organic molecules to harvest the sufficient light. The exciton diffusion length in the organic molecules is about tens of nanometers. If the HTL/LHL/ETL are deposited layer by layer, the exciton does not reach the

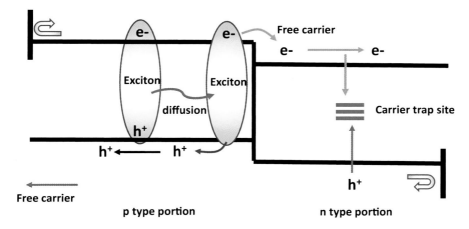

Fig. 3.9. Carrier dynamics in light harvesting layer

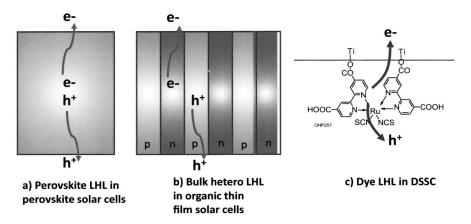

Fig. 3.10. Carrier diffusion in LHL of three solar cells

heterointerface. The bulk hetero layer is composed of interconnected n- and p-type materials (polymers and oligomers) with the tenth of nanometer. This is one of the reasons why the bulk hetero structure is introduced as the LHL. The other reason that the bulk hetero structure is needed is that the organic p-type or n-type material does not carry both of electrons and holes which is needed for the LHL. Figure 3.10 summarizes the carrier generation and carrier diffusion scheme in the LHL for three organic solar cells. Both of electrons and holes can diffuse in the LHL of the perovskite solar cells (Fig. 3.10a). Since organic LHL does not have the bipolar diffusion (electron and hole diffusion) properties, the bulk hetero layer composed of interconnected n- and p-type materials is needed (Fig. 3.10b). In the LHL of the DSSC, electron injection and hole injection occur within one molecule (Fig. 3.10c). The n- and p-type molecules work as the pass of the electron and the hole, respectively. The bulk hetero structure is prepared by coating the precursor solution containing n-type material and p-type material. Both are soluble in a certain solvent as the uniform solution. After the solvent is evaporated, the two materials are not soluble to each other and form the bulk hetero structure with interpenetrated and continuous structure, spontaneously as shown in Fig. 3.11. Polymers and oligomers with n- and p-type natures are used to make the bulk hetero structure. When the polymer is mixed, some phase separation occurs and has various structures such as spherical, lamella, cylinder, and so on. Among them, the spherical phase separation presents independently and is not continuous. Precise material design for both p- and n-type materials is needed to give interpenetrated phase separation automatically. Figure 3.11 shows the whole solar cell structure. On the ITO (indium tin oxide) substrate, HTL, LHL (bulk hetero structure), ETL, and metal electrode are deposited step by step. The bottom HTL and the top ETL work as an electron blocking layer and a hole blocking layer, respectively, because the bulk hetero LHL have the possibility to contact both electrodes.

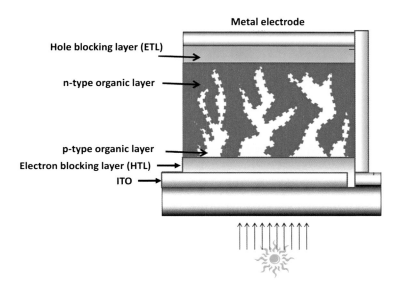

Fig. 3.11. Organic thin film solar cells

Recent drastic efficiency improvement has been brought about by fullerene-free structure. Previously, fullerenes such as C60, PCBM shown in Fig. 3.4, have been used as the n-type materials in LHL. Since the extinction coefficient of these fullerenes is low, the light-harvesting properties were assigned to p-type polymers only. The main role of these fullerenes is carrier path of the electrons in the LHL. Recently, organic n-type materials such as Y6 analogues (Fig. 3.4) have attracted attentions. They have light absorption in the visible light. By coupling the p-type organic polymers and the n-type materials, the mixture can cover the wide range of the visible light region, resulting in the enhancement of Jsc. The example of the organic thin-film solar cells with 19.3% is shown in Fig. 3.12 [9]. D8 and L8-BO show p-type and n-type characters, respectively. In the paper, they realized double-fibril morphology by mixing D8 and PM6 (Fig. 3.4) working as p-type path and L8-BO working as n-type character (Fig. 3.12). The mixture covers the EQE (external quantum efficiency) up to 900 nm, which is almost the same as perovskite solar cells. The double-fibril-type bulk hetero structures enhanced the exciton diffusion length and reduced the recombination rate. The efficiency over 19% has been reported by using the fullerene-free organic thin-film solar cells [10, 11]. These approaches are related to the optimization of the heterointerfaces and show the direction to enhancing the efficiency of the organic thin-film solar cells.

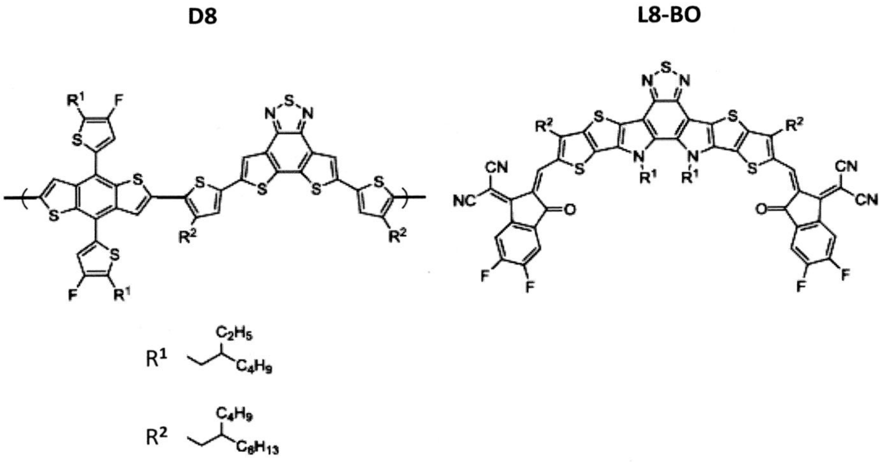

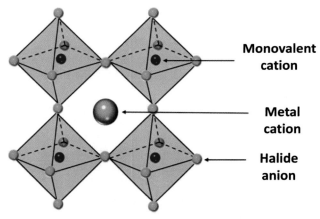

Fig. 3.12. Example of combination of n and p type in LHL [9]. D8: p-type, L8-BO: n-type

ABX$_3$: A (Cation, Cs$^+$, MA$^+$, FA$^+$, etc.), B (Pb^{2+} Sn^{2+}); X (I$^-$, Br$^-$)

MA: methylammonium, FA: Formamidinium

Fig. 3.13. Halide perovskite lattice structure

3.5 Perovskite Solar Cells

The halide perovskite working as LHL of the perovskite solar cells is expressed by ABX$_3$. A is monovalent cations such as methylammonium (MA), formamidinium (FA), Cs$^+$, K$^+$, and so on. B is divalent cations such as Pb^{2+}, Sn^{2+}, and so on. X is halides, such as I$^-$, Br$^-$, and Cl$^-$. Figure 3.13 shows the lattice structure of the representative perovskite. Sn perovskite solar cells follow the Pb perovskite solar

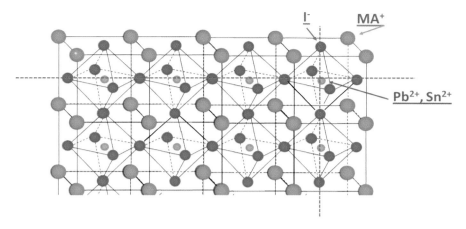

Fig. 3.14. Consecutive halide perovskite lattice unit

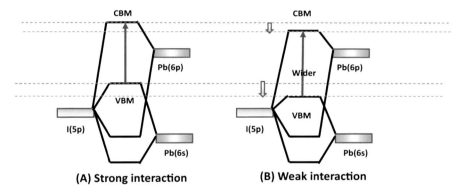

Fig. 3.15. Conduction band and valence band of $APbI_3$

cells. Therefore, in this session, the perovskite solar cell is discussed by comparing Pb perovskite solar cells with Sn perovskite solar cells. The perovskite is composed of octahedron (base unit), where the corner is occupied by these halides. The metal cation is in the center of the octahedron. The space among the octahedron is occupied by A site. The corner of the octahedron unit is occupied by the corner of the octahedron located next the octahedron. Because of this, -Pb-I-Pb-I-Pb- linkages are created in the lattice as shown in Fig. 3.14. This linkage is crucial for explaining the optical properties because the conduction and the valence are associated with these bonding as shown in Fig. 3.15 [12]. The conduction band is the hybrid orbital of I(5p) and Pb(6p). The valence band is the hybrid orbital of I(5p) and Pb(5 s). Both are antibonding orbitals. When the overlapping of I and Pb orbital becomes weaker, the splitting of the hybrid orbital is smaller. Since the overlapping gives larger effects against the I(5p)-Pb(6 s) antibonding level, compared to the I(5p)-Pb(6p)

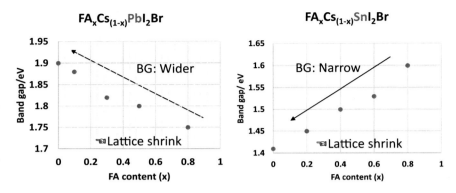

Fig. 3.16. Bandgap shifting depending on A site size [13]

antibonding level, the bandgap becomes wider. For example, when A site of ASnI$_3$ is smaller, the lattice is shrunk with keeping the angle between the octahedron and the next octahedron, resulting in shorter Sn-I bond. Therefore, the bandgap is narrower (From B to A in Fig. 3.15). When the A site of APbI$_3$ is smaller, the angle of the octahedron against the next octahedron is bended (the octahedron is tilted against the next octahedron), and the orbital overlapping was reduced. Therefore, the bandgap becomes wider from A to B in Fig. 3.15. Namely, in ABI$_3$ structure, ASnI$_3$ bandgap is narrower, and APbI$_3$ is wider when A site is smaller. This is experimentally observed as shown in Fig. 3.16 [13]. When FA is replaced with smaller Cs, the bandgap of the Sn perovskite is narrower and that of the Pb perovskite is wider. ABX$_3$ (B:Sn, Pb) with wider bandgap is brought about by replacing I$^-$ with Br$^-$ and Cl$^-$ [14]. Contrary to this, ASnPbI$_3$ (tin-lead alloyed perovskite solar cells) covers the bandgap up to 1.0 eV. The conduction is the antibonding orbital of Pb(p)-I(p), and the valence band is composed of the antibonding of Sn(s)-I(p) which is shallower than that of Pb(s)-I(p). Therefore, the bandgap of the SnPb alloyed perovskite is shallower than that of Pb perovskite and that of Sn perovskite [12]. Figure 3.17 summarizes the relationship between the bandgap and the structure. The bandgap of APbI$_3$ covers from 1.6 eV to 1.5 eV. By replacing I with Br, the bandgap shifts to 2.5 eV. By replacing Br with Cl, the bandgap further shifts to 3.1 eV. ASn$_x$Pb$_{(1-x)}$I$_y$(Br$_{(3-y)}$ covers the narrow bandgap region, the wavelength from 2.1 eV to 1.0 eV by changing x and y [15, 16]. As expected from the structure shown in Fig. 3.13, the stable perovskite structure is obtained by the optimized ion size of A, B, and X in the ABX$_3$ structure. The optimized structure is given by Eq. 3.1 on the tolerance factor. In the case of APbI$_3$, the tolerance factor with 0.8–1.0 gives the perovskite structure shown in Fig. 3.13 at room temperature. Perovskite solar cells have α, β, γ, and δ phase, and so on. The δ phase is transparent and is not appropriate for the solar cell.

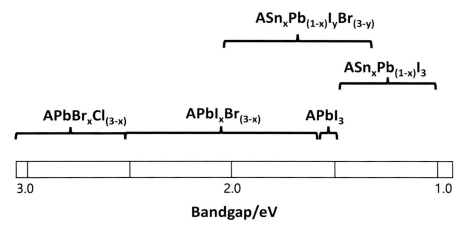

Fig. 3.17. Relationship between bandgap and perovskite structure [12]

$$t = \frac{(R_A + R_X)}{\sqrt{2}(R_B + R_X)} \qquad (3.1)$$

$$\begin{array}{c} \text{AX solution} + \text{BX}_2 \text{ (solution)} \rightarrow \text{ABX}_3 \\ \underline{\text{MeNH}_3{}^+\text{I}^-} + \underline{\text{PbI}_2} \rightarrow \underline{\text{MeNH}_3\text{PbI}_3} \end{array} \qquad (3.2)$$

Precursor solution

$$\underline{3\text{MeNH}_3{}^+\text{I}^- + \text{PbCl}_2} \rightarrow \underline{\text{MeNH}_3{}^+\text{PbI}_3{}^-} + 2\text{MeNH}_3{}^+\text{Cl}^- \qquad (3.3)$$

Perovskite materials   Byproduct

These perovskites are prepared by the methods described in Eq. 3.2 and 3.3. MAPbI$_3$ is taken as the example. Methyl ammonium iodide and PbI$_2$ are dissolved in a certain solvent and dried to give the MAPbI$_3$ crystal (Eq. 3.2). Another way is to use PbCl$_2$ instead of PbI$_2$ (Eq. 3.3). The chloride is replaced by iodide to give the MAPbI$_3$ crystal. The methylammonium chloride is volatile and removed as the by-product. At the beginning of the perovskite solar cell research, Eq. 3.3 was used; however, recently, Eq. 3.2 is often used. The crystal is easy to prepare. However, to prepare the thin and flat film is not easy. To obtain the flat film, antisolvent method is effective [17, 18]. The precursor solution in Eq. 3.2 is spin coated on a substrate, during which poor solvents against the perovskite (chlorobenzene, toluene, ether, and so on) is poured on the wetted film and dried. During the antisolvent process, nucleation occurs under supersaturated conditions, and the crystal grows from the small crystal to give flat perovskite surface with uniform and large grains [19]. Instead of the anti-solvent process, gas flow process is also

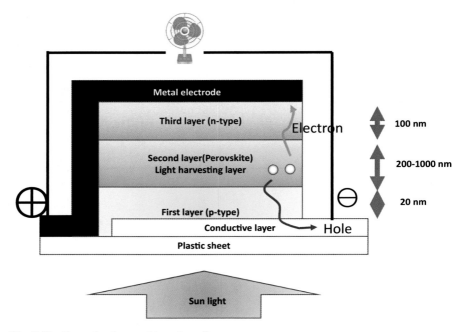

Fig. 3.18. Example of perovskite solar cells

useful to obtain the nucleation [20]. Another important item is dimethyl sulfoxide (DMSO) in the mixed solvent. The DMSO and PbI$_2$ form intercalated structures having 2D structures. After the DMSO is removed by baking, 3D perovskite crystals form [19, 21]. Dimethylformamide (DMF): DMSO [1, 4] is an example of the solvent in the precursor solution. Since the antisolvent process employing much solvent is not appropriate to the mass production, another simple process such as air blow is needed.

The perovskite solar cell is composed of about 20 nm thickness p-type materials, perovskite layer with 200–1000 nm thickness, and n-type materials with around 100 nm thickness as shown in Fig. 3.18. These layers are prepared by spin coating and vapor deposition processes. Figure 3.19 shows the two structures, where the representative materials are added as the example. To fabricate the normal structure, TiO$_2$ working as ETL (n-type), Pb perovskite layer working as LHL, Spiro-MeOTAD (Fig. 3.4) working as HTL (p-type), and metal electrode (Au) are deposited on the ITO or FTO (fluorine-doped tin oxide) substrate, step by step. The inverted structure has a ITO or FTO glass/PEDOT-PSS (Poly3,4-EthyleneDi-OxyThiophene/Poly4-StyreneSulfonate) working as HTL/Pb perovskite/fullerene working as ETL and metal electrode (Ag). Previously, Pb perovskite solar cells with the normal structure gave better efficiency than those with the inverted structure. However, recently, Pb perovskite solar cells with the inverted structure are catching up the normal structure, and 25% efficiency has been reported [22, 23]. The detail is discussed later. On the contrary, the Sn perovskite solar cells with the

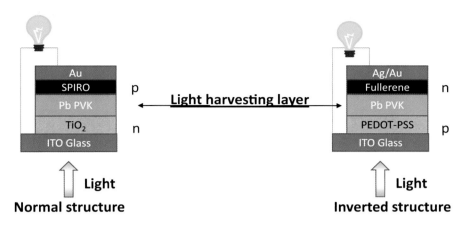

Fig. 3.19 Two types of perovskite solar cells

inverted structure give better efficiency compared to that with the normal structure, because carrier traps form at the interface between metal oxides and Sn perovskite layer [24, 25].

The charge separation mechanism is explained in Fig. 3.3 and 3.8. Solar cell efficiency is discussed in the following items.

- High light absorption
- Low exciton dissociation energy
- Carrier diffusion length of both electron and hole (large carrier mobility and long carrier lifetime)
- Less charge trap density in deep energy level
- Low conduction band offset between LHL and ETL
- Low valence band offset between LHL and HTL

The Pb perovskite film has high absorption coefficient of 10^4–10^5/cm in the wavelength up to 800 nm. The Pb perovskite has low exciton binding energy of 1–5 meV – 32 meV, which is comparable of 17.4 meV of Si, 12.9 meV of CIGS, 4.8 meV of CdTe, smaller than organic materials of 300 meV, and 60 meV of ZnO. Carrier diffusion length reaches 1000 nm. The effective mass is m_h/m_0:0.12 and m_e/m_0:0.15, which is comparable of m_e/m_0:0.26, m_h/m_0:0.39 of Si, m_e/m_0: 0.09, and m_h/m_0:0.72 of CIGS. Therefore, the Pb perovskite can generate carriers with low energy loss, like inorganic solar cells with high efficiency.

One of the items decreasing the efficiency is charge traps. The charge traps are created by the lattice defects such as vacancy, substitution, and interstitial. The vacancy is the lack of ions of A, B, and X in ABX_3 structure. The substitution is created by exchanging the ions with the different ions. For example, X ion is substituted with A or B ion. In the interstitial defect, ions of A, B, and X are located between these ion sites. In each defect, the trap level has been calculated. Carrier traps created at the level shallower than the conduction band (Fig. 3.20a) and deeper

Fig. 3.20. Carrier trap levels in perovskite layer

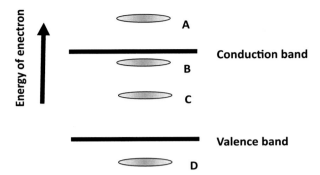

than the valence band (Fig. 3.20d) are not associated with the efficiency losses. Shallow traps close to the conduction band and the valence band do not greatly affect the efficiency decreases. However, serious efficiency losses are brought about by the deep traps shown in Fig. 3.20c. In addition to this, defect formation energy must be considered. Even if a certain defect is calculated to make deep traps, the defect does not affect the serious problems when the defect formation energy is large. According to the calculation [26], the interstitial defect of iodide and the Pb vacancy create the deep traps. Since the defect formation energy of the latter trap is large, the defect does not form a lot. The former defect disappears by reacting with oxygen in the air to form IO_3^- [27, 28]. Therefore, there are no serious defects which affect the efficiency loss. Because of this, Pb perovskite solar cells are called "defect tolerance" solar cells which are one of the items to explain the high efficiency of the Pb perovskite solar cells.

The Sn perovskite, the counterpart of the Pb perovskite solar cells, has different situation. Since the Sn perovskites have shallower conduction and balance than those of the Pb perovskite, iodide vacancy and Sn interstitial make the deep carrier traps [27]. These defect formation energies are not large enough to delete the defect site. In addition, Sn^{2+} defect increases the Sn^{4+} which causes serious efficiency decreases as discussed later [27]. One of the approaches to solve the defect problem is to replace the part of the Sn^{2+} with Pb^{2+} or Ge^{2+} to increase the defect formation energy [29]. The advantage of the alloying is also expected that both carrier traps become shallower because the conduction and the balance band of the SnPb alloyed perovskite are between the Pb perovskite and the Sn perovskite as shown in Fig. 3.21 [26]. Indeed, the efficiency of the SnPb perovskite solar cells has been improved to 23–24% [4, 30].

The presence of Sn^{4+} causes serious efficiency loss. The Sn^{4+} is brought about from the oxidation of Sn^{2+} by oxygen, DMSO [53, 21], and so on, during the storage of the perovskite precursor solutions and device fabrication processes. In addition, the vacancy of Sn^{2+} oxides the Sn^{2+} to Sn^{4+} as discussed just before [26]. The Sn^{4+} works as self-dopants to the Sn perovskite layers and increases the carrier concentration of the layer (in the dark), resulting in decreasing the diode characteristics [13]. To reduce the Sn^{4+} concentration, reducing agents are added. Figure 3.22

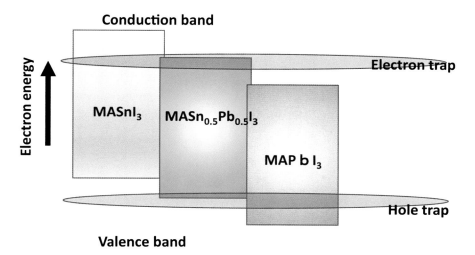

Fig. 3.21. Carrier trap levels in Sn perovskite, Pb perovskite and SnPb alloyed perovskite [26]

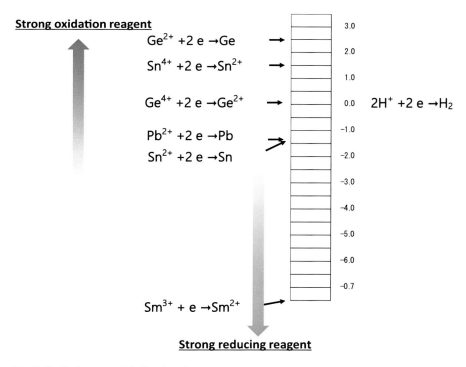

Fig. 3.22 Redox potential of various ions

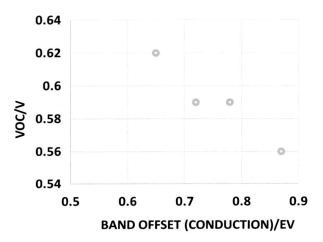

Fig. 3.23. Relationship between band offset and Voc of the Sn perovskite solar cells [13]

shows the redox potential of various ions. These ions are stronger oxidation reagents from the bottom to the top. On the contrary, ions in the bottom are the stronger reducing agent. Sn^{4+} is reduced by the Ge^{2+}, Pb metal, and Sn metal. We have already reported the effectiveness of the Ge^{2+} addition [32, 33]. In addition, the Ge^{4+} formed by reducing the Sn^{4+} is on the heterointerfaces (and probably at the grain boundary) to passivate the surface and decreases the defects at the surfaces [13]. The Sn^{4+} in the precursor ink of the perovskite is reduced by adding Sn metal or Sn metal precursors, followed by filtration to remove the unreacted Sn metal [34–36]. However, in this process, Sn^{4+} formed during the device preparation is not removed. We proposed the direct Sn metal deposition (vacuum deposition) on the perovskite layer to reduce the Sn^{4+} and reported enhance efficiency, because the Sn^{4+} is rich on the surface [37]. SnF_2 [38] and hydrazine salts [39] are effective for suppressing the Sn^{4+} during device fabrication process. Sn^{4+} compounds at the surface of the perovskite layer can be removed as the volatile materials at higher baking temperature [40].

Another item to be solved in the Sn perovskite solar cells is the large conduction band offset of the Sn perovskite and the ETL, because the Sn perovskite has shallow conduction band. One approach is to make the Sn perovskite conduction deeper. We have reported that the introduction of a large A site makes the conduction band deeper, resulting in the efficiency (Voc) as shown in Fig. 3.23 [41–43]. The other approach to reduce the band offset is to use ETL with shallower conduction band than PCBM and C60. ICBA in Fig. 3.4 is one of the candidates. High Voc of about 0.9 V has been reported by using the ICBA [5].

A lot of defects such as I^- vacancy, under-coordinated Sn^{2+} and Pb^{2+}, and so on are in the grain boundary and the heterointerface of HTL/perovskite and perovskite/ETL, because the 3D structure is cut at the edge. When A site of the ABX_3 is bigger, 3D structure is collapsed, and 2D structure forms [44]. The 2D structure does not have cut surfaces at the top and the bottom as shown in Fig. 3.24. To decrease the defect density, the 2D material is added in the perovskite precursor solution, or the 2D material is spin coated on the perovskite layer already prepared. The 2D material

3 Organic Solar Cells 139

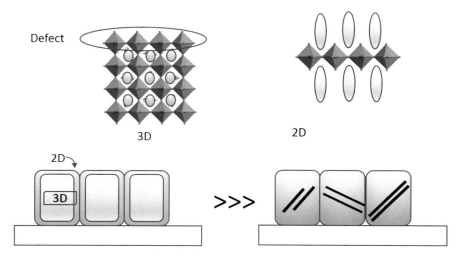

Fig. 3.24. 2D and 3D perovskite structure

must be fabricated vertically so as that the carrier diffusion at the vertical direction would not be disturbed. The representation additives for the 2D material are substituted phenylethyl ammonium cations. The vertical orientation is done by the process conditions such as baking temperature. It is expected that the large A site is adsorbed at a certain facet during the crystal growing process. The highest efficiency of the Pb-free Sn perovskite solar cells with the 2D/3D structure is 14.8% [5]. The structure is ITO/PEDOT-PSS/Sn perovskite (FASnI$_3$)/ICBA/BCP/Al. 4-fluoro-phenethylammonium bromide is added as the additives to for 2D/3D structure. The ICBA with shallower conduction band level is employed as the ETL to decrease the band offset at the interface of the Sn perovskite/ETL.

The passivation of defects at the interface is crucial to enhance the efficiency. The following are examples for the surface passivation to give high efficiency. The highest efficiency of the Pb perovskite solar cells with inverted structure is 25.0% [22]. The structure is ITO/PTAA/Pb perovskite/ferrocenyl-bis-thiophene-2-carboxylate/C60/BCP/Ag. The ferrocenyl-bis-thiophene-2-carboxylate passivates the under-coordinated Pb^{2+} to make Pb-O structure and facilitates the electron injection from the Pb perovskite layer to C60. 3-aminomethylpyridine passivation on Rb$_{0.05}$Cs$_{0.05}$MA$_{0.05}$FA$_{0.85}$Pb[I$_{0.95}$Br$_{0.05}$]$_3$ of ITO/MeO2PACz/perovskite/LiF/C60/BCP/Ag structure gives the efficiency over 25% [23]. The MeO2PACz is discussed later. One of the highest efficiencies of the Pb perovskite solar cells with the normal structure is 25.8%, which is given by the structure of FTO/SnO$_2$/FASnCl$_x$/FAPbI$_3$/SPIRO OMeTAD/Au. SPRO OMeTAD shown in Fig. 3.4. FASnCl$_x$ is the reaction product of SnO$_2$ with FAPbI$_3$ at the heterointerfaces [45]. The bottom line is the insertion of the FASnCl$_x$ layer between SnO$_2$(ETL) and Pb perovskite layer, which reduces the defect concentration at the heterointerfaces and enhances the charge extraction [45]. We have reported the surface passivation with diaminoethane to give

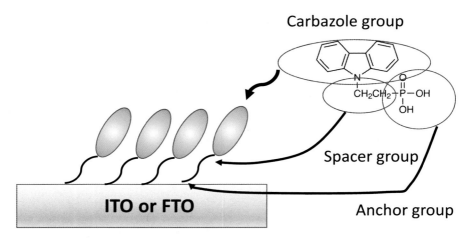

Fig. 3.25. Monolayer of HTL [47]

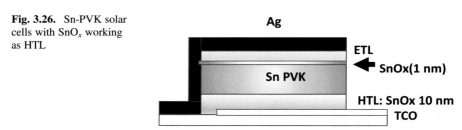

Fig. 3.26. Sn-PVK solar cells with SnO$_x$ working as HTL

the enhanced efficiency of Pb-free Sn perovskite solar cells and SnPb alloyed perovskite solar cells [31, 46].

One of the recent topics is the HTL with monomolecular layer (SAM: self-assemble monolayer). As shown in Fig. 3.25 [47], the SAM molecules are composed of carbazole group (p-type moiety), anchor group (-PO$_3$H$_2$), and spacer group. The anchor group reacts with the $^-$OH of the ITO or FTO to give the monolayer. The spacer group with two carbons, three carbons, and four carbons is discussed. The length of the carbons in the linker group giving the high efficiency depends on the structure of the perovskite layer. In the inverted structure, PEDOT-PSS is employed. However, PEDOT-PSS damages the Sn perovskite layer. Recently, inorganic SnOx ($x = 1.7 - 1.8$) works as the HTL of Sn-PVK [37] and SnPb perovskite solar cells (Fig. 3.26) [48, 49]. The holes are collected by the mid-gap level of the SnO$_x$ as shown in Fig. 3.27 [48, 49].

Fig. 3.27. Hole collection mechanism

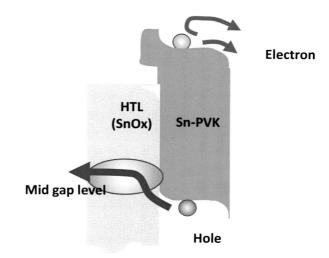

3.6 Perovskite Tandem Solar Cells

The efficiency of the Pb perovskite solar cells is now 25.7% [1]. The research has been shifted to the fabrication of the large modules and stabilities [20, 50]. The theoretical efficiency of the Pb perovskite with the bandgap of 1.55 eV is about 30% [6] (https://github.com/marcus-cmc/Shockley-Queisser-limit/blob/master/SQ%20limit.csv). The perovskite tandem solar cells with the efficiency over 35% which is beyond the theoretical efficiency (33–34%) of the single junction solar cell have attracted attention. The tandem cell is composed of the top cell with the bandgap of 1.6–1.8 eV and the bottom cell with the bandgap of 1.1–1.3 eV as shown in Fig. 3.28. The top cell harvests the visible light, and the bottom cell harvests the near IR light. Supposing that the Voc and Jsc of the bottom cell are Va and Ia and those of the top cell are Vb and Ib, the efficiency of the tandem cell is expressed as shown in Eq. 3.4.

$$\text{Efficiency} = (Va + Vb) \times I \times FF, \text{ where } I = Ia = Ib \quad (3.4)$$

The top cell is connected to the bottom cell with series connection. Therefore, the high voltage of (Va + Vb) is obtained. The Ia of the bottom cells must be matched with the Ib of the top cells. Otherwise, the I is dragged to the lower current side, resulting in large current losses. Because of this current matching, the bandgap of the top and the bottom cells must be optimized. Development of the charge recombination layer located at the interface of the top cell and the bottom cell is crucial, because the holes from the bottom layer must recombine effectively with the electrons from the top cell [51].

Table 3.2 summarizes the efficiency of the tandem cells. The perovskite tandem solar cell consisting of the top Pb perovskite cell (around 1.6 eV)/the bottom Si solar

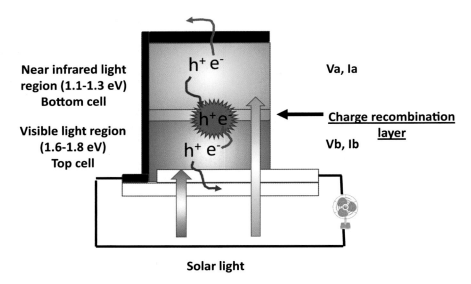

Fig. 3.28. Perovskite tandem solar cell structure

Table 3.2 Efficiency of various perovskite tandem solar cells

Tandem	Efficiency (%)	Area (cm^2)	Institute
Pb PVK(top)/Si(bottom)	31.25 $\Rightarrow$ 32.5		CSEM/EPFL
Pb-PVK/CIGS	24.2	1.045	HZB
Pb-PVK/SnPb-PVK	24.2	1.04	Nanjin Univ.
Pb-PVK/SnPb-PVK	28.0	0.0495	Nanjin Univ.
Si(single) PV	26.7	79	Kanake
Pb PVK PV	25.7	0.096	UNIST

cells (bandgap: around 1.1 eV) gives 32.5%, which is higher than that of the single Si solar cell (26.7%) and that of the single Pb perovskite solar cell (25.7%). The efficiency of the tandem cell consisting of the Pb perovskite top cell/CIGS bottom cell is 24.2%. The all-perovskite solar cells consisting of the Pb perovskite top cell (bandgap: 1.7–1.8 eV)/SnPb alloyed perovskite solar cells with the bandgap of around 1.25 eV give 28% efficiency. The cell is composed of Pb perovskite top cell/atomic layer deposition (ALD)-SnO$_2$/Au/SnPb perovskite bottom cell [52]. 27.4% efficiency of the all-perovskite tandem solar cells with the connection layer of ALD-SnO$_2$/Au/ALD-SnO$_{1.7-1.8}$ as the simplified interlayer is used as the charge recombination [48]. The all-perovskite solar cell has advantage over the other perovskite tandem solar cells in terms of flexible tandem solar cells. To enhance the efficiency of the all-perovskite solar cells, the development of the high-efficiency-bottom cells with the bandgap of 1.0–1.1 eV.

3.7 Conclusion

Organic solar cells are printable solar cells. The advantage of the printable solar cells is that these solar cells can be prepared by the low-temperature process, which make it possible to fabricate lightweight and flexible solar cells, which is not obtained by the other solar cells. The efficiency of these printable solar cells including dye-sensitized solar cells, organic thin-film solar cells, and halide perovskite solar cells has been improved step by step. These solar cells have unique photovoltaic properties with each other, and these application fields would be varied depending on the characteristics. We need various solar cell systems depending on the future needs.

References

1. M.A. Green, E.D. Dunlop, G. Siefer, M. Yoshita, N. Kopidakis, K. Bothe, X. Hao, Prog Photovolt Res Appl **31**, 3–16 (2023) Efficiency Table 61
2. NREL. efficiency chart, https://www.nrel.gov/pv/cell-efficiency.html
3. Y. Ren, D. Zhang, J. Suo, Y. Cao, F.T. Eickemeyer, N. Vlachopoulos, S.M. Zakeeruddin, A. Hagfeldt, M. Grätzel, Nature **613**, 60–65 (2023)
4. H. Shuaifeng, K. Otsuka, R. Murdey, T. Nakamura, M.A. Truong, T. Yamada, T. Handa, K. Matsuda, K. Nakano, A. Sato, K. Marumoto, K. Tajima, Y. Kanemitsu, A. Wakamiya, Energy Environ Sci **15**, 2096–2107 (2022)
5. Y. Bin-Bin, Z. Chen, Y. Zhu, Y. Wang, B. Han, G. Chen, X. Zhang, D. Zheng, Z. He, Adv Mater **33**, 2102055 (2021)
6. S. Ruhle, Sol Energy **130**, 139–147 (2016)
7. B. O'Regan, M. Graetzel, Nature **353**, 737 (1991)
8. J.-M. Ji, H. Zhou, Y.K. Eom, C.H. Kim, H.K. Kim, Adv Energy Mater **10**, 2000124 (2020)
9. L. Zhu, M. Zhang, J. Xinqiu, C. Li, J. Yan, G. Zhou, W. Zhong, T. Hao, J. Song, X. Xue, Z. Zhou, R. Zeng, H. Zhu, C.-C. Chen, R.C.I. MacKenzie, Y. Zou, J. Nelson, Y. Zhang, Y. Sun, F. Liu, Nat Mater **21**, 656–663 (2022)
10. C. Han, J. Wang, S. Zhang, L. Chen, F. Bi, J. Wang, C. Yang, P. Wang, Y. Li, X. Bao, Adv Mater **35**, 2208986 (2023)
11. W. Gao, F. Qi, Z. Peng, F.R. Lin, K. Jiang, C. Zhong, W. Kaminsky, Z. Guan, C.-S. Lee, T.J. Marks, H. Ade, A.K.-Y. Jen, Adv Mater **34**, 2202089 (2022)
12. R. Prasanna, A. Gold-Parker, T. Leijtens, B. Conings, A. Babayigit, H.-G. Boyen, M.F. Toney, M.D. McGehee, Am Chem Soc **139**, 11117–11124 (2017)
13. S. Hayase, Sn-based halide perovskite solar cells., chapter 10, in *Perovskite Photovoltaics and Optoelectronics*, ed. by T. Miyasaka, (Eiley-VCH, 2021), pp. 293–319
14. S. Gholipour, M. Saliba, Chapter 1 - bandgap tuning and compositional exchange for lead halide perovskite materials, in *Characterization Techniques for Perovskite Solar Cell Materials, Micro and Nano Technologies*, (Elsevier, 2020), pp. 1–22
15. K.J. Savill, A.M. Ulatowski, L.M. Herz, Optoelectronic properties of tin–lead halide perovskites. ACS Energy Lett **6**, 2413–2426 (2021)
16. Z. Yang, A. Rajagopal, S.B. Jo, C.-C. Chueh, S. Williams, C.-C. Huang, J.K. Katahara, H.W. Hillhouse, A.K.-Y. Jen, Nano Lett **16**, 7739–7747 (2016)
17. M. Xiao, D.F. Huang, W. Huang, Y. Dkhissi, D.Y. Zhu, P.D.J. Etheridge, D.A. Gray-Weale, P.D.U. Bach, P.D.Y.-B. Cheng, P.D.L. Spiccia, Angew Chem **53**, 9898–9903 (2014)
18. N.J. Jeon, J.H. Noh, Y.C. Kim, W.S. Yang, S. Ryu, S.I. Seok, Nat Mater **13**, 897–903 (2014)

19. M. Bilal Faheem, B. Khan, M. Chao Feng, U. Farooq, F. Raziq, Y. Xiao, Y. Li, All-inorganic perovskite solar cells: energetics, key challenges, and strategies toward commercialization. ACS Energy Lett **5**, 290–320 (2020)
20. D. Li, D. Zhang, K.-S. Lim, H. Yue, Y. Rong, A. Mei, N.-G. Park, H. Han, Adv Funct Mater **31**, 2008621 (2021)
21. Y. Rong, Z. Tang, Y. Zhao, X. Zhong, S. Venkatesan, H. Graham, M. Patton, Y. Jing, A.M. Guloy, Y. Yao, Nanoscale **7**, 10595–10599 (2015)
22. Z. Li, B. Li, W. Xin, S.A. Sheppard, S. Zhang, D. Gao, N.J. Long, Z. Zhu, Science **376**, 416–420 (2022)
23. Q. Jiang, J. Tong, Y. Xian, R.A. Kerner, S.P. Dunfield, C. Xiao, R.A. Scheidt, D. Kuciauskas, X. Wang, M.P. Hautzinger, R. Tirawat, M.C. Beard, D.P. Fenning, J.J. Berry, B.W. Larson, Y. Yan, K. Zhu, Nature **611**, 278–283 (2022)
24. A.K. Baranwal, S. Hayase, Nano **12**(22), 4055 (2022)
25. B. Ajay, S. Shrikant, S. Yoshitaka, K. Gaurav, K.M. Akmal, D. Chao, S.S. Razey, Y. Tomohide, I. Satoshi, S. Qing, M. Koji, H. Shuzi, ACS Appl Energy Mater **5**(8), 9750–9758 (2022)
26. D. Meggiolaro, D. Ricciarelli, A.A. Alasmari, F.A.S. Alasmary, F. De Angelis, J Phys Chem Lett **11**, 3546–3556 (2020)
27. D. Meggiolaro, E. Mosconi, F. De Angelis, ACS Energy Lett **2**(12), 2794–2798 (2017)
28. D. Meggiolaro, S.G. Motti, E. Mosconi, A.J. Barker, J. Ball, C.A.R. Perini, F. Deschler, A. Petrozza, F. De Angelis, Energy Environ Sci **11**, 702–713 (2018)
29. A. Ide, S. Iikubo, K. Yamamoto, Q. Shen, K. Yoshino, T. Minemoto, S. Hayase, Jpn J Appl Phys **61**, 031003 (2022)
30. K. Gaurav, B. Takeru, S. Yoshitaka, S.S. Razey, C. Mengmeng, B. Ajay, L. Dong, S. Yuya, H. Daisuke, N. Daishiro, N. Kohei, K.M. Akmal, S. Qing, S. Hiroshi, H. Shuzi, ACS Energy Lett **7**(3), 966–974 (2022)
31. A.K. Baranwal, K. Nishimura, M.A. Kamarudin, G. Kapil, S. Saini, T. Yabuki, S. Iikubo, T. Minemoto, K. Yoshino, K. Miyazaki, Q. Shen, S. Hayase, ACS Appl Energy Mater **5**(4), 4002–4007 (2022)
32. N. Ito, M.A. Kamarudin, D. Hirotani, Y. Zhang, Q. Shen, Y. Ogomi, S. Iikubo, T. Minemoto, K. Yoshino, S. Hayase, J Phys Chem Lett **9**, 1682–1688 (2018)
33. C.H. Ng, K. Hamada, G. Kapil, M.A. Kamarudin, Z. Wang, S. Likubo, Q. Shen, K. Yoshino, T. Minemoto, S. Hayase, J Mater Chem A **8**, 2962–2968 (2020)
34. M. Chen, M.G. Ju, H.F. Garces, A.D. Carl, L.K. Ono, Z. Hawash, Y. Zhang, T. Shen, Y. Qi, R.L. Grimm, D. Pacifici, X.C. Zeng, Y. Zhou, N.P. Padture, Nat Commun **10**, 16 (2019)
35. T. Nakamura, T. Handa, R. Murdey, Y. Kanemitsu, A. Wakamiya, ACS Appl Electron Mater **2**, 3794–3804 (2020)
36. G. Feidan, S. Ye, Z. Zhao, H. Rao, Z. Liu, Z. Bian, C. Huang, Solar RRRL **2**, 1800136 (2018)
37. W. Liang, C. Mengmeng, Y. Shuzhang, U. Namiki, M. Qingqing, K. Gaurav, B. Ajay, S. Yoshitaka, W. Dandan, L. Dong, M. Tingli, O. Kenichi, S. Takeaki, Z. Zheng, S. Qing, H. Shuzi, ACS Energy Lett **7**(10), 3703–3708 (2022)
38. M.H. Kumar, S. Dharani, W.L. Leong, P.P. Boix, R.R. Prabhakar, T. Baikie, C. Shi, H. Ding, R. Ramesh, M. Asta, M. Graetzel, S.G. Mhaisalkar, N. Mathews, Adv Mater **5**(26), 7122–7127 (2014)
39. C. Wang, G. Feidan, Z. Zhao, H. Rao, Y. Qiu, Z. Cai, G. Zhan, X. Li, B. Sun, Y. Xiao, B. Zhao, Z. Liu, Z. Bian, C. Huang, C. Wang, Adv Mater **32**, 1907623 (2020)
40. J. Zhou, M. Hao, Y. Zhang, X. Ma, J. Dong, F. Lu, J. Wang, N. Wang, Y. Zhou, Matter **5**, 683–693 (2022) Author links open overlay panel
41. Z. Zhang, A.K. Baranwal, S.R. Sahamir, G. Kapil, Y. Sanehira, M. Chen, K. Nishimura, C. Ding, D. Liu, H. Li, Y. Li, M.A. Kamarudin, Q. Shen, T.S. Ripolles, J. Bisquert, S. Hayase, Solar RRL **5**, 2100633 (2021)
42. M.A. Kamarudin, S.R. Sahamir, K. Nishimura, S. Iikubo, K. Yoshino, T. Minemoto, Q. Shen, S. Hayase, ACS Mater Lett **4**(9), 1855–1862 (2022)

43. K. Nishimura, N.A. Kamarudin, D. Hirotani, K. Hamada, Q. Shen, S. Iikubo, T. Minemoto, K. Yoshino, S. Hayase, Nano Energy **74**, 104858 (2020)
44. E.-B. Kim, M.S. Akhtar, H.-S. Shin, S. Ameen, M.K. Nazeeruddin, J Photochem PhotobiolC: Photochem Rev **48**, 100405 (2021)
45. H. Min, D.Y. Lee, J. Kim, G. Kim, K.S. Lee, J. Kim, M.J. Paik, Y.K. Kim, K.S. Kim, M.G. Kim, T.J. Shin, S.I. Seok, Nature **598**, 444–450 (2021)
46. M.A. Kamarudin, D. Hirotani, Z. Wang, K. Hamada, K. Nishimura, Q. Shen, T. Toyoda, S. Iikubo, T. Minemoto, K. Yoshino, S. Hayase, J Phys Chem Lett **10**(17), 5277–5283 (2019)
47. A. Al-Ashouri, V. Magomedov, S.A. Getautis, et al., Energy Environ Sci **12**, 3356 (2019)
48. Y. Zhenhua, Z. Yang, Z. Ni, Y. Shao, B. Chen, Y. Lin, H. Wei, Z.J. Yu, Z. Holman, J. Huang, Nat Energy **5**(9), 1–9 (2020)
49. Y. Zhenhua, J. Wang, M. Bo Chen, A. Uddin, Z. Ni, G. Yang, H. Huang, Adv Mater **34**, 2205769 (2022)
50. Y. Chen, L. Zhang, Y. Zhang, H. Gao, H. Yan, RSC Adv **8**, 10489–10508 (2018)
51. E. Al-Ashouri, B. Köhnen, A. Li, H.H. Magomedov, J.A. Pietro Caprioglio, A.B.M. Márquez, E. Vilches, J.A. Kasparavicius, N. Smith, D. Phung, M. Menzel, L. Grischek, D. Kegelmann, C. Skroblin, T. Gollwitzer, M. Malinauskas, G. Jošt, B. Matič, R. Rech, M. Schlatmann, L. Topič, A. Korte, B. Abate, D. Stannowski, M. Neher, T. Stolterfoht, V. Unold, S.A. Getautis, Science **370**, 1300 (2020)
52. H. Chen, A. Maxwell, C. Li, S. Teale, B. Chen, T. Zhu, E. Ugur, G. Harrison, L. Grater, J. Wang, Z. Wang, L. Zeng, S.M. Park, L. Chen, P. Serles, R.A. Awni, B. Subedi, X. Zheng, C. Xiao, N.J. PodrazaTobin, C.L. Filleter, Y. Yang, J.M. Luther, S. De Wolf, M.G. Kanatzidis, Y. Yan, E.H. Sargent, Nature **613**, s676–s681 (2023)
53. M.I. Saidaminov, I. Spanopoulos, J. Abed, W. Ke, J. Wicks, M.G. Kanatzidis, E.H. Sargent, Conventional solvent oxidizes Sn(II) in perovskite inks. ACS Energy Lett **5**, 1153–1155 (2020)

Chapter 4
Printed Organic Thin-Film Transistors and Integrated Circuits

Hiroyuki Matsui, Kenjiro Fukuda, and Shizuo Tokito

Abstract This chapter focuses on the printed organic thin-film transistor (OTFT) devices and integrated circuits. Section 4.2 describes basic theory for OTFT-based inverters, amplifiers, and feedback circuits. Section 4.3 is devoted to illustrating how to design OTFT-based integrated circuits, including materials, processes, schematic design, circuit simulation, and layout. Section 4.4 picks up key requirements for printed OTFT devices and how to improve the performances. Section 4.5 shows several applications of OTFT integrated circuits such as ultra-flexible circuits, wireless communication, and sensor systems.

Keywords Organic thin-film transistors, OTFT, Printed electronics, Flexible electronics, Integrated circuits, Inkjet

4.1 Introduction

Organic thin-film transistors (OTFT) are the basic elements of integrated circuits, with the functions of switching and amplifying signals [1]. Since p-type (n-type) OTFT devices are turned on only when hole (electron) carriers are accumulated by negative (positive) gate voltage, switching function of OTFT devices would be understood relatively easily. On the other hand, the amplification by OTFT devices needs more detailed consideration. OTFT devices control the drain current by the gate voltage. If we consider the gate voltage as input signal and the drain current as output, the input and output have different unit dimensions, V and A. Hence the meaning of amplification may be difficult to understand. This chapter first introduces how OTFT devices amplify signals in Sect. 4.2.

H. Matsui (✉) · S. Tokito
Research Center for Organic Electronics (ROEL), Yamagata University, Yonezawa, Yamagata, Japan
e-mail: h-matsui@yz.yamagata-u.ac.jp; tokito@yz.yamagata-u.ac.jp

K. Fukuda
Thin-Film Device Laboratory, RIKEN, Wako, Saitama, Japan
e-mail: kenjiro.fukuda@riken.jp

Designing OTFT integrated circuits is one of the challenging parts in the study of OTFT devices. Designing OTFT integrated circuits includes the selection of functional materials, the selection of deposition and patterning method, design rule, circuit design, circuit simulation, and layout. Soluble organic semiconductors and metal nanoparticle inks have enabled integrated circuits to be fabricated by printing methods. Although there are a lot of printable materials these days, it is not always easy to select the materials to be used. Among various printing methods, digital printing such as inkjet can print electronic circuits directly from digital data, making them suitable for rapid prototyping. Design rule is an important communication tool between the process engineers and designers [2]. Designing circuit should always be verified by circuit simulation [3]. Layout of the circuits requires careful consideration to ensure the ease of fabrication and final performance. How to design organic integrated circuits made by digital on-demand printing have rarely been described in literature to our knowledge. Section 4.3 describes how to design OTFT integrated circuits, including practical details, taking an example of complementary organic operational amplifiers (OPAs) [4]. There is a large cultural gap between the researchers in materials, processes, and devices and the researchers in circuit and system designers to date. We hope that this section will help to bridge this gap and make organic integrated circuits more readily accessible in the future.

Section 4.4 picks up key requirements for improving the OTFT performances and how to address them. Surface profile of the printed layers is a unique problem in printed electronics. The most popular phenomenon is so-called a coffee ring effect. Fast operation of OTFT devices is important in many applications since the mobility of OTFT devices is usually not very high. Reducing the channel length and contact resistance is one of the key issues for improving the operation frequency.

Finally in Sect. 4.5, several applications of OTFT integrated circuits are reviewed, including ultra-flexible devices, signal processing, wireless communication, biosensors, and proximity sensors. Some of them are digital circuits, where the voltage signals are assumed to be one of the two values (*high* and *low*), while the others are analog circuits. Regardless of whether they are digital or analog, the basic concept of the circuits is quite similar. Hence the two are explained together in Sect. 4.2.

4.2 Principles of Voltage Amplifiers

This section introduces fundamental voltage amplifiers and feedback circuits. The relation between the input and output voltages of the amplifiers is derived from the characteristics of each OTFT with elementary mathematics. A digital logic gate, inverter, can also be understood in the same way as the amplifiers. In practice, amplifiers are used quite frequently in feedback circuits. The basic principles of feedback with single-end and differential amplifiers are also introduced.

4.2.1 Voltage Transfer Curves of Inverters and Amplifiers

An inverter is the most basic OTFT circuit and a good starting point for understanding the operation of various digital and analog circuits [5]. Figure 4.1a shows a complementary inverter (sometimes referred to as complementary metal-oxide-semiconductor (CMOS) inverter, although it does not actually have an oxide layer) that combines a p-type OTFT and an n-type OTFT. With positively increasing the gate-to-source voltage V_{GS}, the current in a p-type OTFT decreases and that in n-type OTFT increases as illustrated in Fig 4.1c. V_{DD} is a DC power supply and may be, for example, 5 V. When the input terminal voltage V_{IN} is 0 V (*low*), the p-type OTFT has a negative V_{GS} of -5 V and is turned on, and, at the same time, the n-type OTFT has zero V_{GS} and is turned off. As a result, the output terminal V_{OUT} is connected to V_{DD} via the low-resistance channel of the p-type OTFT and has almost the same voltage of V_{DD} (*high*) (Fig. 4.1b). On the other hand, when $V_{IN} = 5$ V (*high*), the p-type OTFT is turned off, and the n-type OTFT is turned on. As a result, V_{OUT} is connected to the ground via the low-resistance channel of the n-type OTFT and becomes 0 V (*low*). Thus this circuit inverts the voltage signal between 0 V (*low*) and 5 V (*high*), which is why this circuit is called inverter.

An inverter is a kind of amplifier, since the V_{OUT} has a sharp transition from *high* to *low* at a certain voltage. The slope of the transition gives a voltage amplification

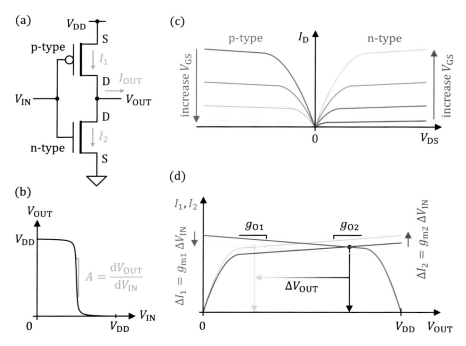

Fig. 4.1 (a) Circuit diagram and (b) voltage transfer curve of a complementary inverter. (c) Output curves of individual OTFTs. (d) How to determine V_{OUT} from the two output curves

factor, the so-called gain: $A \equiv dV_{OUT}/dV_{IN}$. For example, if the slope is -10 mV, 1 mV change in input voltage gives -10 mV change in output voltage. The voltage gain is negative in many cases; this is not problematic and is even preferable in feedback circuit explained in the next section. The relation between the V_{IN} and V_{OUT}, here we call it a voltage transfer curve, can be understood from the characteristic of the two OTFT devices in the following way. Since the drain-to-source voltage V_{DS} of the p-type OTFT is $V_1 = V_{DD} - V_{OUT}$, the current from source to drain of the p-type OTFT, I_1, is a function of V_1. Similarly, the V_{DS} of the n-type OTFT is $V_2 = V_{OUT}$, and the current from drain to source of the n-type OTFT can be written as $I_2(V_2)$. The functions $I_1(V_1)$ and $I_2(V_2)$ are the output ($I_D - V_{DS}$) characteristics of the two OTFT devices. The Kirchhoff's law gives $I_1 = I_2$, assuming that the currents at the input gate terminal and the output terminal are negligible. Therefore, solving

$$I_1(V_{DD} - V_{OUT}) = I_2(V_{OUT}) \tag{4.1}$$

gives the actual V_{OUT}. This equation can be understood by the graph in Fig. 4.1d, where $I_1(V_{DD} - V_{OUT})$ and $I_2(V_{OUT})$ are plotted along the bottom axis of V_{OUT}. Notice that $I_1(V_{DD} - V_{OUT})$ curve is the output characteristics of the p-type OTFT shifted by V_{DD}. Then, the x-coordinate of the cross point of the two curves gives the actual V_{OUT}. If V_{IN} is varied, both the curves of $I_1(V_{DD} - V_{OUT})$ and $I_2(V_{OUT})$ change, and accordingly V_{OUT} changes. The dependence of V_{OUT} on V_{IN} gives the voltage transfer curve. In particular, when the saturation parts of the two curves intersect each other, a small change in V_{IN} causes a large change in V_{OUT}. This provides a high voltage gain.

Here we derivate a formula of the voltage gain from Eq. (4.1). By differentiating Eq. (4.1) by V_{IN}, we obtain

$$\frac{\partial I_1}{\partial V_{IN}} - \frac{\partial I_1}{\partial V_{OUT}} \frac{dV_{OUT}}{dV_{IN}} = \frac{\partial I_2}{\partial V_{IN}} + \frac{\partial I_2}{\partial V_{OUT}} \frac{dV_{OUT}}{dV_{IN}} \tag{4.2}$$

$$A = \frac{dV_{OUT}}{dV_{IN}} = -\frac{-\frac{\partial I_1}{\partial V_{IN}} + \frac{\partial I_2}{\partial V_{IN}}}{\frac{\partial I_1}{\partial V_{OUT}} + \frac{\partial I_2}{\partial V_{OUT}}} = -\frac{g_{m1} + g_{m2}}{g_{o1} + g_{o2}} \tag{4.3}$$

Here g_{m1} and g_{m2} are the transconductance, and g_{o1} and g_{o2} are the output conductance of the two OTFT devices. There are two guidelines for increasing the voltage gain A.

- Increase either g_{m1} or g_{m2}.
- Decrease both g_{o1} and g_{o2}.

The first guideline means sharp switching of the OTFT devices with respect to gate voltage. The second guideline indicates that drain current should be less dependent on drain voltage.

While the current at output terminal have been neglected until now, it is important to understand the effect of non-negligible current at the output terminal I_{OUT} on the output voltage. Considering nonzero I_{OUT}, the Kirchhoff's law gives

$$I_1(V_{DD} - V_{OUT}) = I_2(V_{OUT}) + I_{OUT} \quad (4.4)$$

By differentiating Eq. (4.4) by I_{OUT}, we obtain

$$R_o \equiv \frac{dV_{OUT}}{dI_{OUT}} = -\frac{1}{g_{o1} + g_{o2}} \quad (4.5)$$

R_o is called the output impedance of the circuit and indicates the dependence of the output voltage on the output current. The small output impedance is usually preferable because such circuits can drive high current without affecting the output voltage. However, Eqs. 4.3 and 4.5 indicate that a reduction in output impedance leads to a reduction in gain. Therefore, a two-stage circuit with a high gain stage followed by a low output impedance stage is often used.

Transient characteristics of the amplifiers are also important, since the amplifiers require finite time until the output voltage transit to an expected value. A simplified consideration of the output capacitance C_{OUT} and the output current I_{OUT} helps to understand the transition speed of the output voltage. C_{OUT} is an effective capacitance, representing any loads connected to the output terminal as well as parasitic capacitance. Considering the charge $Q = C_{OUT}V_{OUT}$ at the capacitance, there is a relation of $\frac{dQ}{dt} = C_{OUT}\frac{dV_{OUT}}{dt}$ between the rate of Q and V_{OUT}. The time derivative of charge $\frac{dQ}{dt}$ is the output current required to change the output voltage at a rate of $\frac{dV_{OUT}}{dt}$. Therefore, faster transition speed of $\frac{dV_{OUT}}{dt}$ requires smaller output capacitance C_{OUT} and higher output current. In other words, smaller parasitic capacitance, smaller load capacitance, and higher OTFT current are required for the fast operation of the amplifiers.

Power consumption is another factor to be considered in designing integrated circuits. Power consumption can be estimated by the product of the supply voltage V_{DD} and the current I_1 if I_{OUT} is neglected. Usually, increasing the operation speed and decreasing the output impedance increase the power consumption.

The above concept can also be applied to other types of amplifiers. Figure 4.2a, b show depletion-load (or zero-V_{GS}) and diode-load amplifiers, respectively, and both can be composed of only p-type OTFT devices. Figure 4.2c uses a resistor instead of an OTFT. In these three circuits, the lower elements can be considered simply as two terminal elements, since the gate of OTFT devices are shorted to either source or drain electrodes. The current I_2 through the lower elements does not depend directly on V_{IN} and hence $g_{m2} = 0$. The typical curves of I_2 are shown in Fig. 4.2d. If the output conductance of the upper OTFT is sufficiently low (if the drain voltage dependence of the saturation current is sufficiently small),

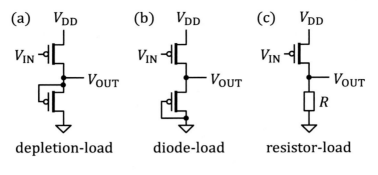

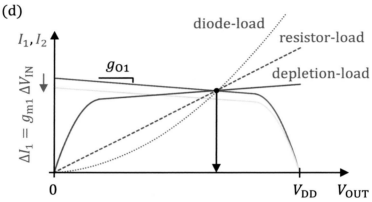

Fig. 4.2 (a) Depletion-load, (b) diode-load, and (c) resistor-load inverters. (d) How to determine V_{OUT} from the characteristics of each component

$$\frac{dV_{OUT}}{dV_{IN}} = -\frac{g_{m1}}{g_{o2}} = -g_{m1}r_{o2} \quad (4.6)$$

where output resistance is defined as $r_{o2} = 1/g_{o2}$. In the resistor-load amplifier, r_{o2} equals to the resistance R. As the curves in Fig. 4.2d show, r_{o2} is typically the largest in depletion-load amplifiers and the smallest in diode-load amplifiers. Therefore, the depletion-load amplifiers have superior gain but inferior output impedance and speed. The diode-load inverters have inferior gain and but superior output impedance and speed. The diode-load inverters also have high linearity (small V_{IN} dependence of gain) and can avoid the distortion of signal waveforms. Resistor-load amplifiers exhibit intermediate characteristics between the depletion-load and diode-load amplifiers. Table 4.1 summarizes the features of the four amplifiers.

Many integrated circuits, not only inverters, have the following points in common.

- DC power is supplied between V_{DD} and the ground. Negative DC power supply V_{SS} can be used instead of the ground to expand the range of output voltage.
- Output voltage is limited between the V_{DD} and ground (or V_{SS}).

Table 4.1 Features of four types of inverters and amplifiers (+: good, −: bad, 0: fair)

	Complementary	Depletion-load	Diode-load	Resistor-load
Gain	+	+	−	−
Output impedance	−	−	+	0
Speed	+	−	+	0
Power consumption	+	0	−	−
Linearity	−	−	+	−

- There are many trade-offs among the performance parameters such as gain, output impedance, speed, and power consumption. Maximizing all the parameters together is impossible. This is why the integrated circuits need to be designed to meet the purpose of each application.
- Device parameters such as channel width and threshold voltage need to be optimized for individual OTFT devices to obtain desired characteristics. Multiple equivalent OTFT devices in an integrated circuit would not help in many cases.

There are several differences between the digital inverters and analog amplifiers. Digital inverters are supposed to output only two values, *high* and *low*. This means that the operation around the transition regime should be avoided. As long as the input voltage is apart from the transition regime, the digital circuits are quite resistant to noise. On the other hand, analog amplifiers need high gain and should be operated around the transition regime.

4.2.2 Feedback Circuits

If an amplifier has a high enough gain, feedback loop can expand the functions of the circuits. The purpose of the feedback can be precise control of gain, high linearity, current sensing, integration, differentiation, frequency filters (low pass, high pass, band pass), etc.

Here we simplify the voltage transfer characteristics of the amplifiers in the previous section as

$$V_{OUT} = -A(V_{IN} - V_M) \qquad (4.7)$$

where V_M is the offset voltage and $A > 0$. Feedback technique usually needs a negative gain, and such an amplifier is depicted by the symbol in Fig. 4.3a, where the V_{DD} and ground terminals are omitted and the small circle indicates the inversion of the signal. Feedback is a technique to connect the output terminal to the input terminal via some circuit element such as resistors and capacitors. Figure 4.3b shows an example of feedback circuits for current sensing [6]. Since the input of the amplifiers has high impedance as gate of OTFT devices, the current I_{IN} flows

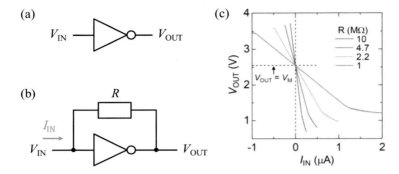

Fig. 4.3 (**a**) Symbol of an inverting amplifier. (**b**) Diagram and (**c**) measured characteristics of a feedback circuit for current sense [6]

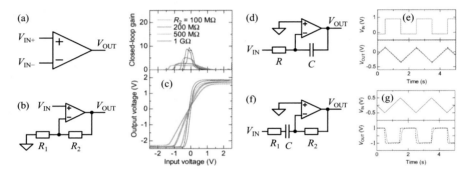

Fig. 4.4 (**a**) Symbol of an operational amplifier. (**b**) Diagram and (**c**) measured characteristics of the non-inverting amplifier. (**d**) Diagram and (**e**) measured characteristics of the integrator. (**f**) Diagram and (**g**) measured characteristics of the differentiator [4]

through the resistor R. The Ohm's law gives $V_{OUT} = -RI_{IN} + V_{IN}$. Since $V_{OUT} = -A(V_{IN} - V_M)$ from Eq. (4.7), we finally obtain

$$V_{OUT} = \frac{A}{1+A}(-RI_{IN} + V_M) \approx -RI_{IN} + V_M \qquad (4.8)$$

Here A is assumed high enough. Since there is a linear relation between I_{IN} and V_{OUT} with a known ratio R, this circuit can be used to measure current by measuring the output voltage with an analog-digital converter (DAC). A problem of this circuit is the offset V_M in the output voltage, because V_M is usually difficult to obtain reproducibly and can depend on temperature. Unexpected changes of V_M cause errors in the measured values.

Differential amplifiers in Fig. 4.4a are more frequently used in practical feedback circuits [4]. The differential amplifiers with extremely high gain (typically 100,000) are called operational amplifiers (OPAs). The voltage transfer characteristics of OPAs are modeled as

$$V_{\text{OUT}} = A(V_{\text{IN}+} - V_{\text{IN}-}) \tag{4.9}$$

where $V_{\text{IN}+}$ and $V_{\text{IN}-}$ are the voltages of positive and negative input terminals. Figure 4.4b shows a non-inverting amplifier with precise control of gain. The voltage transfer characteristics are given by

$$V_{\text{OUT}} = \left(1 + \frac{R_2}{R_1}\right) V_{\text{IN}} \tag{4.10}$$

Here the gain is $1 + (R_2/R_1)$ and can be easily controlled by the ratio of two resistors as in Fig. 4.4c. Figure 4.4d, e show an integrator circuit whose voltage transfer characteristics is given by

$$V_{\text{OUT}}(t) = -\frac{1}{RC} \int_0^t V_{\text{IN}}(t') dt' \tag{4.11}$$

A similar circuit can also be used as a low-pass filter. Figure 4.4f, g show a differentiator whose voltage transfer characteristics are given by

$$V_{\text{OUT}}(t) = -RC \frac{dV_{\text{IN}}}{dt} \tag{4.12}$$

A similar circuit can also be used as a high-pass filter.

4.3 Designing OTFT Circuits

The manufacture of single-crystal silicon devices has required a huge, clean, and rigorously controlled facility environment with extensive use of high temperature, vacuum, and photolithography processes. In contrast, soluble organic semiconductors and metal nanoparticle inks have enabled integrated circuits to be fabricated by printing methods under nearly room temperature and ambient pressure. Among a variety of printing methods, inkjet and dispenser are digital on-demand printing that do not require printing plates. Such printing methods are suitable to rapid prototyping and cost-efficient electronic devices.

This section describes how to design OTFT integrated circuits fabricated by the digital on-demand printing, taking as an example a complementary organic OPA with a layer structure shown in Fig. 4.5. The design flow is as follows:

1. Selecting materials and processes
2. Determining design rule and available device parameters
3. Schematic design and verification by circuit simulation
4. Layout

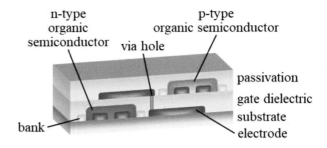

Fig. 4.5 Example of the cross section of OTFT circuits

The first step is to select materials such as metals, semiconductors, and insulators and, at the same time, the processes such as inkjet, dispenser, and spin coat. The second step is to determine the feasible range of dimensional parameters (e.g., line width and space) and device parameters (e.g., mobility), which depend on the selected materials and processes. The collection of acceptable dimensional parameters and guidelines is called a *design rule*. The third step is to design a circuit diagram with schematic symbols of OTFT devices and passive elements. The device parameters such as channel length, channel width, threshold voltage, resistance, and capacitance of passive elements need to be specified in this step, while no positional information of x- and y-coordinates is necessary yet. The designed circuit diagrams are verified by using circuit simulation such as LTspice (Analog Devices, Inc.) and modified if required. Finally, layout design is carried out using computer-aided design (CAD) software within the dimensional parameters of the design rules.

4.3.1 Materials

Material selection is as important as process selection in the fabrication of OTFT devices. The variety of materials for OTFT devices is quite large, and many of them are under development. This section introduces the required properties and representative materials for each layer.

4.3.1.1 Substrates

Substrates are the base for all layers and require various characteristics such as heat resistance, smooth surface, dimensional stability, low cost, and, sometimes, flexibility. Besides rigid substrates such as glass plates and silicon wafers, polyimide (PI), polyethylene naphthalate (PEN), and polyethylene terephthalate (PET) films are often used as flexible substrates these days. Glass is a common rigid substrate material that is inexpensive and readily available and has excellent heat resistance, dimensional stability, and transparency. Silicon wafers are expensive but are often used in basic research because they have excellent surface smoothness, and if the surface is thermally oxidized, the silicon and silicon oxide layer can be used as a gate electrode and dielectric layer, respectively. PI, PEN, and PET are all flexible plastic films and are used to fabricate flexible devices and circuits. Among these, PI is the most expensive and has the highest heat resistance, while PET is the cheapest and has the lowest heat resistance.

Heat resistance is one of the most basic performance requirements for substrates. In the fabrication process of OTFT devices, devices are heated in drying the solvent, sintering the metal nanoparticles, and annealing the organic semiconductor layer. For example, many metal nanoparticles require sintering at around 100–150 °C, making ordinary PET films with a heat resistance temperature of 100 °C unsuitable as substrates.

Dimensional stability is another very important performance requirement for substrates. Plastic films undergo significant dimensional changes with heating and cooling cycles. A 0.1% dimensional change in a 100 mm square substrate would result in a 0.1 mm misalignment at both ends of the substrate. In this case, the layout of different layers must be designed so as to allow for this degree of misalignment.

The thickness and flexibility of substrates have become increasingly important in recent years. A small substrate thickness as well as a small Young's modulus is important for flexible electronic devices because the bending stiffness is proportional to the Young's modulus and the third power of the thickness. Reducing the bending stiffness make the devices more flexible and fit with many curved and soft objects [7].

The surface smoothness of substrates has a significant influence on the quality of the films deposited on them. Large surface roughness of substrates makes film deposition nonuniform, creates more pinholes, and inhibits the crystal growth of organic semiconductor thin films. Ideally, the surface roughness should be less than a few nm. Coating substrates with polymer materials such as polyvinyl phenol (PVP) can improve the surface smoothness [8]. The affinity between the substrate surface and the ink for the next layer is also important when fabricating in wet processes.

Other important factors include cost, as the substrate generally accounts for the largest volume proportion of the entire device, transparency, gas barrier properties, biocompatibility, and biodegradability.

4.3.1.2 Electrodes and Interconnections

Metals are used as electrodes and interconnections. In particular, gold, silver, and copper can be made into nanoparticles and dispersed in a solvent to enable patterning by printing methods [9]. Silver nanoparticle inks are the most readily available, while there are cost issues as with gold nanoparticle inks. Copper nanoparticle inks need to be devised to prevent air oxidation. Other materials such as conductive polymer, poly(3,4-ethlenedioxythiophens) polystyrene sulfonate (PEDOT:PSS), carbon nanotube, and graphene can also be used if the relatively low conductivity is acceptable. They are superior in terms of cost, process temperature, and transparency.

The distance between the source and drain electrodes of OTFT devices requires the most miniaturization in patterning. Therefore, the patterning method of the electrode must have high resolution and positional accuracy. Although photolithography is the most common patterning method, high-resolution printing methods such as inkjet printing, gravure offset printing, and reverse offset printing can also be used [10].

When metallic nanoparticle inks are used, sintering is necessary to fuse the nanoparticles together after printing. Sintering temperatures are typically 100–300 °C, although some sintering can proceed at room temperature [11]. However, there is generally a trade-off between sintering temperature and storage stability; the nanoparticle inks that can be sintered at lower temperatures have shorter storage stability. It is important to select an ink with an appropriate sintering temperature, taking into account the heat resistance of the substrate and other layers.

When a metal and semiconductor are in contact, as in the case of source and drain electrodes in OTFT devices, the work function of the metal has a significant influence on the electrical properties of the interface. Ohmic contact is usually preferable, and Schottky contact should be avoided. Therefore, a metal with a large work function is preferred for p-type organic semiconductors and a metal with a small work function for n-type organic semiconductors. In practice, however, metals with large work functions are costly, such as gold and platinum, while metals with small work functions are unstable in air, such as Ca and Mg. Therefore, the choice of metals available in practice is never large. Silver is most widely used in printing processes, often followed by copper or gold.

4.3.1.3 Carrier Injection Layers

In case it is difficult to form Ohmic contact between the metal and semiconductor, a carrier injection layer may improve the electrical property. There are three types of carrier injection layers. The first is conductive materials with different work functions. As the injection layer is thin, electrical conductivity lower than conventional metals is acceptable [12]. The second is self-assembled monolayers (SAM) with electric dipoles [13]. Self-assembled monolayers are molecular monolayers that exhibit a specific orientation by chemical bonding to a metal surface, and the effective work function of the electrode changes according to the electric dipole of the molecule. The third is a carrier doping layer which increases the carrier density of the semiconductor in the vicinity of the electrode [14]. Acceptor molecules are used for p-type organic semiconductors and donor molecules for n-type organic semiconductors. By increasing the carrier density near the electrode, the thickness of the Schottky barrier can be reduced and electrical properties improved.

4.3.1.4 Gate Dielectrics

Gate dielectric layers have significant impacts on carrier mobility, operating voltage, and device yield. The carrier density n accumulated by gate voltage V_G is $n = C_i(V_G - V_{th})/e = \epsilon(V_G - V_{th})/ed$, where C_i is the capacitance per unit area, V_{th} the threshold voltage, e the elementary charge, ϵ the absolute permittivity, and d the thickness of the dielectric layer. Therefore, the higher the permittivity and the smaller the thickness, the more carriers can be induced at a smaller gate voltage, thus reducing the operating voltage [15]. However, it is important to balance these parameters, as an increase in permittivity leads to decrease mobility and a decrease in film thickness increases the probability of pinholes [16].

Dielectric breakdown occurs when excessive voltage is applied to the gate dielectric layer. The voltage or field strength at which dielectric breakdown occurs is determined by intrinsic factors due to the properties of the dielectric material and extrinsic factors derived from defects such as pinholes. For example, the breakdown electric field of silicon oxide is reported to be 5–10 MV/cm. Parylene, an insulating polymer, also exhibits excellent voltage tolerance, with a breakdown field of around 2 MV/cm. Gate dielectric layers that can be deposited by wet coating include melamine-crosslinked polyvinyl phenol and vinyl cinnamate.

4.3.1.5 Semiconductors

There are p-type and n-type organic semiconductor materials, while the terminology differs from that of impurity-doped inorganic semiconductors such as n-doped silicon. The term p-type organic semiconductor refers to materials with high hole mobility and low ionization energy. Conversely, the term n-type organic semiconductor refers to materials with high electron mobility and high electron affinity. The materials are intrinsic semiconductors in principle, and carriers are accumulated by gate electric field. Crystalline small-molecule semiconductors tend to exhibit high mobility. For example, C8-BTBT, C10-DNBDT-NW, and Ph-BTBT-C10 exhibit mobilities exceeding 10 cm^2/Vs [17–19]. Polymer semiconductors, on the other hand, have better uniformity, and some of them have significant stretchability. Examples of organic semiconductors suitable for OTFT devices are shown in Fig. 4.6.

Important factors in selecting organic semiconductor materials for OTFT devices include mobility, ionization energy or electron affinity, atmospheric stability, thermal stability, solubility, and film uniformity. Mobility is the most important performance indicator for the response speed and on-current of OTFT devices. Ionization energy and electron affinity are closely related to the carrier injection properties from electrodes and the gate threshold voltage of OTFT devices. Density functional theory (DFT) calculations of the highest occupied molecular orbital (HOMO) and the lowest unoccupied molecular orbital (LUMO) give rough estimation of the ionization energy and electron affinity in practice. Chemical and physical stability

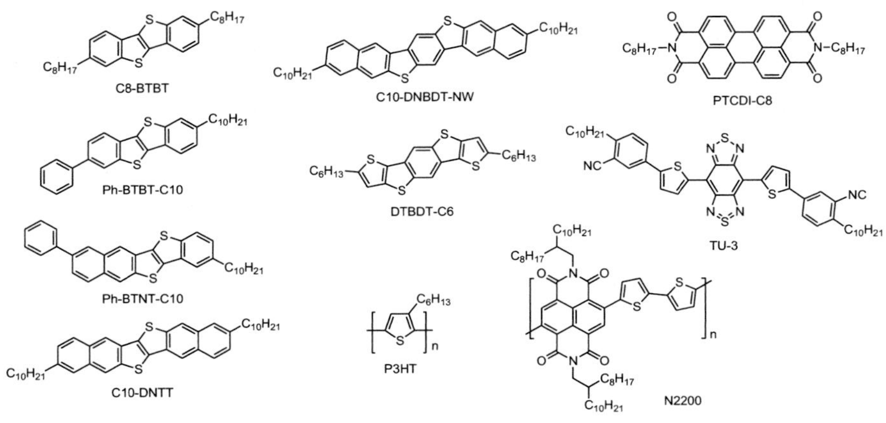

Fig. 4.6 Organic semiconductor materials for OTFTs

against oxygen and moisture in air and thermal stability are also important in practical applications. As a guide to atmospheric stability, the LUMO of n-type organic semiconductors should be more than 4.0 eV, and the HOMO of p-type organic semiconductors should be between 5.0 eV and 5.4 eV [20]. Solubility is important when using printing methods. Film uniformity includes thickness uniformity, film continuity, and crystal domain structures.

4.3.1.6 Other Components

Liquid-repellent materials are often used as banks for patterning other layers. Typical materials are fluorinated polymers such as Teflon and Cytop and fluorinated self-assembled monolayers such as trimethoxy(perfluoroalkyl)silane. Such materials can be directly patterned by printing or partially removed by UV/ozone or plasma after coating.

Via holes are the electrical interconnections between different layers. Holes can be formed by the direct patterning of the insulation layer or by etching with an yttrium-aluminum-garnet (YAG) laser or plasma. Conductive material can be poured into the holes to make electrical interconnection.

Connector pads are terminals for connecting the circuit to external circuits or to instrumentation for testing. As they will have physical contact with the mating metal parts, they should have good adhesion to the substrates and scratch resistance. It is also important that solder erosion does not occur if solder mounting is anticipated.

4.3.2 Deposition and Patterning Methods

Besides vacuum processes such as thermal evaporation and sputtering, wet processes using solutions and dispersions and dry processes using toner can be employed. This chapter focuses on wet processes, of which there are two types: coating, which produces a uniform film without patterning, and printing, which performs direct patterning.

The typical coating method is spin coating. Continuous coatings such as roll-to-roll include blade coating and slit-die coating. Slit-die coating can produce a thin, uniform film but is generally expensive because it requires a very high-precision slit die.

Printing methods include screen printing, gravure offset printing, reverse offset printing, and inkjet printing [10]. Screen printing is widely used in industry and suitable for printing thick films, while it has relatively low resolution with a minimum line width of about 50 µm. Gravure offset printing has higher resolution than screen printing. Reverse offset printing is even finer than gravure offset printing and can print fine lines of 1 µm or less, while it is not suitable for printing thick films. Inkjet printing is a digital on-demand printing and is suitable for rapid prototyping, since printing plates are not required. Inkjet is also more suited to low-viscosity inks than screen printing or gravure offset printing and has the advantage that it does not require the addition of thickening agents, which usually degrade the electric properties of the conductive materials. However, care must be taken to avoid clogging of the nozzle head and the spreading of the ink on the substrate.

4.3.3 Design Rule and Device Parameters

To know the range of permissible device and dimensional parameters in advance is important in designing OTFT integrated circuits. The necessary parameters are device parameters such as mobility and threshold voltage, and dimensional parameters such as line width and space; the former is used for schematic design and the latter for layout design. A summary of the permissible values for dimensional parameters is called a design rule and is an important communication tool between the process engineers and designers [2].

Table 4.2 shows examples of permissible device parameters for p-type OTFT devices that can be fabricated by inkjet printing. In this case, the channel length, channel width, and capacitance per unit area can be freely adjusted within a certain range, while other parameters such as mobility are basically fixed values. Since parameters such as mobility vary depending on the gate insulating material, deposition conditions, etc., even if the semiconductor material is the same, it is necessary to use actual measured values according to the actual fabrication environment.

Figure 4.7 and Table 4.3 show an example of permissible ranges of dimensional parameters (design rules) when inkjet and dispenser are used. In this case, the line

Table 4.2 Example of permissible device parameters for p-type OTFTs

	Minimum	Maximum
Channel length L (μm)	10	100
Channel width W (μm)	100	1000
Specific capacitance C_i (nF/cm^2)	7	20
Mobility μ (cm^2/vs)	1	
Turn-on voltage V_{ON} (V)	0	
Gate threshold voltage V_{TH} (V)	−0.3	
Subthreshold slope SS (V/dec)	0.1	
Source-gate overlap length ΔL_S (μm)	90 - L/2	
Drain-gate overlap length ΔL_D (μm)	90 - L/2	

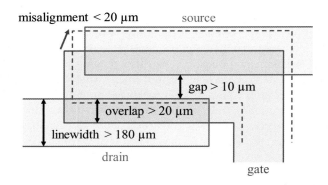

Fig. 4.7 Example of design rule

Table 4.3 Example of permissible dimensional parameters (design rule)

	Minimum	Maximum	Increment
Line width (μm)	180	–	60
Line space (μm)	60	–	60
Source-drain distance (μm)	10	100	–
Source-gate overlap, drain-gate overlap (μm)	20	–	–
Bank width (μm)	300	300	–
Semiconductor area (μm^2)	900 × 900	1500 × 1500	–
Via hole (μm)	300	–	60

width of interconnections needs to be chosen among 180 μm, 240 μm, 300 μm, etc. in 60 μm increments. This restriction on increments is unique to inkjet printing, as inkjet printers usually discharge ink on equally spaced grids in order to draw efficiently using multiple nozzles. In such cases, the centerline of the wiring must also be on the equally spaced grid. The minimum dimensions of the wiring spacing must be set large enough to avoid unwanted short circuits, taking into account the positioning accuracy of the printing device and the spreading of the ink. Source-gate and drain-gate overlaps must be larger than the maximum misalignment to ensure

4 Printed Organic Thin-Film Transistors and Integrated Circuits

the overlap. On the other hand, smaller source-gate and drain-gate overlaps reduce parasitic capacitances and increase the operating speed of circuits.

4.3.4 *Schematic Design and Circuit Simulation*

The schematic design is conducted based on the design rule. Circuit simulations are carried out in parallel with the schematic design in order to verify the circuit operation. Circuit simulation software such as LTspice (Analog Devices, Inc.) usually has a drafting function, so the two tasks, drafting and simulation, can be conducted seamlessly.

When simulating circuits containing OTFT devices using SPICE simulators, device models representing the electrical characteristics of OTFT devices are required. Although metal-oxide-semiconductor field-effect transistor (MOSFET) models based on silicon are readily available in SPICE simulators, OTFT models based on organic semiconductors are rarely available. Here, the procedure for using an OTFT model in LTspice is briefly described. It requires two files: a library file (pOTFT.lib) and a symbol file (pOTFT.asy). A library file is a text file in Fig. 4.8 and defines the characteristics of the OTFT. According to Marinov et al., the drain current in OTFT can be expressed by

```
.subckt pOTFT D G S

.param L = 20u        ; channel length (m)
.param W = 1000u      ; channel width (m)
.param Uo = 1e-4      ; mobility (m^2/(V*s))
.param Vto = -0.3     ; threshold voltage (V)
.param Vss = 0.09     ; subthreshold slope (V/dec)
.param gamma = 0.0    ; mobility enhancement factor
.param Ci = 240u      ; capacitance per unit area (F/m^2)
.param dLs = 80u      ; source-gate overlap length (m)
.param dLd = 80u      ; drain-gate overlap length (m)

Bsd S D I=W/L*Uo*Ci*Vss**(gamma+2)*(
      + ln(1+exp(-(V(G)-Vto-V(S))/Vss))**(gamma+2)
      + -ln(1+exp(-(V(G)-Vto-V(D))/Vss))**(gamma+2)
      + )/(gamma+2)
Cs S G {Ci*W*(dLs+0.5*L)}
Cd D G {Ci*W*(dLd+0.5*L)}

.ends
```

Fig. 4.8 Example of an OTFT library file for SPICE simulation

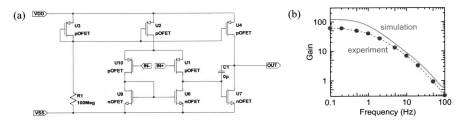

Fig. 4.9 (**a**) Circuit diagram and (**b**) frequency dependence of gain of an OTFT-based operational amplifier [4]

$$I = \frac{W}{L}\mu_o C_i V_{SS}^{\gamma+2} \frac{\left\{\ln\left[1+\exp\left(\frac{V_G-V_T-V_S}{V_{SS}}\right)\right]\right\}^{\gamma+2} - \left\{\ln\left[1+\exp\left(\frac{V_G-V_T-V_D}{V_{SS}}\right)\right]\right\}^{\gamma+2}}{\gamma+2}$$

(4.13)

where μ_o is the mobility, V_{SS} the subthreshold slope, and γ the exponent factor for gate-voltage-dependent mobility [21]. The lines beginning with param define various device parameters such as L and W. The lines beginning with Bsd represent the expression for the current flowing between source and drain. The lines beginning with Cs and Cd represent the source-gate and drain-gate capacitance, which are not found in the reference but are necessary for simulating dynamic characteristics. Contact resistance and channel length modulation effects were ignored in the example. Symbol file can be copied from the existing MOSFET symbol file and renamed. These two files enable the circuit simulation with OTFT devices.

Figure 4.9a shows a circuit diagram of a complementary organic OPA combining five p-type OTFT devices (M1-M5) and three n-type OTFT devices (M6-M8) [4]. Although the diagram only shows the connections between the OTFT devices, the channel length and channel width of each OTFT need to be specified individually to obtain the desired characteristics in practice. Figure 4.9b shows the results of the circuit simulation (dashed line) and the measured values after the circuit was actually fabricated (solid line). The circuit simulation adequately predicts the characteristics of the actual OTFT integrated circuit.

More advanced simulations may include adding contact resistance and channel length modulation effects, or performing Monte Carlo analysis to account for variations in device parameters. Contact resistance and channel length modulation effects are particularly important when simulating OTFT devices with small channel lengths. Monte Carlo analysis is important for predicting variations in circuit characteristics such as gain and offset, as well as the yield of integrated circuits.

4.3.5 Layout

After schematic design, layout of each layer is designed by using CAD software. Figure 4.10 shows an example of layout and optical micrograph of a complementary organic OPA fabricated by inkjet and dispenser printing. Although the circuit was printed approximately as per the design drawing, minor differences can be seen. For example, the line width of the interconnections in the horizontal direction was 180 µm, while the line width of the interconnections in the vertical direction was 120 µm, indicating the dependence of the line width on the printing direction. This is due to the scan direction in inkjet printing being the horizontal direction. For the same reason, the vertical line of the metal layer is not smooth, and therefore the horizontal line of the metal layer should be used as the source and drain electrodes of the OTFT devices. The corners of the metal layer in actual circuits are rounded, and the line width of the metal layer is not perfectly constant but slightly thinner at the edges. Therefore, the channel length of the OTFT devices is slightly larger at both ends of the channel than in the middle of the channel.

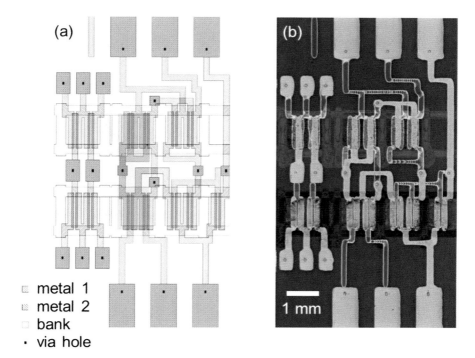

Fig. 4.10 (a) Layout and (b) optical microscope image of the OTFT-based operational amplifier [4]

4.3.6 Fabrication and Evaluation

Finally, the actual fabrication method of the OTFT integrated circuit is briefly introduced. The complementary organic OPAs presented here have a layer structure as shown in Fig. 4.5. The metal layer was printed with silver nanoparticle ink (NPS-JL, Harima Kasei) using an inkjet printer (DMP-2831, Fujifilm). The p-type organic semiconductor layer was printed with a mesitylene solution of 2,8-difluoro-5,11-bis(triethylsilylethynyl)anthradithiophene (diF-TES-ADT) by inkjet. The n-type organic semiconductor layer was printed with a 1-methylnaphthalene solution of TU-3 (Future Ink) using a dispenser (IMAGEMASTER 350, Musashi Engineering). Interlayer connections through via holes were made by inkjet printing silver nanoparticle ink after laser etching.

The characteristics of the complementary organic OPA are summarized in Table 4.4 alongside those of a commercially available ultralow power OPA, LPV821 (Texas Instruments). It can be seen that the organic OPAs are inferior in terms of gain and response speed, but not significantly different in terms of current consumption and supply voltage. These characteristics make on-demand printed organic OPAs suitable for sensor applications requiring relatively low speed and low power consumption, especially for wearable healthcare devices.

4.4 Requirements for Printed OTFT Circuits

4.4.1 Flat and Thin Electrodes

An OTFT consists of three kinds of electrodes (viz., gate, source, and drain), insulator, and semiconductor. As shown in Fig. 4.11, an OTFT has stacked layers; thereby each layer requires thin thickness and uniform profiles. Especially, the thickness of bottom electrodes and insulators affect both the operation voltage and yield of OTFT devices.

When printed ink dries on the surface of a substrate, the solute is generally transported from the center to the edge, and the resulting solute film forms a nonuniform ring-like profile. Deegan et al. studied this phenomenon, known as the

Table 4.4 The performances of the printed organic OPA [4] and a commercial low-power OPA (LPV821, Texas Instruments)

	Printed organic OPA	Commercial low-power OPA
Power consumption (nA)	30	650
Supply voltage (V)	5	1.7–3.6
Open-loop gain (dB)	36	135
Gain bandwidth product (Hz)	50	8000
Through rate (V/ms)	0.1–0.3	3.3
Input offset voltage (mV)	100	0.01

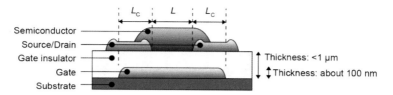

Fig. 4.11 Schematic illustration of an organic thin-film transistor. L means channel length and L_C means contact length

coffee ring effect, for colloidal suspension systems [22]. The nonuniform profile in the cross-section for printed electrodes originating from this effect is the major issue to be solved for the devices with stacked layers such as capacitors and OTFT. The thicker edges of bottom electrodes interfere with the flatness and uniformity of overlying dielectric layers. As a result, fully solution-processed OTFT devices have difficulty operating at high voltages due to the potential for electrical shorts between the lower and upper electrodes [23, 24]. In order to achieve low operation voltage less than 10 V, the thickness of gate insulators should be less than 500 nm because the relative permittivity of printable gate insulating materials is usually ranging from 1 to 5. For this demand, thin (less than 100 nm) and uniform electrodes are required for electronic devices which have stacked layers.

4.4.2 Fast Operation

A cutoff frequency in saturation mode can be described by

$$f_T \approx \frac{\mu_{\text{eff}}(V_{\text{GS}} - V_{\text{th}})^2}{2\pi L(L + 2L_C)} \quad (4.14)$$

with the effective charge carrier mobility μ_{eff}, the gate–source voltage V_{GS}, the threshold voltage V_{th}, the channel length L, and the contact length L_C [25]. Equation 4.14 clearly shows that the downscaling of channel length is the most important for improving the operation speed of the circuits.

We should also make consideration of the effect of contact resistance (R_C, Fig. 4.12). When the current flow from source to drain electrodes through a semiconducting layer, the total resistance (R_{total}) is divided into channel resistance (R_{ch}) and R_C:

$$R_{\text{total}} = R_{\text{ch}} + R_C \quad (4.15)$$

The existence of R_C causes the decrease of μ_{eff} from intrinsic mobility. μ_{eff} is calculated using

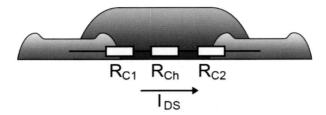

Fig. 4.12 Total resistance of the OTFT. The total resistance (R_{total}) is divided into channel resistance (R_{ch}) and R_C. The R_C is divided into the resistance between source and semiconducting layer (R_{C1}) and drain and semiconducting layer (R_{C2})

$$\mu_{eff} \approx \mu_0 \left[1 - \left(\frac{\mu_0 C_i R_C W |V_{GS} - V_{th}|}{L + \mu_0 C_i R_C W |V_{GS} - V_{th}|} \right)^2 \right] \quad (4.16)$$

with the intrinsic mobility μ_0, the gate dielectric capacitance per unit area C_i, the channel width W, and threshold voltage V_{th}. Limitations by contact resistance are becoming increasingly crucial [26], and finding ways to reduce these limitations has become a key issue for high-speed operation of printed OTFT devices and circuits as indicated by Eq. (4.14).

4.4.3 Profile Control of Inkjet-Printed Silver Electrodes

To achieve thin (less than 1 μm) electrodes, low viscosity inks are favorable. Inkjet print requires low viscosity (about 10 mN/m); thereby the inkjet-printed electrodes are suitable for the use of OTFT devices. However, the low-viscosity inks tend to cause the coffee ring effect. Okuzono et al. have proposed a simple model that predicts a final shape of a dried thin film [27]. According to the model, solvent evaporation rate (J_s) and diffusion coefficient (D) affect the final shape of the film (Fig. 4.13a). The smaller value J_s or the larger value D more readily induces a convex shape. This is because diffusion tends to homogenize the concentration field contrary to the outward flow. In order to control the final shape of printed silver electrodes, the J_s of the silver-nanoparticle ink should be suppressed. Silver nanoparticles dispersed in a water-based solvent (DIC Corp. Japan, JAGLT) were used. Since both environmental temperature and humidity decide the J_s of water, the drying conditions were controlled, and the dependencies of the ambient humidity and drying time on the profiles were assessed [28]. The silver nanoparticle ink was patterned with an inkjet printer (Fujifilm Dimatix, DMP2800) onto the cross-linked poly-4-vinylphenol (PVP) layers using a printhead with 10 pl nozzles. After the printing, the substrates were stored in an environmental test chamber following the printing process (espec, SH-221) in order to evaporate the solvents from printed ink. Temperature in the chamber was held at 30 °C, and relative humidity was changed

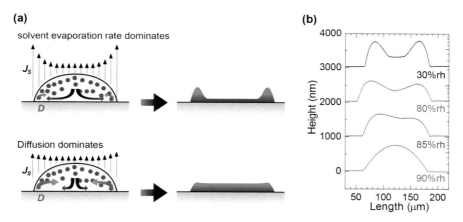

Fig. 4.13 Profile control of printed electrodes. (**a**) The correlation between the evaporation speed of the solvent and the final shape of the printed electrodes. When solvent evaporation rate (J_S) dominates, the final shape of the solute tends to be concave. On the other hand, when diffusion (D) dominates, convex shape can be obtained. (**b**) Controlling the shape of inkjet-printed silver electrodes by changing the ambient humidity. Profiles of the electrodes with different drying humidity between 30%RH and 90%RH. The profiles were obtained from laser microscopic images. The temperature of the chamber and drying time was 30 °C and 30 min [29]

from 30%RH to 90%RH, while the storage time was fixed to 30 min. After the drying process, the substrates were heated at 140 °C for 1 h to sinter the silver nanoparticles.

Figure 4.13b shows line profiles (cross-sectional view) of printed silver electrodes dried at various humidity levels. The electrode profiles varied widely with the ambient humidity in the chamber. The cross-sectional profile for a line with ambient humidity of 30%RH was concave. These nonuniformities in silver electrode thickness are a result of the *coffee-ring effect*. The ratio of thickness between the edge and center of the profile (t_e/t_c) is 3.0. This concave shape was suppressed by increasing the ambient humidity from 30%RH to 80%RH ($t_e/t_c = 2.1$), such that a nearly trapezoidal shape was observed at an ambient humidity level of 85%RH ($t_e/t_c = 1.3$). Furthermore, for ambient humidity levels of 90%RH, the silver electrodes formed a convex shape ($t_e/t_c < 1$). These results clearly show that the silver electrode profiles were very sensitive to the ambient humidity levels during the drying process.

How the flatness of the electrodes affected the functionality and performance of electronic devices with stacked layer constructions was also investigated. Thin-film capacitors were fabricated. Two types of silver layers with different drying conditions were prepared as bottom electrodes: (1) 30 °C, 30%RH for 30 min (concave), and (2) 30 °C, 85%RH for 30 min (almost flat). After forming these electrodes, a solution of cross-linked PVP was spin-coated and baked to form 210-nm thick dielectric layers. Silver nanoparticle ink was then applied using inkjet printing to form the upper source/drain electrodes. Figure 4.14 shows histograms for the electrical breakdown voltage results of the fabricated capacitors with two different

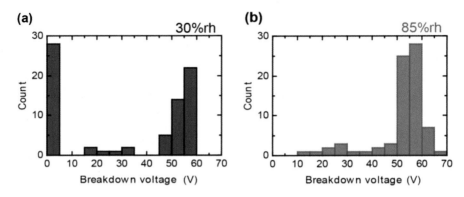

Fig. 4.14 Breakdown voltage histograms for the fabricated capacitors with different drying conditions, at 30 °C, 30%RH for 30 min (**a**) and at 30 °C, 85%RH for 30 min (**b**). The total number of counts was 75 for each condition [29]

lower silver electrodes prepared at drying conditions of 30 °C, 30%RH for 30 min (a) and 30 °C, 85%RH for 30 min (b). The capacitors with concave-shaped electrodes did not have sufficient insulating properties; 37% of the capacitors exhibited the breakdown voltages of less than 5 V, as shown in Fig. 4.14a, which indicates that the upper and lower electrodes had shorted. The peaks of the lower electrodes pose nonuniformity of the dielectric layers and/or the increase of the electric field at the edge of the lower electrodes, which causes the shorted capacitors. On the other hand, no capacitor electrodes shorted when the shaped electrodes were used for the lower electrodes. Additionally, the average breakdown voltage improved from 33 V to 52 V by using trapezoidal-shaped electrodes. A breakdown voltage of 52 V corresponds to 2.54 MV/cm in electric field strength, which was comparable to that for cross-linked PVP used as dielectric layers and evaporated metal used as lower gate electrodes [29].

The methods for controlling the shape of printed electrodes can increase freedom in printing conditions and could help in the practical realization of printed electronics.

4.4.4 Improvement of Field-Effect Mobility for Printed OTFT Devices

As shown in Eq. (4.14), not the intrinsic mobility but the effective mobility of semiconducting layer decides the operation speed of the integrated circuits. This requires that both R_{ch} and R_C should be decreased for the printed OTFT devices. As several previous reports show, the R_C is decreased when the energy barrier between the work function of source/drain electrodes and highest occupied molecular orbital (HOMO) level of p-type organic semiconducting layer is suppressed by the carrier

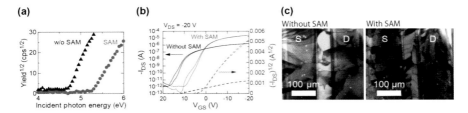

Fig. 4.15 Effect of source-drain electrode modification by SAM treatment on transistor characteristics. (**a**) Square root of the yield as a function of incident photon energy from photoemission spectroscopy. The black dots represent the untreated electrodes, and the red dots represent the treated electrodes. (**b**) Transfer characteristics of fabricated TFTs. The black lines represent the transfer curve for the device without the SAM treatment, and the red lines those with the SAM treatment. The SAM modification process improved the transistor electrical characteristics dramatically, whereby on-current increased from 1.6 mA to 27 mA, and the estimated mobility in saturation regime increased from 0.02 cm^2 V^{-1} s^{-1} to 0.9 cm^2 V^{-1} s^{-1}. (**c**) Polarization microscope images of channel region of fabricated TFTs with untreated and with treated electrodes. A mesitylene-based formulation of a soluble small-molecule organic semiconducting layer (Merck, lisicon® S1200) with a deep ionization potential of 5.4 eV was used as organic semiconducting layer [34]

injection layer. For printed electrodes, self-assembled monolayer (SAM) can change the work function of the electrodes, which causes the reduction of energy barrier between source/drain electrodes and semiconducting layer. Figure 4.15a shows how the work function of printed silver electrodes is changed by the SAM layer. The SAM layer changed the work function of the printed silver electrodes from 4.7 eV to 5.3 eV. How the SAM treatment affected the transistor characteristics was also revealed [30]. A mesitylene-based formulation of a soluble small-molecule material with a deep ionization potential of 5.4 eV was used as organic semiconducting layer (Merck, lisicon® S1200) [31]. Figure 4.15b shows the transfer characteristics of the fabricated OTFT devices, having the same W/L ratio of 50, with and without applying a SAM treatment to the source-drain electrodes. The SAM modification process improved the transistor electrical characteristics dramatically, whereby on-current increased from 1.6 μA to 27 μA and the estimated mobility in saturation regime increased from 0.02 cm^2 V^{-1} s^{-1} to 0.9 cm^2 V^{-1} s^{-1}. The crystallinity of semiconducting layer between source/drain electrodes was observed with a polarization microscope, as shown in Fig. 4.15c. Both devices had nearly same crystalline domains, even though there were large differences in mobility between the devices with the SAM treatment and those without it. These results indicate that the SAM modification layer reduces only the R_C of the printed OTFT devices, which cause dramatic improvement of the transistor characteristics.

The R_C of the fabricated OTFT devices was estimated using a transfer-line method. Figure 4.16a plots the channel width-normalized total on-resistance (R_{Total}) as a function of channel length. R_C was obtained by extrapolating the linear fit to a channel length of zero and plotted as a function of gate-source voltage (V_{GS}) (Fig. 4.16b). R_C decreases with increasing gate-source voltage, likely due to an

Fig. 4.16 Estimation of contact resistance. The contact resistance of the TFT devices with treated source-drain electrodes was estimated by using the transfer-line method. (**a**) Channel width-normalized total on resistance (R_{Total}) as a function of channel length. (**b**) Width-normalized contact resistance as a function of gate-source voltage (V_{GS}) [34]

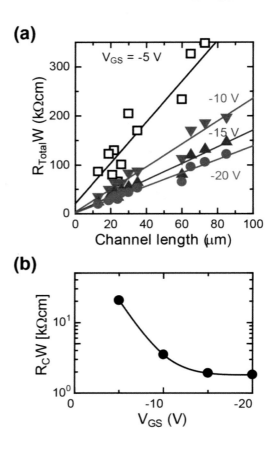

increase in carrier density in the channel and near the contacts. R_C decreased to a value as low as 1.83 k Ω cm, a remarkably low contact resistance value for fully solution-processed OTFT devices, which is attributed to there being a low-energy barrier between the printed organic semiconducting layer and source/drain electrodes.

4.5 Examples of Printed OTFT Circuits

4.5.1 Ultra-Flexible and Large-Area Circuits

Thin and ultra-flexible devices that can be manufactured in a process that covers a large area will be essential to realizing low-cost, wearable electronic applications including foldable displays and medical sensors. The evolution from rigid, heavy, and thick electronics to new flexible electronics has reached the point whereby electronics can be attached to curved and moving surfaces such as the skin of the

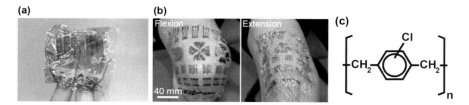

Fig. 4.17 Fully printed organic circuits fabricated on ultrathin substrate. (**a**) A photograph of organic TFT devices on 1-μm-thick parylene-C films. (**b**) Organic device films conforming to a human knee. Our devices were fabricated entirely by printing, enabling them to be easily fabricated on a large scale, which can cover whole area of a human skin as shown in (**b**). (**c**) The chemical structure of parylene-C [34]

human body without any concern to the wearer [32, 33]. Achieving flexible organic electronics based on OTFT devices fabricated with fully printed processes will be essential for realizing wearable electronic applications that are low in cost and environmentally friendly. For this demand, OTFT devices with excellent electrical characteristics and mechanical stability that were fully printed on ultra-flexible polymer films were fabricated [30].

The devices were fabricated entirely by printing, enabling them to be easily fabricated on a large scale (Fig. 4.17a). The polychloro-p-xylylene (parylene-C, Fig. 4.17c) films with thickness of only 1 μm were used as base substrates. The parylene-C films were formed by chemical vapor deposition onto the supporting glass plates with release layer (fluoropolymer layer). The parylene-C films are attached to a release layer with weak adhesive strength (13 mN), so that the fabricated devices can be safely peeled off the supporting plates. The fabricated organic devices are extremely thin and ultra-flexible; their total thickness is less than 2 μm, and total weight is only 2 g m^{-2}. An example is the printed device conforming to a human knee. Figure 4.17b illustrates their potential in healthcare and monitoring applications; they can be gently attached to the skin or wrapped around limbs without the wearer perceiving any discomfort. They can be bent or even wrinkled so that they conform to the movements of the human body, which has uneven surfaces and a large range of motion.

The fabricated devices exhibited remarkable mechanical stability. Figure 4.18a shows a photograph of a fabricated OTFT device tightly wrapped around a copper wire with bending radius of 140 μm. The transfer characteristics were measured in ambient air with and without strain due to bending, as shown in Fig. 4.18b, such that there was no discernible change in the electrical characteristics during the bending. The change in the on-current was 3.9% and the change in the mobility was 1.6% with bending and the on/off ratio remained more than 10^5. The ultrathin film substrates help reduce the applied strain in the OTFT devices, whereby the calculated strain was roughly 0.5% or less, even when the devices were bent to a radius of 140 μm [35, 36]. As several research groups have already reported, such a small degree of strain does not significantly change the electrical performance of OTFT devices [37].

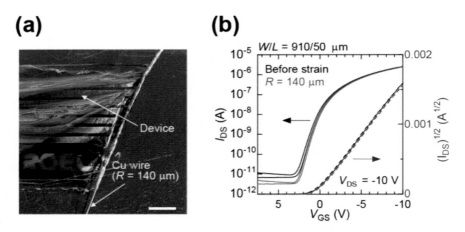

Fig. 4.18 (a) Photograph of fully printed OTFT devices on 1-μm-thick parylene-C films wrapped around a copper wire with a radius of 140 μm. Scale bar, 5 mm. (b) The transfer characteristics of the OTFT devices were measured in the bent and unbent states, with no discernible changes in the characteristics due to bending [34]

The mechanical stability of unipolar diode-load inverters fabricated on thin parylene-C films was evaluated. Figure 4.19a shows a photograph of the inverter circuit, and Fig. 4.19b shows a circuit diagram for the inverter. The ratio between the channel widths of the drive and load OTFT devices is 2.6:1. Both OTFT devices had channel lengths of 20 μm. Figure 4.19c plots static transfer characteristics measured in ambient air of a diode-load inverter under no compressive strain and under 50% compressive strain. The output voltage (V_{OUT}) and small-signal gain were plotted as a function of input voltage (V_{IN}). There was no significant change in the electrical characteristics during bending, and the inverter functioned properly even at a driving voltage (V_{DD}) of 2 V. The trip point when the inverter was not under compression was 5.92 V at $V_{DD} = 10$ V and 0.95 V at $V_{DD} = 2$ V. The small-signal gain of the inverter without compression was 1.57 at $V_{DD} = 10$ V and 0.97 at $V_{DD} = 2$ V. The trip point of the inverter under 50% compressive strain was 5.84 V at $V_{DD} = 10$ V and 0.92 V at $V_{DD} = 2$ V. The small-signal gain of the inverter under 50% compression was 1.61 at $V_{DD} = 10$ V and 1.00 at $V_{DD} = 2$ V. The changes in the trip point and gain in the case of 50% compressive strain were only less than 5% of their values in the case of no compressive strain. Figure 4.19d shows the dynamic response of the inverter circuit. A continuous V_{DD} of 10 V and an AC input voltage with 100-Hz rectangular waveform ranging from 0 V to 10 V were applied to the inverter, and the output voltage was monitored using a digital oscilloscope. The measured rising and falling time in the case of no compression were 427 μs and 691 μs, which correspond to a total delay time of 1.12 ms. This delay is quite short for fully printed organic circuits [38, 39]. The operating speeds remained stable even when compressive strain was applied to the films and the circuits crumpled. The measured rising and falling time of the inverter under 50% compressive strain were

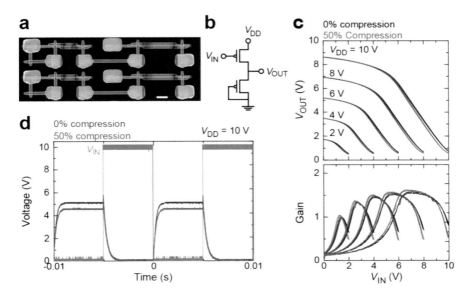

Fig. 4.19 Fast operating of printed organic circuits. (**a**) Photograph of fabricated unipolar organic diode-load inverter circuits. Scale bar, 1 mm. (**b**) Circuit diagram of the inverter device. (**c**) Static transfer characteristics of the inverter and small-signal gain as a function of input voltage (V_{IN}). The black solid line indicates the characteristics without strain, and the red solid lines indicate those of circuits under 50% compressive strain. (**d**) Dynamic operation of the inverter circuit. A continuous driving voltage (V_{DD}) of 10 V and an AC input voltage with a 100 Hz rectangular waveform from 0 V to 10 V (gray) were applied to the inverter, and the output voltage was monitored using a digital oscilloscope. The blue line indicates the output voltage without strain, and the red line indicates the output voltage under 50% compressive strain [34]

495 µs and 705 µs, which correspond to a total delay of 1.20 ms and only a 7% change from the initial total delay. These results exemplify the outstanding mechanical stability of fast-operating printed organic circuits fabricated on thin films.

4.5.2 Digital Logic Gates

NAND and NOR logic gates (Fig. 4.20a, b) can be designed in a similar way as inverters. NAND and NOR mean NOT-AND and NOT-OR, respectively, and their truth tables are shown in Table 4.5. Since NAND has p-type OTFT devices in parallel and n-type OTFT devices in series, A = B = *high* is the only case that the output is connected to the ground. On the other hand, NOR has p-type OTFT devices in series and n-type OTFT devices in parallel, and its output is *high* only when both inputs A and B are *low*. NAND and NOR gates are the basic building blocks of logic operation circuits because any logic gates can be implemented by a combination of NAND gates (or NOR gates).

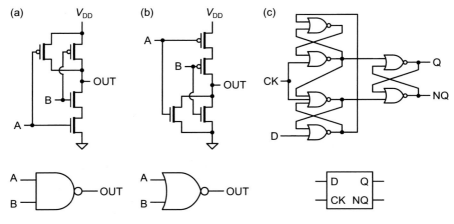

Fig. 4.20 (a) NAND, (b) NOR, and (c) D flip-flop

Table 4.5 Truth tables of inverter, NAND, and NOR. "0" denotes *low* voltage and "1" denotes *high* voltage

A	B	OUT (NAND)	OUT (NOR)
0	0	1	1
0	1	1	0
1	0	1	0
1	1	0	0

Table 4.6 Truth tables of D flip-flop

CK	D	Q	NQ
Rising edge	0	0	1
Rising edge	1	1	0
Non-rising	Any	Keep previous state	Keep previous state

D flip-flop (D-FF, Fig. 4.20c) is another important digital circuit, since it has a 1-bit memory function [40]. The truth table of D-FF is shown in Table 4.6. A D-FF receives 1-bit data from the data terminal D at the rising (or falling) edge of the clock terminal CK and keeps to output the data at the output terminal Q until the next rising edge of clock. The inverted output NQ is always the opposite to Q. A D-FF can be composed of eight NAND gates or six NOR gates (one of six has three inputs). D-FF can be used for composing shift registers and counters.

4.5.3 Radio-Frequency Identification (RFID) Tags

Figure 4.21 shows a prototype of radio-frequency identification (RFID) tags using OTFT devices and a temperature sensor [40]. Here, two D-FF circuits compose a 2-bit counter. The counter increments the output in binary format at every clock

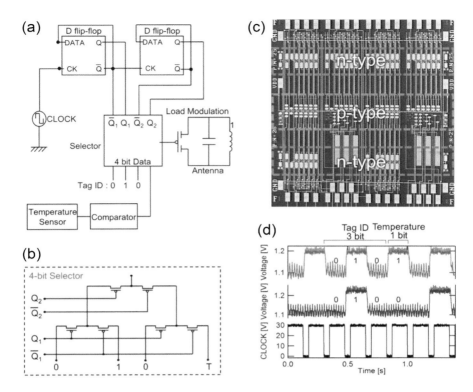

Fig. 4.21 Radio-frequency identification (RFID) tag using OTFTs and a temperature sensor. Block diagrams of (**a**) a whole circuit and (**b**) a 4-bit selector. (**c**) Optical microscope image of the OTFT integrated circuits. (**d**) Signals measured by a wireless receiver and a clock signal. Red: temperature is high. Blue: temperature is low [40]

signal. A 4-bit selector, or multiplexer (MUX), selects one of the 4-bit data (3-bit tag ID and 1-bit data from a temperature sensor). The selection depends on the signal from the counter. Thus, the counter and selector acquire 4-bit data one by one and transmit the data through the antenna with a load modulation. This allows multiple bits of digital information to be transmitted by a 13.56 MHz radio wave. A temperature sensor is connected to a comparator to check whether the temperature is higher than a certain threshold. Figure 4.21d shows the actual transmitted data when the temperature is high (red) and low (blue) along with the clock signal.

4.5.4 Pseudo-CMOS Logic Circuits

Complementary organic circuits consist of both p- and n-type OTFT devices. Although soluble n-type semiconductor materials have been developed in recent

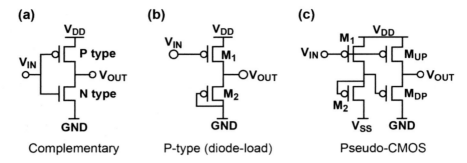

Fig. 4.22 Circuit diagrams (**a**) complementary, (**b**) diode-load, and (**c**) pseudo-CMOS inverters

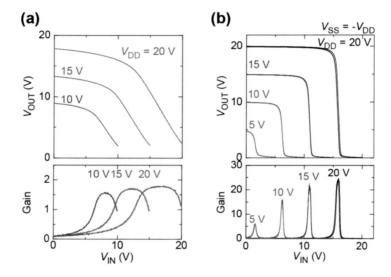

Fig. 4.23 (**a**) Voltage transfer curves and gain of fully printed diode-load inverter. (**b**) Voltage transfer curves and gain of fully printed pseudo-CMOS inverter

years [41], their performance in OTFT devices remains lower than those of p-type semiconductor materials. As a result, operation speed of the complementary organic circuits is determined by response of n-type OTFT devices. In order to solve the problem, pseudo-CMOS logic design was suggested by Huang et al. [42]. Pseudo-CMOS inverters comprise four p-type OTFT devices as shown in Fig. 4.22. Excellent voltage transfer characteristics with high gain and fast speed were reported using pseudo-CMOS logic design [43, 44]. Fully printed pseudo-CMOS integrated circuits exhibited good static and dynamic characteristics.

Figure 4.23 shows the voltage transfer characteristics of a fabricated fully printed diode-load and pseudo-CMOS inverters [30, 45]. The diode-load inverter exhibited low gain and small static noise margin. On the other hand, the pseudo-CMOS

Table 4.7 Signal gain of the fully printed diode-load and pseudo-CMOS inverters at each operating voltage V_{DD}

V_{DD} (V)	Diode-load	Pseudo-CMOS
20	−2.9	−24
15	−1.9	−23
10	−1.1	−16
5	–	−6.1

inverter exhibited much better characteristics. The inverter was operated successfully at small operation voltage ($V_{DD} = -V_{SS} = 5$ V). Table 4.7 summarizes signal gains of fully printed inverter circuits at each operating voltage V_{DD}. The gain was 24 at $V_{DD} = 20$ V, which was 14 times larger than that of diode-load inverter. These static results clearly show that the fully printed pseudo-CMOS circuits can be comparable with conventional complementary organic integrated circuits.

In order to demonstrate the applicability of the fully printed pseudo-CMOS circuits to a logic circuit, RS flip-flop (RS-FF) was fabricated [46]. The fabricated RS-FF comprises two pseudo-NAND circuits (Figure 4.24a). A block diagram and a truth table of the RS-FF are shown in Figure 4.24b. Figure 4.24c shows a photograph of a fabricated RS-FF. The RS-FF comprises 12 OTFT devices. Figure 4.24d shows the voltage transfer characteristics of the NAND circuits when input voltage V_{inA} was changed from 0 V to 15 V at supply voltages of 15 V and tuning voltage $V_{SS} = -V_{DD}$. The NAND was operated successfully as the truth table; when the V_{inB} was fixed at 15 V, the fabricated NAND device exhibited good switching characteristics with a signal gain 23 at a trip point of $V_{inA} = 7.24$ V, while the V_{out} remained *high* voltage when the V_{inB} was fixed at 0 V. As shown in Fig. 4.24e, the RS-FF circuit exhibited switching characteristics in accordance with truth table at operating voltage of 10 V. The output (Q) of the fabricated RS-FF changed its state at the falling edge of the input signal ($\overline{S}$ or $\overline{R}$), and the rise of either $\overline{S}$ or $\overline{R}$ signal did not affect Q value. The delay time of the RS-FF circuit was also evaluated. The delay time is the sum of rising and falling times, which are defined as the time difference between 10% and 90% of the output signal for transient changes in the circuit input from a logically *low* to *high* level (rising time) and *high* to *low* level (falling time). A measured total delay time of the RS-FF was 6.4 ms at 10 V, which was quite fast speed among the fully printed integrated circuits.

4.5.5 Wearable Biosensors

Wearable biosensors enable real-time health monitoring without disturbing the user's activity. Figure 4.25 shows a wearable lactate sensor system using pseudo-CMOS OTFT amplifiers [6]. The working electrode has immobilized lactate oxidase on its surface and generates small current depending on the concentration of lactate. The small current is amplified by the detection unit composed of an OTFT amplifier and a resistor into a large output voltage. Another amplifier is used as a feedback control unit, which helps to keep the electrochemical potential constant with respect

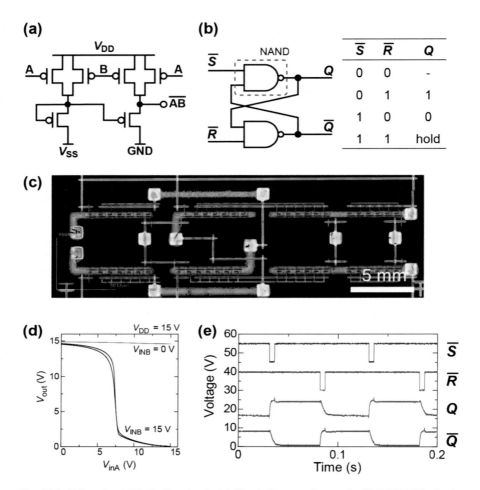

Fig. 4.24 Fully printed RS flip-flop circuit. (**a**) Circuit diagram of a pseudo-CMOS NAND circuit. (**b**) Block diagram and the truth table of the RS flip-flop circuit. (**c**) An optical image of the fabricated pseudo-CMOS NAND-based RS flip-flop circuit. (**d**) Voltage transfer characteristics of the pseudo-NAND circuit. (**e**) The input-output characteristics of fabricated RS flip-flop circuit at a supply voltage (V_{DD}) of 10 V [46]

to the reference electrode as so-called a potentiostat. The two units were printed on a flexible substrate as shown in Fig. 4.25b. Figure 4.25c, d show the results of the lactate detection with the printed lactate sensor system. The output voltage exhibited a linear response to the lactate concentration up to 0.5 mM. Different kinds of working electrodes such as different enzymes, ionophores, antibodies, and aptamers enable such sensor systems to be used for a variety of biosensing [47].

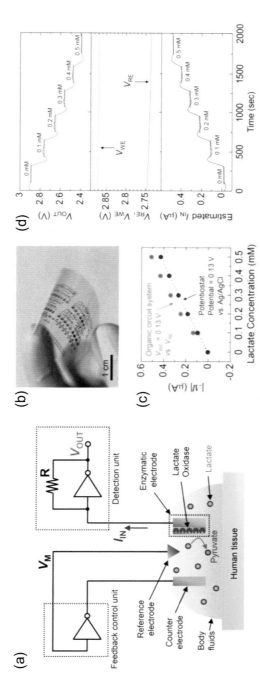

Fig. 4.25 Wearable lactate sensor system using pseudo-CMOS OTFT amplifiers. (**a**) Schematics and (**b**) photograph of the sensor system. (**c**) Current at the working electrode at different lactate concentrations. (**d**) Output voltage, reference, and working electrode voltages and estimated working electrode current at different lactate concentrations [6]

4.5.6 Electrostatic Proximity Sensor Matrix

Floating extended-gate OTFT devices can detect electrostatic field with high sensitivity (Fig. 4.26) [48]. The floating extended gate has a larger size than a usual gate electrode is electrically floating. This structure works like an antenna, receives electrostatic field, and modulates the drain current in OTFT devices (Fig. 4.26b). If the target object has electrostatic charges, the device can detect the proximity of the object. Figure 4.26c–e show an ultra-flexible electrostatic proximity sensor passive matrix array composed of 4 × 4 floating extended-gate OTFT devices. The sensor sheet has a small thickness of 2 μm and can detect a human hand at a several centimeter distance. The multiple sensor pixels provide a rough imaging of the position and shape of the hand. Similar system can also be used to image static electricity on a variety of films and substrates, which may cause problems in manufacturing.

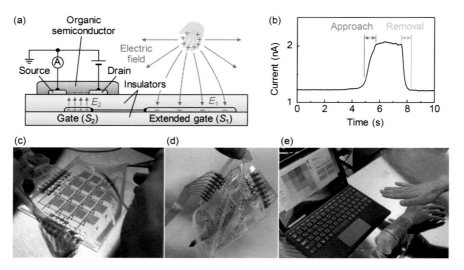

Fig. 4.26 4 × 4 proximity sensor array using floating extended-gate OTFTs. (**a**) Operation principle. (**b**) The change in drain current when human hand approached the sensor. Photographs of the devices (**c**) during peeling off from a substrate, (**d**) after peeling off, and (**e**) under operation [48]

References

1. H. Matsui, Y. Takeda, S. Tokito, Org Electron **75**, 105432 (2019)
2. J. Kwon, S. Baek, Y. Lee, S. Tokito, S. Jung, Langmuir **37**, 10692 (2021)
3. S. Jung, J. Kwon, S. Tokito, G. Horowitz, Y. Bonnassieux, S. Jung, J Phys D Appl Phys **52**, 444005 (2019)
4. H. Matsui, K. Hayasaka, Y. Takeda, R. Shiwaku, J. Kwon, S. Tokito, Sci Rep **8**, 8980 (2018)
5. T. Leydecker, Z.M. Wang, F. Torricelli, E. Orgiu, Chem Soc Rev **49**, 7627 (2020)
6. R. Shiwaku, H. Matsui, K. Nagamine, M. Uematsu, T. Mano, Y. Maruyama, A. Nomura, K. Tsuchiya, K. Hayasaka, Y. Takeda, T. Fukuda, D. Kumaki, S. Tokito, Sci Rep **8**, 6368 (2018)
7. S. Cheng, Z. Lou, L. Zhang, H. Guo, Z. Wang, C. Guo, K. Fukuda, S. Ma, G. Wang, T. Someya, H.-M. Cheng, X. Xu, Adv Mater **35**, 2206793 (2023)
8. P. Vicca, S. Steudel, S. Smout, A. Raats, J. Genoe, P. Heremans, Thin Solid Films **519**, 391 (2010)
9. A. Kamyshny, S. Magdassi, Small **10**, 3425 (2014)
10. K. Fukuda, T. Someya, Adv Mater **29**, 1602736 (2016)
11. K. Fukuda, T. Sekine, Y. Kobayashi, Y. Takeda, M. Shimizu, N. Yamashita, D. Kumaki, M. Itoh, M. Nagaoka, T. Toda, S. Saito, M. Kurihara, M. Sakamoto, S. Tokito, Org Electron **13**, 3296 (2012)
12. D. Kumaki, Y. Fujisaki, S. Tokito, Org Electron **14**, 475 (2013)
13. D.J. Gundlach, J.E. Royer, S.K. Park, S. Subramanian, O.D. Jurchescu, B.H. Hamadani, A.J. Moad, R.J. Kline, L.C. Teague, O. Kirillov, C.A. Richter, J.G. Kushmerick, L.J. Richter, S.R. Parkin, T.N. Jackson, J.E. Anthony, Nat Mater **7**, 216 (2008)
14. T. Minari, T. Miyadera, K. Tsukagoshi, Y. Aoyagi, H. Ito, Appl Phys Lett **91**, 053508 (2007)
15. M. Halik, H. Klauk, U. Zschieschang, G. Schmid, C. Dehm, M. Schütz, S. Maisch, F. Effenberger, M. Brunnbauer, F. Stellacci, Nature **431**, 963 (2004)
16. A.F. Stassen, R.W.I. de Boer, N.N. Iosad, A.F. Morpurgo, Appl Phys Lett **85**, 3899 (2004)
17. H. Minemawari, T. Yamada, H. Matsui, J. Tsutsumi, S. Haas, R. Chiba, R. Kumai, T. Hasegawa, Nature **475**, 364 (2011)
18. H. Iino, T. Usui, J. Hanna, Nat Commun **6**, 6828 (2015)
19. C. Mitsui, T. Okamoto, M. Yamagishi, J. Tsurumi, K. Yoshimoto, K. Nakahara, J. Soeda, Y. Hirose, H. Sato, A. Yamano, T. Uemura, J. Takeya, Adv Mater **26**, 4546 (2014)
20. Y. Zhao, Y. Guo, Y. Liu, Adv Mater **25**, 5372 (2013)
21. O. Marinov, M.J. Deen, U. Zschieschang, H. Klauk, IEEE Trans Electron Devices **56**, 2952 (2009)
22. R.D. Deegan, O. Bakajin, T.F. Dupont, G. Huber, S.R. Nagel, T.A. Witten, Nature **389**, 827–829 (1997)
23. S. Chung, S.O. Kim, S.K. Kwon, C. Lee, Y. Hong, IEEE Electron Device Lett **32**, 1134–1136 (2011)
24. M. Hambsch, K. Reuter, M. Stanel, G. Schmidt, H. Kempa, U. Fügmann, U. Hahn, A.C. Hübler, Mater Sci Eng B **170**, 93–98 (2010)
25. H. Klauk, Chem Soc Rev **39**, 2643–2666 (2010)
26. F. Ante, D. Kälblein, T. Zaki, U. Zschieschang, K. Takimiya, M. Ikeda, T. Sekitani, T. Someya, J.N. Burghartz, K. Kern, H. Klauk, Small **8**, 73–79 (2012)
27. T. Okuzono, M. Kobayashi, M. Doi, Phys Rev E **80**, 021603 (2009)
28. K. Fukuda, T. Sekine, D. Kumaki, S. Tokito, ACS Appl Mater Interfaces **5**, 3916–3920 (2013)
29. M.-H. Yoon, H. Yan, A. Facchetti, T.J. Marks, J Am Chem Soc **127**, 10388–10395 (2005)
30. K. Fukuda, Y. Takeda, M. Mizukami, D. Kumaki, S. Tokito, Sci Rep **4**, 3947 (2014)
31. G. Lloyd, T. Backlund, P. Brookes, L.W. Tan, P. Wierzchowiec, J.Y. Lee, S. Bain, M. James, J. Canisius, S. Tierney, K. Kawamata, T. Wakimoto, Proc IDW **10**, 469 (2010)
32. M. Kaltenbrunner, T. Sekitani, J. Reeder, T. Yokota, K. Kuribara, T. Tokuhara, M. Drack, R. Schwödiauer, I. Graz, S. Bauer-Gogonea, S. Bauer, T. Someya, Nature **499**, 458–463 (2013)

33. D.H. Kim, N. Lu, R. Ma, Y.S. Kim, R.H. Kim, S. Wang, J. Wu, S.M. Won, H. Tao, A. Islam, K.J. Yu, T. Kim, R. Chowdhury, M. Ying, L. Xu, M. Li, H.J. Chung, H. Keum, M. McCormick, P. Liu, Y.W. Zhang, F.G. Omenetto, Y. Huang, T. Coleman, J.A. Rogers, Science **333**, 838–843 (2011)
34. K. Fukuda, Y. Takeda, Y. Yoshimura, R. Shiwaku, L.T. Tran, T. Sekine, M. Mizukami, D. Kumaki, S. Tokito, Nat Commun **5**, 4147 (2014)
35. H. Gleskova, S. Wagner, W. Soboyejo, Z. Suo, J Appl Phys **92**, 6224–6229 (2002)
36. T. Sekitani, Y. Kato, S. Iba, H. Shinaoka, T. Someya, T. Sakurai, S. Takagi, Appl Phys Lett **86**, 073511 (2005)
37. P. Cosseddu, G. Tiddia, S. Milita, A. Bonfiglio, Org Electron **14**, 206–211 (2013)
38. M. Hambsch, K. Reuter, H. Kempa, A.C. Hübler, Org Electron **13**, 1989–1995 (2012)
39. K. Suzuki, K. Yutani, M. Nakashima, A. Onodera, S. Mizukami, M. Kato, T. Tano, H. Tomono, M. Yanagisawa, K. Kameyama, Proc IDW **09**, 1581–1584 (2009)
40. A. Yamamura, H. Matsui, M. Uno, N. Isahaya, Y. Tanaka, M. Kudo, M. Ito, C. Mitsui, T. Okamoto, J. Takeya, Adv Electron Mater **3**, 1600456 (2017)
41. S. Chen, Y. Zhao, A. Bolag, J. Nishida, Y. Liu, Y. Yamashita, A.C.S. Appl, Mater Interfaces **4**, 3994–4000 (2012)
42. T.C. Huang, K. Fukuda, C.M. Lo, Y.H. Yeh, T. Sekitani, T. Someya, K.T. Cheng, IEEE Trans Electron Devices **58**, 141–150 (2011)
43. K. Fukuda, T. Sekitani, T. Yokota, K. Kuribara, T.C. Huang, T. Sakurai, U. Zschieschang, H. Klauk, M. Ikeda, H. Kuwabara, T. Yamamoto, K. Takimiya, K.T. Cheng, T. Someya, IEEE Electron Device Lett **32**, 1448–1450 (2011)
44. T. Yokota, T. Sekitani, T. Tokuhara, N. Take, U. Zschieschang, H. Klauk, K. Takimiya, T.C. Huang, M. Takamiya, T. Sakurai, T. Someya, IEEE Trans Electron Devices **59**, 3434–3441 (2012)
45. Y. Takeda, Y. Yoshimura, Y. Kobayashi, D. Kumaki, K. Fukuda, S. Tokito, Org Electron **14**, 3362–3370 (2013)
46. Y. Takeda, Y. Yoshimura, F.A.E.B. Adib, D. Kumaki, K. Fukuda, S. Tokito, Jpn J Appl Phys **54**, 04DK03 (2015)
47. K. Nagamine, A. Nomura, Y. Ichimura, R. Izawa, S. Sasaki, H. Furusawa, H. Matsui, S. Tokito, Anal Sci **36**, 291 (2020)
48. I. Shoji, H. Wada, K. Uto, Y. Takeda, T. Sugimoto, H. Matsui, Adv Mater Technol **6**, 2100723 (2021)

Chapter 5
Ultra-Flexible Organic Electronics

Tomoyuki Yokota

Abstract Flexible electronics have much attracted to realize biomedical applications for their flexibility and conformability. To improve these characteristics, reducing Young's modulus of material itself or thickness of the device is very effective. In recent years, it has been reported that devices can be made thinner than a few µm to achieve high skin conformability. In this chapter, we introduce ultra-flexible organic electronics which were fabricated on the 1 µm-thick plastic substrate. The ultra-flexible organic electronics do not show any degradation after crumpling test. These ultra-flexible organic electronics can be applied to the stretchable electronics by integrating the pre-stretched elastomer. The detail of the fabrication process and device structure is also explained. In addition, we introduce the bio-signal measurement applications of ultra-flexible organic electronics. Finally, the breathable nanomesh sensor will be introduced. The breathability of the device is very important for long-term continuous measurement of bio-signal. We will introduce the detail of the nanomesh sensor and explain the biocompatibility of the sensor.

Keywords OTFT · OLED · OPD · PPG sensor · Image sensor

5.1 Introduction

The richness of our lives has improved with the advancement of electronic technology over time. In the early postwar period, the development of electronics centered on home appliances, but in recent years, the forms of electronics have become more diverse. A decade ago, it was common for a family to have one television set, telephone, and other household appliances; however, in today's society, each individual owns one or more personal computers and telephones. In fact, the number of electronics around us is growing every year: in 1990, each person owned roughly

T. Yokota (✉)
The University of Tokyo, Tokyo, Japan
e-mail: yokota@ntech.t.u-tokyo.ac.jp

© The Author(s), under exclusive license to Springer Nature Japan KK 2024
S. Ogawa (ed.), *Organic Electronics Materials and Devices*,
https://doi.org/10.1007/978-4-431-56936-7_5

one processor; today, this number has increased more than 100-fold. This number is expected to increase in the future, and an era of owning countless electronics will arrive in the near future. However, it is very difficult to use all electronics with constant attention, and it is expected to become a form in which we use electronics without being aware of it.

As a form of such electronics, wearable electronics have attracted attention in recent years. Wearable electronics have been developed based on the health boom against the backdrop of an aging population, which is a social problem today. A variety of shapes and forms have been commercialized, including ring- and bracelet-shaped rings, contact lenses, and textiles. However, these devices incorporate rigid electronics into conventional worn objects. Therefore, the number of units that can be fitted is limited, and the problem is that they inevitably feel worn. The ultimate solution to these problems is the development of wear-free electronics.

To realize the wear-free electronics described above, it is essential to develop new electronics that are different from silicon-based electronics of the past. Flexible electronics, in which electronics are fabricated on plastic film substrates, have attracted attention as representatives of such electronics. Flexible electronics are lightweight, thin, and mechanically flexible, because they are fabricated on plastic film substrates. In addition, devices can be fabricated by printing processes and thus have the potential to be fabricated over large areas at low cost. Flexible electronics is a field that has developed with the idea of developing and realizing displays that can be bent [1, 2]. In recent years, many companies have developed flexible displays, and the time will soon come when such displays become commonplace in our daily lives.

Flexible electronics are expected to have many applications other than in displays. Large-area sensors that can be bent or attached to curved surfaces [3], solar cells [4], radio-frequency identification (RFID) tags [5], and other integrated circuits can be used to fabricate large-area flexible devices. Such flexible electronics will become even more compatible with our daily lives by adding the mechanical property of elasticity and not just bendability. The environment in which people live, especially the objects with which they come into contact, has a large number of free-form surfaces, and elasticity is essential for covering these surfaces. As such an electronics field, stretchable electronics has also been actively studied in recent years, and stretchable displays [6], sensors [7], and solar cells [8] have been reported (Fig. 5.1).

In recent years, devices for clinical and biomedical applications have attracted attention as applications that focus on the softness of flexible and stretchable electronics [9–11]. Conventional devices made of silicon are rigid and do not exhibit good compatibility with soft living organisms. On the other hand, flexible electronics are soft and flexible and are thus considered to have a high affinity for soft living organisms. Our group previously reported an electronic artificial skin [12, 13] with temperature and pressure sensors integrated on a film substrate. Recently, a group at Stanford University reported a pressure sensor with higher sensitivity than the human skin [14]; a group at National Institute of Advanced Industrial Science and Technology (AIST) reported a carbon nanotube (CNT)-based strain sensor [15]; and

5 Ultra-Flexible Organic Electronics

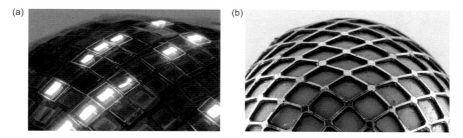

Fig. 5.1 Stretchable electronics. (**a**) Stretchable display. Reproduced with permission [6], copyright 2009, Springer Nature. (**b**) Stretchable pressure and temperature sensor. Reproduced with permission [13], copyright 2005, National Academy of Sciences

a group at the University of Illinois reported an epidermal device [16] in which various sensors were integrated. In addition to devices that sense from outside the human body, applications in medical devices such as pressure-sensing catheters [17] and stretchable balloon catheters [18] have also been studied. Furthermore, reports on flexible electrode matrices [19] and brain probes made of soft materials [20] for measuring biological signals in direct contact with organs in the body have been published, and further research is expected to progress in the future.

In this chapter, we introduce some of our recent work in clinical and biomedical applications by increasing the softness of the device.

5.2 Ultra-Flexible Electronics

Adhesion is an important property in the development of these devices for biological and medical applications. Poor adhesion of the device to the living body makes it difficult to measure the signal at the same location over a long period of time, and signal quality is easily affected by noise and other factors. Therefore, high adhesion between the organism and the film is essential for the long-term measurement of biological signals. A group at the University of Illinois experimentally demonstrated that thinning the substrate improves its adhesion to objects with complex shapes, as shown in Fig. 5.2 [21, 22]. Therefore, the realization of electronics on such thin-film substrates would be a great advantage for biological and medical device applications. In this section, we introduce various organic electronic devices fabricated on flexible substrates.

5.2.1 Ultra-Flexible Organic Integrated Circuits

Our group succeeded in realizing organic transistors and integrated circuits on substrates as thick as 1 μm [23]. Figure 5.3 shows the fabricated ultra-flexible

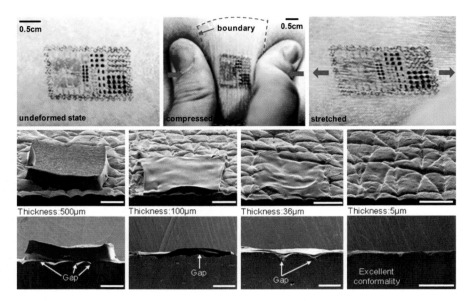

Fig. 5.2 Conformability of the varying thickness device. The thin device shows the good conformability compared with thicker devices. Reproduced with permission [21], copyright 2010, Springer Nature. Reproduced with permission [22], copyright 2013, Wiley-VCH

organic integrated circuit. The fabricated device had a total thickness of 2 µm, including the encapsulating film, and thus adhered very well to the skin. It is extremely light, weighing only 3 g/m^2, which reduces the feeling of wearing it on the body. Such devices fabricated on thin-film substrates can dramatically improve not only the adhesion but also the flexibility. The fabricated device showed almost no change in electrical properties even after crumpling, confirming its remarkable flexibility (Fig. 5.4). This is because the magnitude of the strain on the device is approximately proportional to the thickness of the substrate [24]. In addition, our device placed an organic integrated circuit in the middle of the entire device by depositing an encapsulating film that was almost as thick as the substrate [25]. This structure reduces the strain applied to the device when it is bent and is thought to be responsible for the remarkable flexibility of the device.

Furthermore, we can add stretchability to the ultra-flexible device by combining it with stretchable substrates (Fig. 5.5). First, our ultra-flexible device is attached to a pre-stretched soft substrate. If the stretchable substrate returns to its original state, a regular wrinkle structure can be fabricated, as shown in Fig. 5.5, because the device cannot shrink. This wrinkle structure flattens when the device is stretched and forms a wrinkle structure when it shrinks. Such a structure is mechanically very stable, and the device characteristics do not exhibit a significant change, even after more than 100 stretchable cycle tests [23].

Next, a sensor application using the fabricated ultra-flexible electronics is described. Figure 5.6 shows an example of a sensor application using organic

Fig. 5.3 Ultra-flexible organic circuits. Ultra-flexible organic circuits with organic transistors directly attached to the skin. Reproduced with permission [23], copyright 2013, Springer Nature

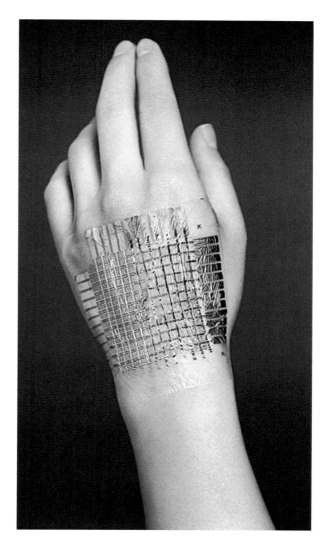

transistors. The device integrates an organic transistor active matrix and tactile sensor with a comb structure. The comb electrodes do not conduct when they are not touched, but when touched by a conductive object or material such as a finger or metal, they conduct. When a metal ring was placed on this tactile sensor matrix, a current was observed in its shape (Fig. 5.7).

Furthermore, we fabricated more complex organic amplified circuits by integrating electrical elements such as organic transistors [26]. The fabricated organic amplifier circuit is based on a pseudo-complementary metal-oxide-semiconductor (CMOS) circuit consisting of only p-type organic transistors and operates at a very low voltage of 2 V or less. In addition, an extremely high amplification gain of more

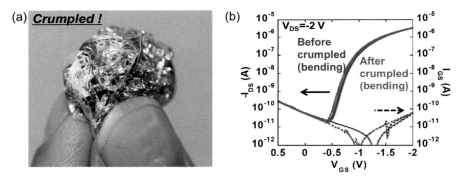

Fig. 5.4 Mechanical durability of the ultra-flexible organic transistors. (**a**) Picture of the ultra-flexible organic transistor after crumpling test. (**b**) The electrical characteristics of the ultra-flexible organic transistors before and after crumpling test. Reproduced with permission [23], copyright 2013, Springer Nature

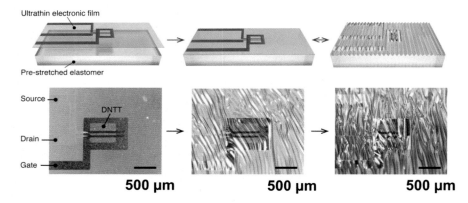

Fig. 5.5 Stretch-compatible ultra-flexible organic transistors. Ultra-flexible organic transistor was attached to the pre-stretched elastomer. The surface wrinkle structure shows the good stretchability. Reproduced with permission [23], copyright 2013, Springer Nature

than 50 dB was achieved in the low-frequency range. Such organic amplification circuits, with their high amplification gain in the low-frequency range, are well suited for amplifying biological signals. Therefore, we amplified rat electrocardiograms (ECGs) using the developed organic amplification circuit. As shown in Fig. 5.8, the magnitude of the biological signal was amplified approximately 100 times, and the signal-to-noise ratio was successfully improved more than 10 times by connecting an organic amplifier circuit [26].

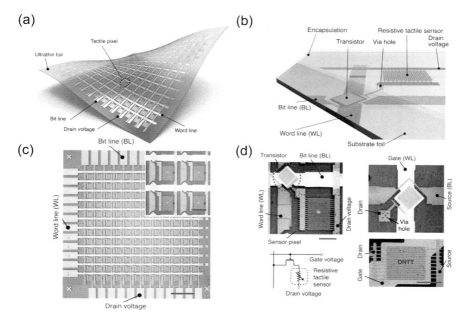

Fig. 5.6 The application of the ultra-flexible organic circuits. (**a**) Schematic illustration of the ultra-flexible tactile sensor system. (**b**) Schematic illustration of one pixel of the tactile sensor. (**c**) Picture of the ultra-flexible tactile sensor (**d**) picture of one pixel of the tactile sensor. Reproduced with permission [23], copyright 2013, Springer Nature

5.2.2 Ultra-Flexible Organic Light-Emitting Diodes

In addition to electrical measurements, the measurement of various biological signals using light has attracted attention. Unlike electrical measurements, the measurement of biological signals using light is superior in terms of the resolution and sensitivity. Furthermore, noninvasive photosensors measure various biological information in the body from living surfaces such as the skin and is used in a variety of medical devices [27–29]. We have successfully fabricated ultra-flexible optical devices that follow the skin surface by fabricating organic light-emitting diodes and photodetectors on ultra-flexible substrates [30]. Furthermore, by integrating the fabricated organic light-emitting diode and photodetectors, we succeeded in developing a flexible pulse oximeter that could be wrapped around a finger. The newly developed ultra-flexible organic light-emitting diodes are expected to be applied not only to display and sensor applications but also to a wide range of other fields, such as devices for light stimulation.

Figure 5.9 shows the device structure and photograph of the fabricated organic light-emitting diode (OLED). This flexible OLED was extremely thin, with a substrate thickness of 1 μm. Therefore, glass substrates are used as support substrates during the fabrication process to improve handling. First, a fluoropolymer was

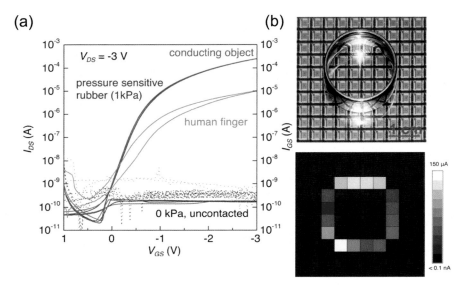

Fig. 5.7 Demonstration of the ultra-flexible tactile sensor system. (**a**) The electrical characteristics of the tactile sensor. If the conducting object attach to the sensor, the current was increased. (**b**) Top-view photograph of a metallic ring placed on the tactile sensor system. (**c**) Corresponding drain current of a metallic ring placed on the sensing sheet. Reproduced with permission [23], copyright 2013, Springer Nature

formed as a release layer on the surface of a glass substrate, which was used as a support substrate via spin coating. Next, 1 μm-thick parylene was deposited on the glass substrates by chemical vapor deposition. After the deposition of parylene, 500-nm polyimide was deposited by spin coating as a smooth layer. ITO was then deposited as a transparent electrode via sputtering at 70 nm. The ITO electrode thus formed is amorphous and flexible. Following ITO deposition, the hole injection layer (HIL) and hole transport layer (HTL) were deposited in air via spin coating. Each layer was deposited by spin coating and then annealed in air. Next, a polymer material was deposited as the emission layer via spin coating. After deposition, the film was annealed in a nitrogen atmosphere, and sodium fluoride and aluminum were deposited as cathodes by vacuum evaporation. Finally, 1 μm of parylene was deposited as an encapsulating film by chemical vapor deposition. The fabricated OLED can be used as a freestanding device by peeling it off from the support substrate after fabrication. The fabricated OLEDs have a total thickness of less than 3 μm, indicating that they are highly flexible and do not break even when crumpled. As shown in Fig. 5.10, light emission was observed even when the device was placed on a 100-μm thick razor blade and bent without breaking.

The characteristics of the fabricated device are shown in Fig. 5.11. The fabricated OLEDs showed almost the same characteristics as the devices fabricated on glass substrates, with external quantum efficiencies of more than 10% for the red and green OLEDs and higher than 5% for the blue OLED (Fig. 5.11). One of the factors

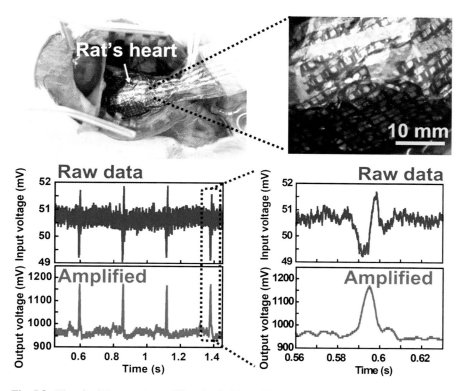

Fig. 5.8 Ultra-flexible organic amplifier circuit. The ECG signal of rat was amplified by the ultra-flexible organic amplifier circuit. Reproduced with permission [26], copyright 2016, Springer Nature

that contributed to these properties was the deposition of a smoothing layer on the parylene substrate. Parylene substrates have higher surface roughness than commercially available flexible film substrates. This factor greatly reduced the device yield when fabricating vertical devices such as OLEDs. Therefore, we succeeded in improving the surface flatness by one order of magnitude by depositing a polyimide film on a parylene substrate. As mentioned earlier, the fabricated OLEDs must be peeled off from the support substrate, and there is a possibility that the device may be damaged during the process. However, the fabricated OLED had high flexibility, and it was confirmed that its electrical and optical properties remained almost unchanged when it was peeled off from the glass support substrate. Furthermore, by combining the fabricated OLED with a rubber substrate, we successfully added elasticity to the device. First, the rubber substrate was stretched, and an ultra-flexible OLED was attached to it. The pre-stretched rubber can then be restored to its original state, thereby adding stretchability to the OLED and organic transistors. Figure 5.12 shows a photograph of the OLED as it was stretched. Stretchability was determined by the

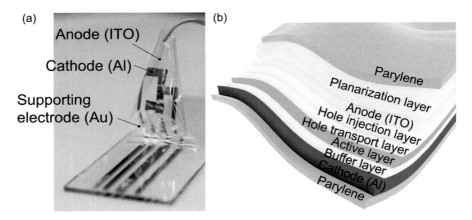

Fig. 5.9 Ultra-flexible organic light-emitting diode (OLED). (**a**) The photograph of the ultra-flexible OLED (**b**) the structure of the ultra-flexible OLED. Reproduced with permission [31], copyright 2020, National Academy of Sciences

Fig. 5.10 Flexibility of the ultra-flexible OLED. The ultra-flexible OLED was put on the 100 μm-thick blade. Reproduced with permission [31], copyright 2020, National Academy of Sciences

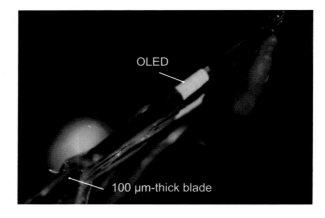

pre-stretched value of the rubber substrate, and in this case, 200% stretchability was successfully achieved.

Furthermore, ultra-flexible OLEDs can be used for the optical stimulation of nerves, taking advantage of their flexibility. The fabricated blue ultra-flexible OLEDs were successfully used to conduct experiments on the light stimulation of rat nerves [31]. Ultra-flexible OLEDs have a total thickness of 3 μm and are extremely thin, allowing them to follow the surface topography of the skin and muscle. This allows for more efficient light stimulation compared with conventional solid LEDs and other devices. Figure 5.13 shows an experiment in which rats were photo-stimulated. The fabricated OLED has a light-emitting area of 4 mm^2, and by using gold as the supporting electrode, a high output power of 0.5 mW/mm^2 at a drive voltage of less than 10 V was successfully achieved. Figure 5.14 shows the myoelectric potential of the rat when the rat nerves were stimulated by the OLED. In

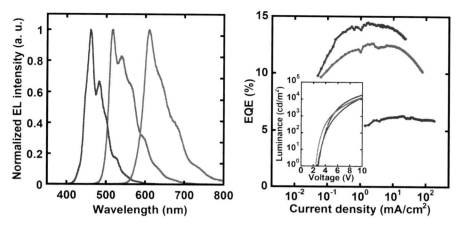

Fig. 5.11 The electrical characteristics of the ultra-flexible OLED. The ultra-flexible OLED shows the almost same characteristics as the devices fabricated on glass substrates. Reproduced with permission [30], copyright 2016, American Association for the Advancement of Science

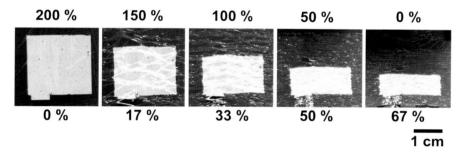

Fig. 5.12 Stretch-compatible ultra-flexible OLED. The ultra-flexible OLED shows stretchability of over 200%. Reproduced with permission [30], copyright 2016, American Association for the Advancement of Science

Fig. 5.13 Optical stimulation by the ultra-flexible OLED. The ultra-flexible blue OLED was put on the gracilis muscle of rat hind limb and optical stimulated of gracilis muscle nerve. Reproduced with permission [30], copyright 2020, National Academy of Sciences

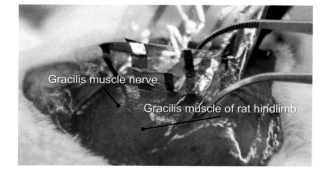

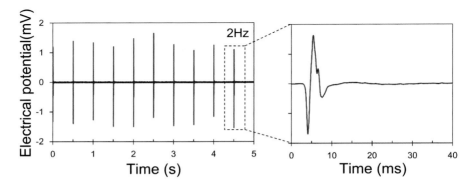

Fig. 5.14 The myoelectric potential of the rat when the rat nerves were stimulated by the OLED. Light stimulation was performed using a very short pulse of 5 ms at a period of 2 Hz. The myopotentials were generated immediately after photo-stimulation. Reproduced with permission [31], copyright 2020, National Academy of Sciences

the present study, light stimulation was performed using a very short pulse of 5 ms at a period of 2 Hz. The blue lines in Fig. 5.14 show the period during which light stimulation was performed. The results showed that myopotentials were generated immediately after photo-stimulation. Furthermore, the EMGs showed a period of 2 Hz, similar to that of light stimulation, confirming that the flexible OLED can be used for proper optical stimulation of the nerves. Similar evoked potentials were observed when the frequency was increased to 10 Hz.

5.2.3 Ultra-Flexible Organic Photodetectors

An organic photodiode (OPD) can be realized using a process similar to that used for OLEDs. An organic semiconductor with high sensitivity in the near-infrared region (850 nm) was used as the active layer of an ultra-flexible OPD. ITO was used as the transparent electrode, and a zinc oxide (ZnO) layer was formed on its surface to form an electron transport layer. PMDPP3T was used as the donor material, $PC_{61}BM$ was used as the acceptor material, and a bulk heterostructure containing a mixture of these two materials was used as the active layer. Poly(3,4-ethlenedioxythiophens) polystyrene sulfonate (PEDOT:PSS) and silver were used as the hole transport layer (HTL) and the top electrodes, respectively. Finally, a 1-μm thick polymer film was deposited as the encapsulating film (Fig. 5.15). Flexible organic photodiodes responsive to near-infrared light, which have been reported in the past, have problems such as low detection sensitivity owing to high dark current [32, 33]. Organic photodiodes with low dark currents have also been reported, but their low photosensitivity makes them difficult to use in biological imaging [33–35]. By optimizing the mixing ratio of the donor and acceptor materials and the film thickness of the active layer, we have succeeded in developing a flexible OPD that simultaneously achieves a low

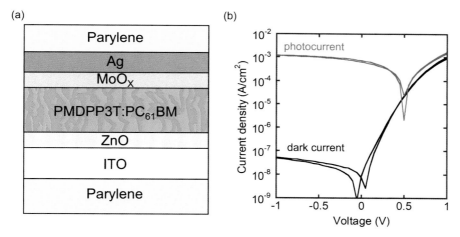

Fig. 5.15 Ultra-flexible organic photo diode (OPD). (**a**) Structure of the ultra-flexible OPD (**b**) electrical characteristics of the ultra-flexible OPD. Reproduced with permission [38], copyright 2020, Springer Nature

dark current and high optical sensitivity in the near-infrared region. Photosensitivity was 0.57 A/W at the 850 nm wavelength used for vein authentication and other applications. This sensitivity is comparable to that of silicon photodiodes used in conventional imagers. The dark current density of the fabricated OPD was as low as 10^{-7} A/cm^2 or less, and the photocurrent density was 1.7×10^{-3} A/cm^2 when irradiated with near-infrared light at a wavelength of 850 nm and an output of 2.9 mW/cm^2 (Fig. 5.15).

5.2.4 Sensor Applications of Ultra-Flexible Organic Optical Devices

Next, the sensor applications of ultra-flexible OLEDs and OPDs are introduced. We succeeded in developing an ultra-flexible pulse oximeter that can be wrapped around a finger. Figure 5.16 shows a photograph of the developed ultra-flexible pulse oximetry. Pulse oximetry consists of a two-color OLED and OPD. Hemoglobin in the blood has different absorption coefficients depending on the wavelength of light. Therefore, we can calculate the ratio of hemoglobin to oxidized hemoglobin using two different wavelengths and detecting the light scattered and reflected inside the finger with the OPD. From this calculated ratio, we estimated the blood oxygen level [36]. Figure 5.17 shows the photoplethysmogram (PPG) signals measured using this pulse oximetry. In the experiment, measurements were conducted on two subjects with different blood oxygen levels (99% and 90%). In subjects with a 99% blood-oxygen ratio, the amplitudes of the pulse waves measured with green and red light were almost of the same magnitude. In contrast, in subjects with a 90%

Fig. 5.16 Ultra-flexible organic pulse oximetry. The thickness of each component is 3 μm. Reproduced with permission [30], copyright 2016, American Association for the Advancement of Science

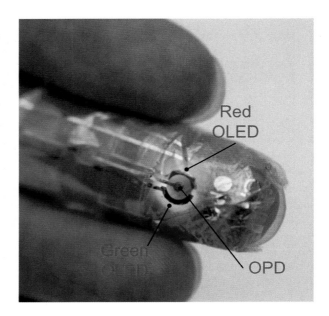

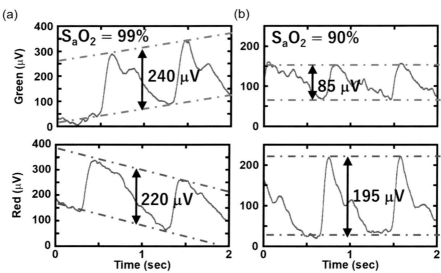

Fig. 5.17 Measurement of PPG signal (**a**) Output signal from OPD with 99% oxygenation of blood. The green and red lines represent the signals when the green and red PLEDs, respectively, were operated. (**b**) Output signal from OPD with 90% of oxygenation of blood. The green and red lines represent the signals when the green and red PLEDs, respectively, were operated. Reproduced with permission [30], copyright 2016, American Association for the Advancement of Science

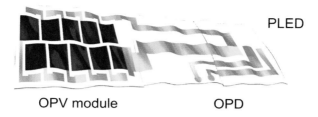

Fig. 5.18 Ultra-flexible self-powered PPG sensor system. The ultra-flexible self-powered PPG sensor contains ultra-flexible OLED, OPD, and organic solar cell (OPV). Reproduced with permission [37], copyright 2021, Springer Nature

blood-oxygen ratio, the amplitude of the pulse wave measured with red light was approximately two to three times greater than that measured with the green light-emitting device. The results showed a similar trend to previously reported commercial pulse oximetry, meaning that differences in blood oxygen levels can be measured with an ultra-flexible organic optical sensor. To achieve further measurement accuracy, the number of subjects should be increased, and a calibration curve should be produced.

One of the remaining challenges in ultra-flexible pulse oximetry was the rigid power supply. Although the sensor itself was extremely thin, the power supply to drive the sensor was not sufficiently thin, and the system as a whole was not sufficiently flexible. Therefore, our group developed a self-powered pulse wave sensor by forming organic thin-film solar cells (OPVs) on an ultra-flexible substrate [37]. Figure 5.18 shows a schematic of the developed self-powered sensor. The newly developed self-powered sensor integrated three types of devices: OLED, OPD, and OPV. The voltage required to drive the OLED was achieved by connecting several OPVs in series. The fabricated OPV modules had a total device area of 1.2 cm^2 and power conversion efficiency of 5.8%. This efficiency was almost the same as that of devices fabricated on glass substrates. Each device was fabricated on a separate substrate and connected using anisotropic conductive tape. The fabricated self-powered pulse-wave sensor can be operated by generating electricity with sunlight to drive the OLED. The light from the OLED was incident inside the finger, and the OPD received the reflected light to detect the pulse wave. In fact, when the developed sensor was attached to a hand and driven using a solar simulator, it successfully detected pulse waves, as shown in Fig. 5.19.

5.3 Sheet-Type Image Sensor

Other important applications of organic optical devices are image sensors. We developed flexible organic image sensors by combining organic photodiodes and backplane technology [38]. Figure 5.20 shows the device structure and device photograph of the sheet-type organic image sensor. Sheet-type organic image

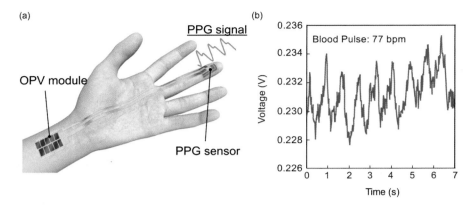

Fig. 5.19 Demonstration of the PPG measurement by the ultra-flexible self-powered sensor. (**a**) Schematic illustration of the PPG measurement (**b**) Output voltage characteristics of OPD with PPG measurement. Reproduced with permission [37], copyright 2021, Springer Nature

sensors use organic photodiodes with high sensitivity to near-infrared light as the optical sensor and low-temperature polysilicon (LTPS) thin-film transistors as the backplane for cell selection. The flexible organic image sensor has an effective total photosensitive area of 12.6 × 12.8 mm^2, a cell pitch of 50 μm, and a resolution of 508 dpi, for a total pixel count of 252 × 256. The organic photodiode had maximum sensitivity at 850 nm in the near-infrared region. In addition, the organic photodiode uses an inverse structure that is atmospherically stable, achieving high reliability under atmospheric conditions. When light is irradiated on a flexible organic image sensor, the organic semiconductor layer absorbs the light, which generates a photocurrent through photoelectric conversion, and the charge is stored in the capacitor of the sensor element in each cell. Imaging can be performed by measuring the accumulated charge as a signal. The circuit part that performs this readout and signal processing is mounted on a flexible cable, making the entire organic imager system highly flexible.

The newly developed image sensor uses LTPS transistors with high mobility in the backplane to achieve high-resolution and high-speed operation. In fact, LTPS transistors have a mobility of more than 10 cm^2/Vs, compared to the average mobility of 1 cm^2/Vs for organic and amorphous silicon thin-film transistors. In addition, the on/off ratio of the transistor is also large, 10^7 or higher, compared to organic transistors and amorphous silicon transistors, which range from 10^4 to 10^6. Therefore, the LTPS transistor backplane fabricated in this study could read out one cell as fast as 60 μs. This is because the mobility is 5–10 times higher than that of previously reported transistors using organic transistors, IGZO, and amorphous silicon [39–41]. In addition, the high mobility of the transistors allows for a finer cell size, and the sensor was designed with a resolution of 508 dpi, which is higher than that of conventional flexible image sensors. The integration of this

5 Ultra-Flexible Organic Electronics

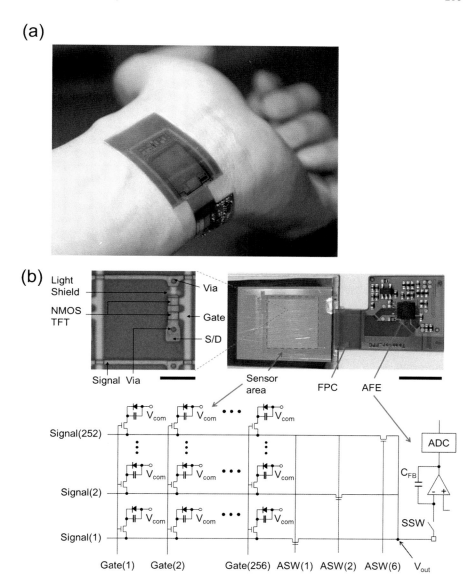

Fig. 5.20 Vain imaging by the sheet-type image sensor. (**a**) Photograph of the sheet-type organic image sensor, which can easily attach to the curved surface. (**b**) Picture of the image sensor pixel and circuit diagram of the sheet-type organic image sensor. Reproduced with permission [38], copyright 2021, Springer Nature

high-resolution, fast-operating LTPS transistor with a highly sensitive organic photodiode enables the high-speed detection of extremely weak light at 4.7 $\mu W/cm^2$.

Fig. 5.21 Fingerprint imaging by sheet-type image sensor. The finger was directly put on the sheet-type image sensor and capture the fingerprint image. Reproduced with permission [38], copyright 2021, Springer Nature

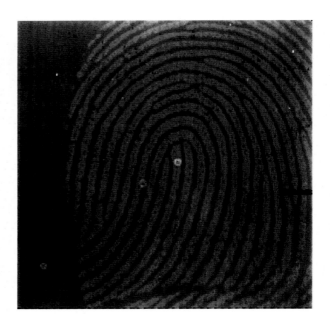

One of the features of a sheet-type organic image sensor is that it can capture fingerprints without using an optical system. Figure 5.21 shows the image of the fingerprints taken by the image sensor directly in contact with the skin. The captured fingerprint image can detect not only feature points such as the number of ridges, branching points, and end points but also the location of sweat glands, thus capturing fingerprint information that is important for biometric authentication. As mentioned earlier, the newly developed image sensor can capture high-resolution fingerprint images without the use of an optical system. Therefore, the entire system can easily be made smaller and more flexible.

Furthermore, vein image can be captured by combining a flexible organic image sensor with an optical lens. Figure 5.22 shows the results of finger-vein imaging using a flexible organic image sensor. To obtain a vein image, light was irradiated from the top of the finger using a near-infrared LED, and the light transmitted through the finger was focused using a lens. The vein images captured in this study clearly show characteristic information, such as branching points, branching angles, and number of veins, which are generally used for vein authentication. In addition, the vein images captured with the organic image sensor were compared to those captured using the CMOS image sensor, and the images were almost equivalent.

As mentioned above, the sheet-type organic image sensor can also capture images at a high speed. Therefore, static biometric data such as fingerprints and veins can be obtained, and dynamic biometric data such as photoelectric volumetric pulse waves can be measured. Figure 5.23 shows the results of a photoelectric volume pulse wave measurement using a flexible organic image sensor. A comparison of the results

Fig. 5.22 Vain imaging by the sheet-type image sensor. The vain image was captured using optical lens. Reproduced with permission [38], copyright 2021, Springer Nature

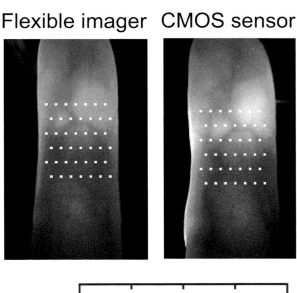

Fig. 5.23 Measurement of the PPG signal by the sheet-type image sensor. The finger was directly put on the sheet-type image sensor and pulse wave signal was measured. Reproduced with permission [38], copyright 2021, Springer Nature

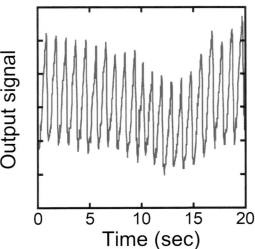

measured with the developed sheet-type image sensor with those measured with a commercially available pulse wave sensor showed that both sensors had the same heart rate of 77 beats per min. Furthermore, when the average time difference of the peak position of the photoelectric volume pulse wave was evaluated, it showed a very small value of 30 ms compared to a commercially available pulse-wave sensor, confirming that it can measure the same performance as a commercially available pulse-wave sensor.

The almost flexible pulse-wave sensor reported to date has only one sensor point. Therefore, when measuring the photoelectric volumetric pulse wave at the wrist, the

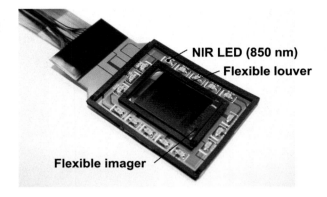

Fig. 5.24 Wearable-type image sensor. Photograph of the wearable-type image sensor. The wearable image sensor contains the sheet-type image sensor, flexible louver and near-infrared LEDs. Reproduced with permission [38], copyright 2021, Springer Nature

signal intensity of the photoelectric volumetric pulse wave can vary significantly depending on the position at which the sensor is attached. This is because when the sensor is located near blood vessels, the photoelectric volume pulse wave can be measured stably, whereas in the absence of blood vessels, the light reflected and scattered inside the body is small, resulting in a smaller signal intensity. On the other hand, sheet-type organic image sensors have high resolution and high speed and can measure photoelectric volumetric pulse waves cell by cell. Therefore, even when the sensor attachment point is shifted, the amplitude intensity of the pulse photoelectric volume pulse wave is measured for each cell, and the cell that can most stably measure the photoelectric volume pulse wave is selected to enable a stable photoelectric volume pulse wave measurement.

The sheet-type organic image sensors described above do not have integrated optical systems, such as light sources and lenses, and to use them as wearable devices, it was necessary to integrate these elements into the image sensor. Therefore, we developed a wearable image sensor, as shown in Fig. 5.24. The wearable image sensor integrated 16 near-infrared LED chips as light sources around a flexible organic image sensor. In addition, a flexible louver film was integrated into an image sensor as an optical system for vein imaging. LED chips can be mounted on flexible substrates, allowing them to bend and follow curved surfaces such as arms.

The results of the vein imaging using this wearable image sensor are shown in Fig. 5.25. Wearable image sensors are flexible as a whole. For example, a wearable image sensor can be wrapped around the wrist area to image veins. Compared with sheet-type organic image sensors, wearable image sensors increase the size of the sensor portion to 3 cm × 4 cm and use a sensor with reduced resolution. Therefore, compared with the vein image shown in Fig. 5.22, the vein boundaries are slightly blurred. However, the branching shape of the vein was captured well, and the vein image was confirmed to be sufficient for biometric authentication.

One of the applications of the developed wearable image sensor is the visualization of the blood flow. The visualization of blood flow was verified by analyzing vein images taken with the developed flexible image sensor using a method called singular value decomposition (SVD). SVD is a technique that has long been used in

5 Ultra-Flexible Organic Electronics

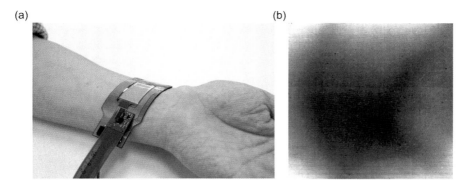

Fig. 5.25 Vain imaging by the wearable-type image sensor. (**a**) Photograph of the measurement setup of the vein image. (**b**) Vain image capture by the wearable-type image sensor. Reproduced with permission [38], copyright 2021, Springer Nature

image analysis to extract features, such as the time variation of images. Figure 5.26 shows a schematic of the SVD analysis. First, an image sensor was used to obtain time-varying images of the finger veins. The images used in this analysis were 1000 vein images taken in 18 s. The image data were converted into a two-dimensional matrix, and SVD analysis was performed based on this matrix. The obtained images were found to contain different information for each singular value. For example, some singular values can be found that contain photoplethysmography volume pulse wave information or body motion noise. By retransforming the image using only singular values containing biometric information, an image with a reduced noise signal can be obtained. By analyzing these images, we succeeded in obtaining images with time-varying contrast in the vein [42].

5.4 Flexible Temperature Sensor

Many factors are considered important for health monitoring. Body temperature is an important vital parameter. Body temperature is very useful biometric information as a vital sign for health status and early detection of disease, and its constant monitoring in daily life is considered very important. Measuring and controlling temperature is very important not only for health monitoring but also for industrial purposes. For this reason, research and development have been conducted for decades, and various types of temperature sensors have been widely put into practical use. Thermocouples and resistance thermometers are commonly used in the industry as temperature sensors. These temperature sensors have a sensitivity of 0.1 °C or less and can measure temperature with high accuracy. On the other hand, to read out these temperature sensors with high accuracy, complex circuits for readout are required, and considering multipoint measurement, there is a need to line up single temperature sensors. In conventional applications, these points have not been

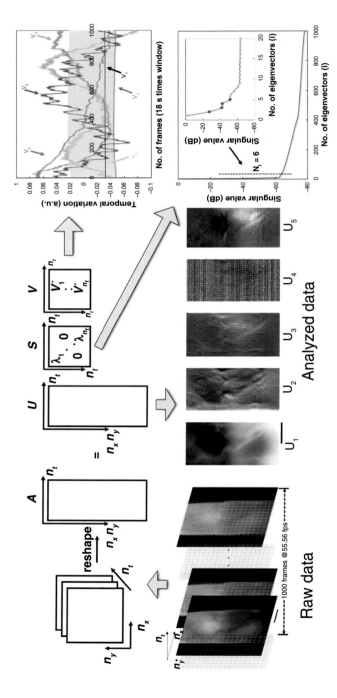

Fig. 5.26 Concept of singular value decomposition (SVD) analysis. 3D matrix hologram is reshaped into 2D matrix, and then, decomposed into three matrices of spatial (U) and temporal (V) eigenvectors as well as a diagonal matrix of singular values. Reproduced with permission [42], copyright 2022, Optica Publishing Group

a major issue, but when considering wearable devices, it is very important to make them as small, simple, and weightless as possible to ensure comfort. In particular, epidermal body temperature differs from place to place, making multiple-point measurements important for monitoring health status. In addition, considering monitoring in daily life, there is a need to accurately measure the temperature of a moving living body, which requires high adhesion between the temperature sensor and the living body and reduces damage to the living body. These are difficult to achieve with conventional hard sensors, and it is important to develop soft temperature sensors that are gentle to the living body.

5.4.1 Conventional Flexible Temperature Sensor

One approach is to fabricate a hard temperature sensor such as a thermocouple or a resistance thermometer on a flexible substrate and use it as a temperature sensor that is less invasive to the living body. Such flexible temperature sensors have been reported by several researchers worldwide. For example, Professor John Rogers and his colleagues at the University of Illinois successfully measured the temperature distribution of biological tissue by arranging multiple resistance thermometers on a flexible substrate and affixing the device directly to biological tissue [43]. However, although such flexible temperature sensors can measure with high accuracy, the amount of output change (resistance or voltage) in response to temperature change is very small, and a highly accurate readout circuit is essential. Therefore, it is difficult to make the readout circuit part softer and smaller, which makes it incompatible with wearable devices and implantable temperature sensors.

Thermography is a typical example of other methods capable of imaging temperature distributions. Because thermographs measure the temperature by detecting infrared emission energy, a certain distance is required between the object to be measured and the sensor, making it difficult to miniaturize the system. In addition, the thermograph itself is a rigid element, making it incompatible with flexible, wearable electronics, and its practical resolution is limited to approximately 0.5 ° C. Furthermore, one of the biggest problems is the inability to measure the temperature of the body surface while wearing overclothing.

5.4.2 Polymer Positive Temperature Coefficient (PTC) Type Temperature Sensor

Printability is imperative for wearable and flexible electronics. This is because the printing process allows the devices to be fabricated directly on textile fabrics and flexible substrates. One printable temperature sensor material is the polymer PTC. Polymer PTCs consist of conductive filler materials, including graphite and silver

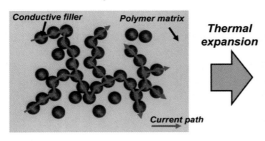
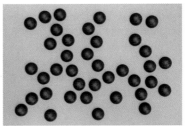

Fig. 5.27 The operation mechanism of polymer PTC sensor. At low temperatures, the conductive fillers are in contact with each other, and there are many conductive paths within the polymer. On the other hand, as the temperature increased, the distance between conductive fillers increased owing to the gradual expansion of the polymer volume

nanoparticles, dispersed in a polymer and can be fabricated by printing. The principle of operation of the polymer PTC is very simple: At low temperatures, the conductive fillers are in contact with each other, and there are many conductive paths within the polymer, resulting in low resistance. On the other hand, as the temperature increased, the distance between conductive fillers increased owing to the gradual expansion of the polymer volume. As a result, the final conductive path is reduced, and the resistance becomes very high (Fig. 5.27). The amount of resistance change in polymer PTCs varies depending on the polymer material; however, some devices using polyethylene and other materials have been reported to show resistance changes of six digits or more [44]. One of the characteristics of polymer PTCs is their rapid change in resistance. This temperature change occurs near the melting point or glass transition temperature of the polymer itself, and many have been reported to change the resistance by several orders of magnitude. Owing to this characteristic of a large resistance change at a specific temperature, they have been used in industrial applications as protection circuits for devices and thermal fuses. On the other hand, although conventionally used polymer PTCs exhibit excellent resistance changes, their extremely high reaction temperature of approximately 100 °C, low durability for repeated operation, and low flexibility owing to the thickness of the device (1 mm or more) are some of the obstacles for their application.

5.4.3 Polymer PTC with Sensitivity Near Body Temperature

Most of the polymer PTCs reported thus far are temperature sensors that show a resistance change above 100 °C because of the high melting point of the polymer itself [45, 46]. Therefore, although it can be used in protection circuits for electronics, it is difficult to use it for monitoring biological temperatures, and a reduction in

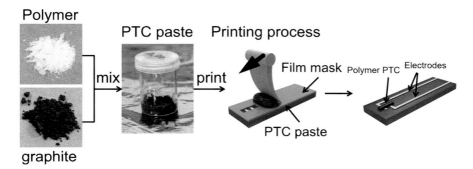

Fig. 5.28 The fabrication process of the polymer PTC sensor. The polymer PTC paste was made by mixing the synthesized semicrystalline acrylate polymer and conductive filler. The paste shows the high viscosity; we can easily print the paste by the printing method. Reproduced with permission [48], copyright 2015, National Academy of Sciences

the reaction temperature to near body temperature has long been awaited. Recently, Professor Zhenan Bao and his colleagues at Stanford University have developed a temperature sensor that achieves a high repeat durability of over 100 cycles and a large resistance change near body temperature by mixing two types of polymers, one with a low melting point and the other with a high melting point [47]. They designed a polymer PTC material by mixing two polymer materials, polyethylene and polyethylene oxide, and a nickel filler as the conductive material. Polyethylene oxide has a melting point near 45 °C, which is close to body temperature, while polyethylene has a melting point near 95 °C. By mixing two polymer materials with different melting points, they succeeded in realizing a temperature sensor that achieves a resistance change near the body temperature and yet has high durability for repeated use. However, this temperature sensor is more than 1-mm thick and lacks flexibility. In addition, because the reaction temperature is controlled by the molecular weight of the polyethylene oxide, it is very difficult to control the reaction temperature, which is a drawback because the product can only be used for certain applications.

A technique to address the issues with polymer PTC materials described earlier involves synthesizing a polymer material with a melting point near the body temperature by photopolymerizing acrylic monomer materials with two different alkyl chain lengths. By mixing 25 wt% graphite with the synthesized polymer material, it can be used as a polymer PTC material that exhibits resistance changes near the body temperature [48]. This material has a rigid state at room temperature, but when temperatures of approximately 40 °C are applied, the viscosity increases and the material becomes a printable paste (Fig. 5.28). The temperature sensor could be patterned using a metal mask or film mask, and the thickness could be reduced to 12 μm by reducing the thickness of the metal mask. Thus, compared with conventional temperature sensors using polymer PTCs, the film thickness can be reduced, enabling temperature sensors with high flexibility that do not break even when bent. Furthermore, by integrating this temperature sensor with an active matrix of organic transistors, a large-area multipoint temperature sensor can be attained.

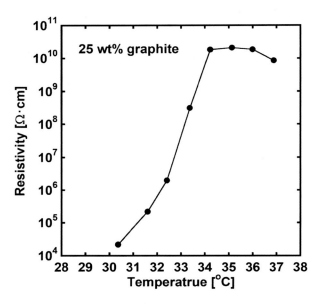

Fig. 5.29 The electrical characteristics of the temperature sensor. The resistivity was dramatically changed around skin temperature. Reproduced with permission [48], copyright 2015, National Academy of Sciences

Figure 5.29 shows the electrical characteristics of the fabricated temperature sensor using this polymer PTC material. The fabricated temperature sensor exhibited a low resistivity of approximately 10^4 Ω cm at approximately 36 °C, while the resistivity gradually increased as the temperature increased. It increases to 10^{10} Ω cm at approximately 40 °C, indicating a very large resistance change of approximately six orders of magnitude at 4 °C, which is near the body temperature. In addition, the polymer PTC fabricated in this study showed very high repeatability compared with temperature sensors using conventional polymer PTC materials. Even after applying a temperature change history of approximately 2000 cycles, a large resistance change of approximately five to six orders of magnitude was observed. Furthermore, because the thickness of the temperature sensor can be as low as 12 μm, the total thickness can be reduced to 25 μm by reducing the substrate thickness. By reducing the total thickness, the temperature sensor exhibited high flexibility and could be bent to a radius of curvature of less than 1 mm without breaking.

The polymer was synthesized by photopolymerization of octadecyl acrylate and butyl acrylate. Each material is a solid and liquid substance at room temperature, and the melting point of the synthesized polymer can be controlled by changing the ratio of these polymerizations. Consequently, the sensitive temperature can be controlled by constructing polymer PTC materials with different monomer material ratios. Figure 5.30 shows the characteristics of the temperature sensors with different proportions of monomers during synthesis. As the percentage of octadecyl acrylate increased from 60 wt% to 100 wt%, the temperature at which the resistivity of the temperature sensor changed shifted toward a higher temperature. The sensitive temperature could be controlled from 25 °C to approximately 50 °C, indicating a very wide range of controllability.

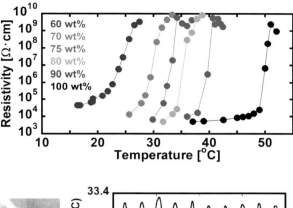
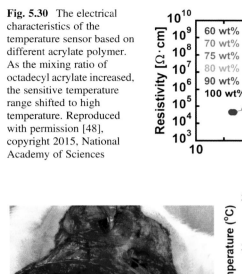
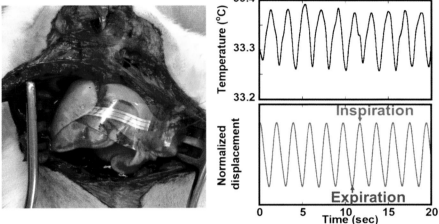

Fig. 5.30 The electrical characteristics of the temperature sensor based on different acrylate polymer. As the mixing ratio of octadecyl acrylate increased, the sensitive temperature range shifted to high temperature. Reproduced with permission [48], copyright 2015, National Academy of Sciences

Fig. 5.31 Measurement of rat lung temperature during respiration. The lung temperature periodically fluctuated due to the heat exchange in synchronization with breathing. Reproduced with permission [48], copyright 2015, National Academy of Sciences

5.4.4 Measurement of Biological Temperature

Temperature sensors using this polymer PTC material exhibited a very large resistance change, similar to temperature sensors using conventional polymer PTCs. Therefore, very high sensitivity can be obtained near the sensitive temperature region. The sensitivity of the temperature sensor was measured near the sensitive temperature region and was found to be very high (less than 0.1 °C). Such high sensitivity and flexibility may be effective for measuring minute temperature changes at soft biological temperatures. One example is the measurement of temperature changes in the lungs of rats owing to respiration. Temperature changes over time were measured by attaching a flexible polymer PTC sensor directly to the lungs of rats. Figure 5.31 shows the measured temperature changes in the lungs of the rats.

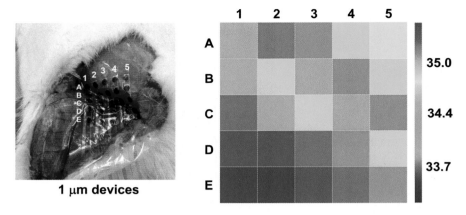

Fig. 5.32 Ultra-flexible temperature sensor matrix. Temperature mapping measurement of the rat heart by using a 5 × 5 array of ultra-flexible temperature sensors. The substrate thickness of the sensor is 1 μm

Rat lungs were shown to undergo periodic temperature changes of approximately 0.1 °C. Simultaneously with the measurement of the temperature change, the displacement of the lungs can be measured using a laser displacement meter to confirm the displacement at approximately the same period as the temperature change. This means that as the rat breathes, it draws cooler air from the atmosphere into its lungs, reducing the lung temperature. This high sensitivity and flexibility make it possible to detect very small temperature changes at biological temperatures. Furthermore, because this temperature sensor is printable, as mentioned earlier, it can be easily deposited over a large area. As shown in Fig. 5.32, a 5 × 5 temperature sensor matrix was fabricated to measure the temperature distribution in the heart of rats, and the results were similar to those obtained using an infrared camera, confirming the feasibility of applying this technology to large-area devices.

5.4.5 Application for Wearable Devices

The developed polymer PTCs exhibited a large resistance change near the body temperature, making them suitable for a variety of wearable device applications when combined with organic integrated circuits. The device shown in Fig. 5.33 is a wearable temperature-sensing system that integrates a solar cell, organic integrated circuit, and temperature sensor. The system uses a solar cell as the power supply. The power generated by the solar cells was used to operate the organic circuits. In this system, when the body temperature is normal (approximately 36 °C), the resistance of the temperature sensor is low, and the ring oscillator integrated into the organic circuit does not operate [48]. However, when the body temperature increased to approximately 38 °C, the resistance of the temperature sensor increased, and the ring

5 Ultra-Flexible Organic Electronics

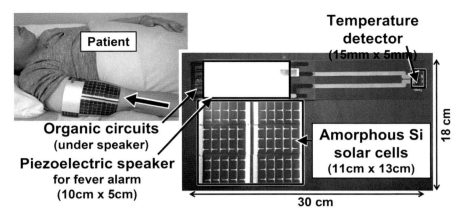

Fig. 5.33 Fever Alarm Armband system. The fever alarm armband system contains three components. Flexible polymer PTC sensor, piezoelectric speaker, and solar cell. Reproduced with permission [49], copyright 2015, IEEE

oscillator was activated. The ring oscillator oscillates at a high frequency of several kilohertz, which applies a voltage to the PVDF sheet, a piezoelectric element, and generates sound waves of a certain frequency. By detecting these sound waves, the system can detect an increase in a patient's body temperature. Thus, a system that could wirelessly detect an increase in body temperature would be an extremely useful wearable device for heat stroke prevention and other applications.

5.5 Breathable Nanomesh Sensors

The skin performs a wide variety of functions, including protection from external stimuli and shocks, temperature regulation, sensory perception, respiration, and excretion, and is extremely important for measuring or stimulating biological information. It can also electrically measure myoelectric, cardiac, electroencephalography (EEG), and other internal body information through the skin, and capture movement from skin deformation. For these biometric data, sensors can be placed in close contact with the skin surface to enable highly accurate and precise measurements and stimulation under normal activities.

However, the problem with conventional sensors is that wearing them restricts movement and activity, causing the measurement to have some reaction to the object, and the stiffness of the sensor causes incomplete adhesion to the living body, resulting in missing or noisy data and reduced reliability. Furthermore, when applied in the fields of medicine and sports, continuous measurements over a long period of time are often required. Therefore, the thin-film or rubber sheet-type devices commonly used have a low gas permeability, which inhibits the secretion of sweat and other substances from the skin, and the safety of long-term use has not

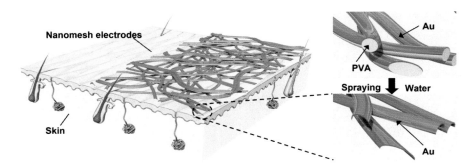

Fig. 5.34 On-skin nanomesh sensor. A schematic of the nanomesh sensor. The process of laminating an Au nanomesh onto skin is as follows: first, Au is evaporated onto electrospun PVA nanofibers; PVA meshes are then dissolved by spraying water; after PVA removal, nanomesh sensor adheres to the skin. Reproduced with permission [50], copyright 2017, Springer Nature

been proven from a dermatological standpoint. In this section, we introduce the nanomesh sensor, a skin-attachable electrode that is both breathable and stretchable, as one method to solve this problem [50].

Figure 5.34 shows the structure of the nanomesh sensor. Nanomesh sensors are fabricated by forming nanofibers of a biocompatible polymer material (polyvinyl alcohol (PVA)) using an electrospinning device and depositing 70–100 nm of gold on the surface of the nanofibers. Since PVA is highly soluble in water, the fabricated nanomesh sensor can be attached to the skin by spraying a small amount of water after the sensor is attached to the skin. Furthermore, the nanomesh sensor is so soft that it sticks to the fingerprints and along the microscopic irregularities of the skin, as shown in Fig. 5.35.

The sensor is ultrathin and ultra-lightweight; therefore, the user does not even feel that they are wearing it, and there is no discomfort when wearing it. Furthermore, a skin irritation test confirmed that the product did not cause an inflammatory reaction, even after being applied to the skin for 1 week [50]. For the skin irritation test, the nanomesh sensor and two comparison samples (silicone and parylene films) were applied to the forearms of 20 subjects for 7 days, and skin reactions at the application site were compared. Seven days after application, two dermatologists conducted a patch test to determine the presence or absence of allergic reactions and primary irritation reactions on the skin of each application site according to the patch test criteria of the International Contact Dermatitis Research Group (ICDRG). In the post-application dermatitis evaluation, all nanomesh sensors showed negative, and weakly positive reactions (+?) were not observed. In contrast, weakly positive reactions were observed for silicon and parylene films in a few cases. Thus, nanomesh sensors are less susceptible to stimulus-response effects (Fig. 5.36).

Furthermore, myoelectric and electrocardiographic measurements were performed by attaching the sensor to the skin as an application of the nanomesh sensor. Myoelectric measurements were performed by attaching a nanomesh sensor to the skin of the forearm and using a commercially available wireless module. The

5 Ultra-Flexible Organic Electronics

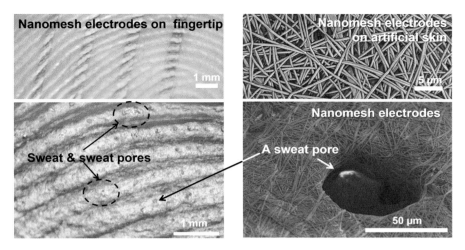

Fig. 5.35 Conformability of a nanomesh sensor. A picture of a nanomesh sensor attached to a fingertip. It shows a high level of conformability and adherence to the skin. Reproduced with permission [50], copyright 2017, Springer Nature

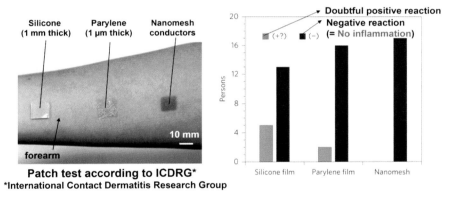

Fig. 5.36 Biocompatibility test of nanomesh sensor. A depiction of the biocompatibility test. Films of silicone (1 mm thick), parylene (1 μm thick), and nanomesh sensors were attached to the forearm for a week. Dermatitis evaluations were compared for different substrates according to the patch test criteria of the ICDRG. "+?" and "−" stand for doubtful positive reaction and negative reaction, respectively. Reproduced with permission [50], copyright 2017, Springer Nature

measured results showed that the noise level was almost equivalent to that measured using a commonly used gel electrode. To measure electrocardiograms, a nanomesh sensor was attached to the chest and connected to a wireless module. The connection between the nanomesh sensor and wireless module was made using a belly button textile with electrode wiring formed on it (Fig. 5.37).

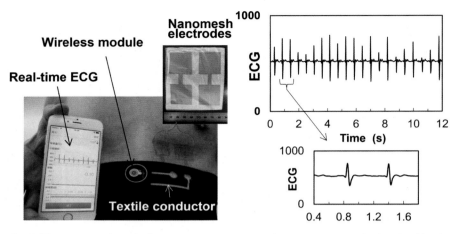

Fig. 5.37 Demonstration of ECG measurement. Nanomesh sensors were attached to the skin; the wireless module and nanomesh sensors were connected by the textile-based wire. The ECG signal can be monitored by smartphone

5.6 Summary

Ultra-flexible electronics have high adhesion to the skin and are expected to contribute significantly to the realization of wearable electronics without the feeling of being worn. If research and development on power supply and wireless communication, which are the remaining issues, are advanced, biological signals may be monitored naturally in our daily lives in the near future.

References

1. G.H. Gelinck, H.E.A. Huitema, E. van Veenendaal, E. Cantatore, L. Schrijnemakers, J.B.P.H. van der Putten, T.C.T. Geuns, M. Beenhakkers, J.B. Giesbers, B.-H. Huisman, E.J. Meijer, E.M. Benito, F.J. Touwslager, A.W. Marsman, B.J.E. van Rens, D.M. de Leeuw, Flexible active-matrix displays and shift registers based on solution-processed organic transistors. Nat Mater **3**, 106–110 (2004)
2. L. Zhou, A. Wanga, W. Sheng-Chu, J. Sun, S. Park, T.N. Jackson, All-organic active matrix flexible display. Appl Phys Lett **88**, 083502 (2006)
3. A.C. Arias, J. Devin MacKenzie, I. McCulloch, J. Rivnay, A. Salleo, Materials and applications for large area electronics: Solution-based approaches. Chem Rev **110**, 3–24 (2010)
4. M. Pagliaro, R. Ciriminna, G. Palmisano, Flexible Solar Cells. ChemSusChem **1**, 880–891 (2008)
5. A. Rida, L. Yang, R. Vyas, M.M. Tentzeris, Conductive inkjet-printed antennas on flexible low-cost paper-based substrates for RFID and WSN applications. IEEE Antennas Propag Mag **51**, 13–23 (2009)
6. T. Sekitani, H. Nakajima, H. Maeda, T. Fukushima, T. Aida, K. Hata, T. Someya, Stretchable active-matrix organic light-emitting diode display using printable elastic conductors. Nat Mater **8**, 494–499 (2009)

7. Q. Hua, J. Sun, H. Liu, R. Bao, Y. Ruomeng, J. Zhai, C. Pan, Z.L. Wang, Skin-inspired highly stretchable and conformable matrix networks for multifunctional sensing. Nat Commun **9**, 244 (2018)
8. D.J. Lipomi, B.C.-K. Tee, M. Vosgueritchian, Z. Bao, Stretchable organic solar cells. Adv Mater **23**, 1771–1775 (2011)
9. W. Gao, S. Emaminejad, H.Y.Y. Nyein, S. Challa, K. Chen, A. Peck, H.M. Fahad, H. Ota, H. Shiraki, D. Kiriya, D.-H. Lien, G.A. Brooks, R.W. Davis, A. Javey, Fully integrated wearable sensor arrays for multiplexed in situ perspiration analysis. Nature **529**, 509–514 (2016)
10. C.M. Boutry, Y. Kaizawa, B.C. Schroeder, A. Chortos, A. Legrand, Z. Wang, J. Chang, P. Fox, Z. Bao, A stretchable and biodegradable strain and pressure sensor for orthopaedic application. Nat Electron **1**, 314–321 (2018)
11. J.T. Reeder, J. Choi, Y. Xue, P. Gutruf, J. Hanson, M. Liu, T. Ray, A.J. Bandodkar, R. Avila, W. Xia, S. Krishnan, X. Shuai, K. Barnes, M. Pahnke, R. Ghaffari, Y. Huang, J.A. Rogers, Waterproof, electronics-enabled, epidermal microfluidicdevices for sweat collection, biomarker analysis, and thermography in aquatic settings. Sci Adv **5**, eaau6356 (2019)
12. T. Someya, T. Sekitani, S. Iba, Y. Kato, H. Kawaguchi, T. Sakurai, A large-area, flexible pressure sensor matrix with organic field-effect transistors for artificial skin applications. PNAS **101**, 9966–9970 (2004)
13. T. Someya, Y. Kato, T. Sekitani, S. Iba, Y. Noguchi, Y. Murase, H. Kawaguchi, T. Sakurai, Conformable, flexible, large-area networks of pressure and thermal sensors with organic transistor active matrixes. PNAS **102**, 12321–12325 (2005)
14. S.C.B. Mannsfeld, B.C.-K. Tee, R.M. Stoltenberg, C.V.H.-H. Chen, S. Barman, B.V.O. Muir, A.N. Sokolov, C. Reese, Z. Bao, Nat Mater **9**, 859 (2010)
15. T. Yamada, Y. Hayamizu, Y. Yamamoto, Y. Yomogida, A.I. Najafabadi, D.N. Futaba, K. Hata, Nat Nanotechnol **6**, 296 (2011)
16. D.H. Kim, N. Lu, R. Ma, Y.S. Kim, R.H. Kim, S. Wang, J. Wu, S.M. Won, H. Tao, A. Islam, K.J. Yu, T. Kim, R. Chowdhury, M. Ying, L. Xu, M. Li, H.J. Chung, H. Keum, M. McCormick, P. Liu, Y.W. Zhang, F.G. Omenetto, Y. Huang, T. Coleman, J.A. Rogers, Science **333**, 838 (2011)
17. T. Sekitani, U. Zschieschang, H. Klauk, T. Someya, Nat Mater **9**, 1015 (2010)
18. D.H. Kim, N. Lu, R. Ghaffari, Y.S. Kim, S.P. Lee, L. Xu, J. Wu, R.H. Kim, J. Song, Z. Liu, J. Viventi, B. Graff, B. Elolampi, M. Mansour, M.J. Slepian, S. Hwang, J.D. Moss, S.M. Won, Y. Huang, B. Litt, J.A. Rogers, Nat Mater **10**, 316 (2011)
19. D. Khodagholy, T. Doublet, M. Gurfinkel, P. Quilichini, E. Ismailova, P. Leleux, T. Herve, S. Sanaur, C. Bernard, G.G. Malliaras, Adv Mater **23**, H268 (2011)
20. S. Takeuchi, T. Suzuki, K. Mabuchi, H. Fujita, J Micromech Microeng **14**, 104 (2004)
21. D.H. Kim, J. Viventi, J.J. Amsden, J. Xiao, L. Vigeland, Y.S. Kim, J.A. Blanco, B. Panilaitis, E.S. Frechette, D. Contreras, D.L. Kaplan, F.G. Omenetto, Y. Huang, K.C. Hwang, M.R. Zakin, B. Litt, J.A. Rogers, Nat Mater **9**, 511 (2010)
22. J.-W. Jeong, W.-H. Yeo, A. Akhtar, J.J.S. Norton, Y.-J. Kwack, S. Li, S.-Y. Jung, S. Yewang, W. Lee, J. Xia, H. Cheng, Y. Huang, W.-S. Choi, T. Bretl, J.A. Rogers, Adv Mater **25**, 6839 (2013)
23. M. Kaltenbrunner, T. Sekitani, J. Reeder, T. Yokota, K. Kuribara, T. Tokuhara, M. Drack, R. Schwödiauer, I. Graz, S.B. Gogonea, S. Bauer, T. Someya, Nature **499**, 458–463 (2013)
24. Z. Suo, E.Y. Ma, H. Gleskova, S. Wagner, Mechanics of rollable and foldable film-on-foil electronics. Appl Phys Lett **74**, 1177 (1999)
25. T. Sekitani, S. Iba, Y. Kato, Y. Noguchi, T. Someya, T. Sakurai, Ultraflexible organic field-effect transistors embedded at a neutral strain position. Appl Phys Lett **87**, 173502 (2005)
26. T. Sekitani, T. Yokota, K. Kuribara, M. Kaltenbrunner, T. Fukushima, Y. Inoue, M. Sekino, T. Isoyama, Y. Abe, H. Onodera, T. Someya, Nat Commun **7**, 11425 (2016)
27. A. Boss, S. Bisdas, A. Kolb, M. Hofmann, U. Ernemann, C.D. Claussen, C. Pfannenberg, B.J. Pichler, M. Reimold, L. Stegger, J Nucl Med **51**, 1198 (2010)

28. S. Miwa, T. Otsuka, J Orthop Sci **22**, 391 (2017)
29. D. Guhathakurta, A. Dutta, Front Neurosci **10**, 261 (2016)
30. T. Yokota, P. Zalar, M. Kaltenbrunner, H. Jinno, N. Matsuhisa, H. Kitanosako, Y. Tachibana, W. Yukita, M. Koizumi, H. Jinno, T. Someya, Ultraflexible organic photonic skins. Sci Adv **2**, e1501856 (2016)
31. D. Kim, T. Yokota, T. Suzuki, S. Lee, T. Woo, W. Yukita, M. Koizumi, Y. Tachibana, H. Yawo, H. Onodera, M. Sekino, T. Someya, Ultraflexible organic light-emitting diodes for optogenetic nerve stimulation. Proc Natl Acad Sci U S A **117**, 21138–21146 (2020)
32. S. Park, K. Fukuda, M. Wang, C. Lee, T. Yokota, H. Jin, H. Jinno, H. Kimura, P. Zalar, N. Matsuhisa, S. Umezu, G.C. Bazan, T. Someya, Adv Mater **30**, 1802359 (2018)
33. Z. Wu, W. Yao, A.E. London, J.D. Azoulay, T.N. Ng, ACS Appl Mater Interfaces **9**, 1654–1660 (2017)
34. B. Siegmund, A. Mischok, J. Benduhn, O. Zeika, S. Ullbrich, F. Nehm, et al., Nat Commun **8**, 15421 (2017)
35. T. Rauch, M. Böberl, S.F. Tedde, J. Fürst, M.V. Kovalenko, G. Hesser, et al., Nat Photonics **3**, 332–336 (2009)
36. W.G. Zijlstra, A. Buursma, W.P. Meeuwsen-van der Roest, Absorption spectra of human fetal and adult oxyhemoglobin, de-oxyhemoglobin, carboxyhemoglobin, and methemoglobin. Clin Chem **37**, 1633–1638 (1991)
37. H. Jinno, T. Yokota, M. Koizumi, W. Yukita, M. Saito, I. Osaka, K. Fukuda, T. Someya, Self-powered ultraflexible photonic skin for continuous bio-signal detection via air-operation-stable polymer light-emitting diodes. Nat Commun **12**, 2234 (2021)
38. T. Yokota, T. Nakamura, H. Kato, M. Mochizuki, M. Tada, M. Uchida, S. Lee, M. Koizumi, W. Yukita, A. Takimoto, T. Someya, A conformable imager for biometric authentication and vital sign measurement. Nat Electron **3**, 113–121 (2020)
39. B. Kang, W.H. Lee, K. Cho, Recent advances in organic transistor printing processes. ACS Appl Mater Interfaces **5**, 2302–2315 (2013)
40. H. Akkerman, B. Peeters, A. van Breemen, S. Shanmugam, D. Tordera, J.-L. van der Steen, A.J. Kronemeijer, P. Malinowski, F. De Roose, D. Cheyns, J. Genoe, W. Dehaene, P. Heremans, G. Gelinck, Printed organic photodetector arrays and their use in palmprint scanners. SID Symposium Digest of Technical Papers **49**, 494–497 (2018)
41. M.J. Powell, The physics of amorphous-silicon thin-film transistors. IEEE Trans Electron Devices **36**, 2753–2763 (1989)
42. D. Cheng, J. Wang, T. Yokota, T. Someya, Spatiotemporal processing in photoplethysmography for skin microcirculatory perfusion imaging. Biomed Opt Express **13**, 838–849 (2022)
43. X. Lizhi, S.R. Gutbrod, A.P. Bonifas, S. Yewang, M.S. Sulkin, L. Nanshu, H.-J. Chung, K.-I. Jang, Z. Liu, M. Ying, L. Chi, R. Chad Webb, J.-S. Kim, J.I. Laughner, H. Cheng, Y. Liu, A. Ameen, J.-W. Jeong, G.-T. Kim, Y. Huang, I.R. Efimov, J.A. Rogers, "3D multifunctional integumentary membranes for spatiotemporal cardiac measurements and stimulation across the entire epicardium", nature. Communications **5**, 3329 (2014)
44. Y. Luo, G. Wang, B. Zhang, Z. Zhang, The influence of crystalline and aggregate structure on PTC characteristic of conductive polyethylene/carbon black composite. Eur Polym J **34**, 1221–1227 (1998)
45. X. Xiang-Bin, Z.-M. Li, K. Dai, M.-B. Yang, Anomalous attenuation of the positive temperature coefficient of resistivity in a carbon-black-filled polymer composite with electrically conductive in situ microfibrils. Appl Phys Lett **89**, 032105 (2006)
46. R. Strümpler, Polymer composite thermistors for temperature and current sensors. J Appl Phys **80**, 6091 (1996)
47. J. Jeon, H.-B.-R. Lee, Z. Bao, Flexible wireless temperature sensors based on Ni microparticle-filled binary polymer composites. Adv Mater **25**, 850–855 (2013)
48. T. Yokota, Y. Inoue, Y. Terakawa, J. Reeder, M. Kaltenbrunner, T. Ware, K. Yang, K. Mabuchi, T. Murakawa, M. Sekino, W. Voit, T. Sekitani, T. Someya, Ultraflexible, large-

area, physiological temperature sensors for multipoint measurements. PNAS **112**, 14533–14538 (2015)
49. H. Fuketa, M. Hamamatsu, T. Yokota, W. Yukita, T. Someya, T. Sekitani, M. Takamiya, T. Someya, T. Sakurai, 16.4 Energy-autonomous fever alarm armband integrating fully flexible solar cells, piezoelectric speaker, temperature detector, and 12V organic complementary FET circuits, in *2015 IEEE International Solid-State Circuits Conference-(ISSCC) Digest of Technical Papers*, (IEEE, 2015)
50. A. Miyamoto, S. Lee, N.F. Cooray, S. Lee, M. Mori, N. Matsuhisa, H. Jin, L. Yoda, T. Yokota, A. Itoh, M. Sekino, H. Kawasaki, T. Ebihara, M. Amagai, T. Someya, Inflammation-free, gas-permeable, lightweight, stretchable on-skin electronics with nanomeshes. Nat Nanotechnol **12**, 907–913 (2017)

Chapter 6
Polymer Nanosheets with Printed Electronics for Wearable and Implantable Devices

Tatsuhiro Horii and Toshinori Fujie

Abstract Polymer nanosheets are ultrathin films with thicknesses of tens to hundreds of nanometers, which can be fabricated and functionalized by using the wet process, such as spin coating, molecular adsorption methods, inkjet printing, and gravure coating. In this chapter, we explain physical properties (e.g., adhesiveness and elastic modulus) and applications (e.g., medical and electronic) of the polymer nanosheets.

Keywords Polymer nanosheets · Characteristics · Spin coating · Langmuir–Blodgett assembly · Layer-by-layer deposition · Sol-gel method · Polysaccharides · Polylactic acid (PLA) · Chitosan · Alginate · Wound dressing material · Conductive nanosheets · Dielectric elastomer actuators (DEAs) · Implantable medical devices

6.1 Polymer Nanosheets

6.1.1 What Is Polymer Nanosheets?

Before reviewing examples of polymer nanosheets applied to wearable and implantable devices, we introduce polymer nanosheets and discuss their characteristics. In general, polymer nanosheets comprise polymeric ultrathin films with thicknesses of tens to hundreds of nanometers and areas of several square centimeters (i.e., size aspect ratio greater than 10^6) [1, 2]. Spin coating [3, 4], Langmuir–Blodgett assembly [5], layer-by-layer (LbL) deposition [6], and a sol-gel method involving interpenetrating networks comprising organic and inorganic materials [7–10] enable

T. Horii
School of Life Science and Technology, Tokyo Institute of Technology, Yokohama, Japan

T. Fujie (✉)
School of Life Science and Technology, Tokyo Institute of Technology, Yokohama, Japan

Living Systems Materialogy Research Group, International Research Frontiers Initiative, Tokyo Institute of Technology, Yokohama, Japan
e-mail: t_fujie@bio.titech.ac.jp

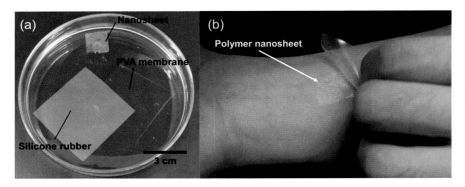

Fig. 6.1 (a) Photograph of three kinds of freestanding sheets in acetone. The surface of the polysaccharide nanosheet was modified with luminescent pigment. (b) Photograph of the polysaccharide nanosheet detached from the silicone rubber sheet. The image was captured in the dark (Adapted from ref. [6])

the fabrication of polymer nanosheets with unique mechanical properties, such as tunable flexibility, non-covalent adhesion to human skin, and high transparency, as shown in Fig. 6.1.

6.1.2 Polymer Nanosheets and Their Characteristics

Polysaccharides and polylactic acid (PLA), shown in Fig. 6.2, are representative polymer building blocks of nanosheets for medical applications. Alternating layers of oppositely charged chitosan (polycation) and alginate (polyanion) (Fig. 6.2b, c) on a substrate produces nanosheets composed of polysaccharides whose film thickness can be controlled by adjusting the number of layers [6]. In addition, nanosheets consisting of a single layer of PLA, a hydrophobic polymer, can be obtained using spin coating. The film thickness can be modulated by controlling the concentration of the PLA solution and rotation speed.

The nanosheet can be detached from a substrate by peeling a tape frame affixed to the surface of a support layer (e.g., polyvinyl alcohol (PVA)) covering the nanosheet fabricated on the substrate. Dissolving the water-soluble PVA layer with water then yields a self-supporting nanosheet. Alternatively, self-supporting nanosheets can be fabricated by precoating the substrate with a PVA layer in advance, forming the nanosheet on top of the PVA layer, and then immersing the whole assembly in water to dissolve the PVA. Collecting the peeled nanosheets from the substrate allows the nanosheets can be characterized as two-dimensional molecular assemblies composed of polymers. Nanosheets can be recovered with a soft nylon mesh and then attached to skin or organ surfaces without the adhesives required for bandages.

Adhesion measurements of nanosheets using the micro-scratch method showed that the adhesion force of nanosheets, represented by the normailized clitical

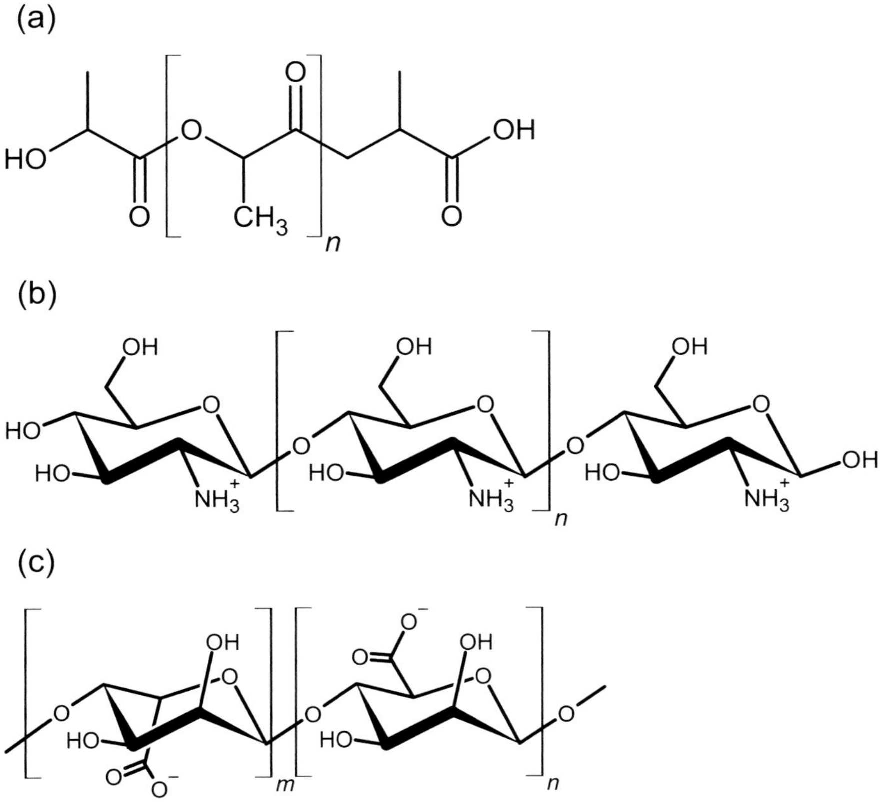

Fig. 6.2 Chemical structures of (**a**) poly lactic acid, (**b**) chitosan, and (**c**) alginate

loading, to substrates increased with decreasing nanosheet thickness and significantly increased at thicknesses below 200 nm, as shown in Fig. 6.3a [11]. In addition, the mechanical properties of the nanosheets were evaluated using the bulge test, which indicated that the flexibility of the nanosheets increased with decreasing thickness, as shown in Fig. 6.3b. The elastic modulus specific to the material increased with decreasing thickness, which was attributed to the proportional decrease in the bulk layer due to the ultrathin nature of the film and the increase in the degrees of freedom of the polymer chains within the nanosheets.

We investigated the application of chitosan and alginate nanosheets prepared by the LbL method as wound dressing materials. A nanosheet (thickness: 75 nm, area: 4 cm^2) supported by a PVA membrane was applied to a lung pleural defect ($\varnothing$ = 6 mm) in a beagle dog, and then the PVA was dissolved with saline solution, as shown in Fig. 6.4. The nanosheets showed adhesion and mechanical strength to withstand normal respiratory pressure at 5 min after application. At 3 h, the pressure resistance (60 cm H_2O) was comparable to that of a fibrin sheet (collagen mesh with

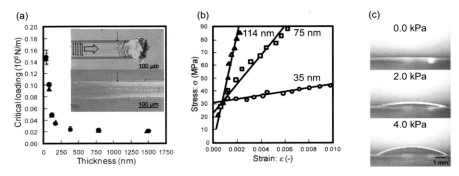

Fig. 6.3 (a) Thickness dependency of the critical load of the polysaccharide nanosheet before (filled circle) and after (filled triangle) detachment. Inset: Photographs of the polysaccharide nanosheets with thicknesses of 1482 nm (upper) and 77 nm (lower) after the micro-scratch test. Black arrows indicate the direction of the stylus on the nanosheet and dashed arrows show points where the nanosheets detached from the SiO_2 substrate. (b) Stress-strain curves for various thicknesses of polysaccharide nanosheet obtained using a 1 mm diameter circular hole in bulge tests. (c) Macroscopic cross-section photographs of the deflected 75 nm polysaccharide nanosheet during the bulge test (hole diameter: 6 mm) (Adapted from ref. [11])

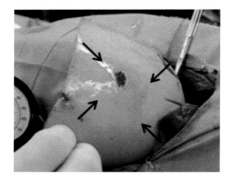

Fig. 6.4 Photograph of a 75-nm thick polysaccharide nanosheet securing a defect 3 h after repair, having withstood pressure over 50 cm H_2O. The region indicated by arrows shows the nanosheet-sealed area (Adapted from ref. [11])

fibrin glue), a conventional wound dressing material. At 7 days after application, the injured area had healed and formed a smooth surface similar to that before the injury. The nanosheets showed negligible adhesion to other tissues, while the fibrin sheet showed strong adhesion to the chest wall tissue side due to an inflammatory response. These results indicate that the nanosheets are a suitable medical material with wound covering and anti-adhesion properties. The nanosheet surface can also be loaded with drugs or cells for drug release or cell transplantation in vivo.

6.2 Development of Biological Tissue-Compatible Electronics Based on Polymer Nanosheets Functionalized Using Printing Technology

Biological organisms have hierarchical structures. Organs consist of tissues, composed of cells that are bound together by the extracellular matrix (ECM). The ECM, in particular the basement membrane, is a large two-dimensional molecular assembly composed of biomolecules that convey dynamic information (adhesion, migration, proliferation, differentiation) of adjacent cell clusters by changing their mass diffusion and mechanical properties. Using polymer self-assembly and microfabrication techniques, we can produce molecular assemblies using polymer nanosheets (film thickness: tens to hundreds of nm), which are comparable to those of basement membranes in artificial systems [12]. The nanosheets have the advantage that they can be attached to surfaces in vivo without the use of adhesives, due to their compatibility resulting from their flexibility. The nanosheet can be considered as an ultrathin substrate with a high surface area, and various substances (e.g., drugs, cells, nanoparticles, electronic elements) can be loaded onto the nanosheet surface using various printing techniques, resulting in a functional nanosheet that can be applied to biological tissues. We present the developments and applications of nanosheet electronics using printed electronics.

6.2.1 Inkjet Printing

Nanosheets composed of styrene-butadiene copolymer (SBS), which has excellent stretchability, have been developed for application to circuit substrates and encapsulation membranes [13]. Silver ink (particle size: 15 nm) was inkjet printed to form conductive circuits (resistivity: 5.6×10^{-5} Ω cm) on the surface of an ultrathin layer of acrylic-copolymer (thickness: 115 nm) coated on SBS nanosheets (thickness: 383 nm), as shown in Fig. 6.5a, b. The water contact angle was reduced by coating the surface of SBS nanosheets with cationic acrylic copolymer using the roll-to-roll method, which resulted in the surface properties appropriate for inkjet printing. A surface-mounted chip LED was placed on the electronic circuit, and an SBS nanosheet was used to cover the circuit to attach the LED without soldering. Observing the SBS nanosheet from the back, it was found that the silver wiring flexibly followed the electrode part of the LED, as shown in Fig. 6.5d. The 3 V battery successfully powered the LED on the skin surface (Fig. 6.5e, f). The nano-ink-printed nanosheet is a useful circuit in which electronic devices can be driven on the body. It is expected to be applied in skin-contact electronics.

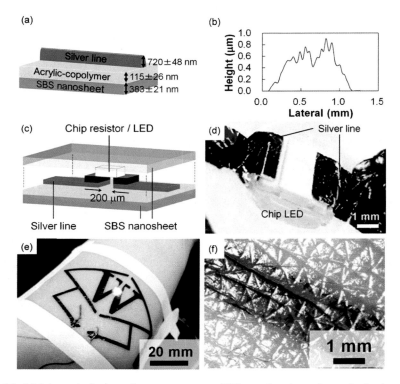

Fig. 6.5 (**a**) Schematic diagram of components on an SBS nanosheet in an electronic circuit and (**b**) cross-section profile of the silver line. (**c**) Schematic diagram showing the lamination of an electronic element, such as a chip resistor or a chip LED, using an SBS nanosheet with silver lines and another SBS nanosheet. (**d**) Micrograph of the reverse side of silver lines connected to the electrodes of a chip LED. (**e**) Photograph of an SBS nanosheet with an LED lighting circuit attached to the skin surface. (**f**) Micrograph of an SBS nanosheet with inkjet-printed silver wire on skin (Adapted from ref. [13])

6.2.2 Conductive Nanosheets

The roll-to-roll gravure coating system (Fig. 6.6) enables the production of nanosheets that are several meters long with thicknesses of less than 1 μm by controlling the solution concentrations and the amount of coating quantity. Conductive nanosheet coated with poly(3,4-ethilenedioxythiophene) doped with poly (4-styrenesulfonate acid) (referred to as PEDOT:PSS) with a thickness of 120 nm showed conductivity (450–500 S/cm) [14]. We will now introduce the applications of conductive nanosheets for electronic devices.

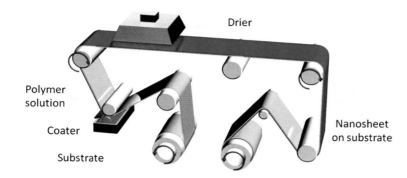

Fig. 6.6 Schematic diagram of the roll-to-roll gravure coating system (Adapted from ref. [14])

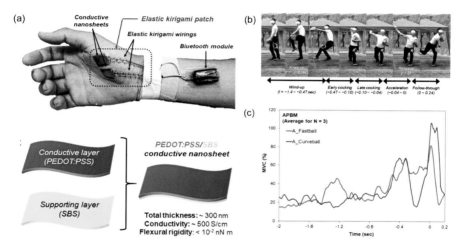

Fig. 6.7 (a) Image of conductive nanosheets connected to stretchable wires (i.e., elastic kirigami patch) and a sEMG measuring unit (upper). Schematic diagram of the construction of the conductive nanosheet (lower). (b) Photographs of the pitching motion separated into five phases. (c) The mean sEMG signals, normalized to the maximum voluntary contraction (MVC), for two different kinds of pitching trial (Adapted from ref. [15])

6.2.2.1 Bio-electrodes for Humans and Plants

Conventional bio-electrodes consist of rigid metal wiring and electronic elements, which can cause discomfort due to the mechanical mismatch between the human skin and the electrodes. To address the limitations of conventional bio-electrodes, flexible thin-film electrodes are being developed for skin-contact electronics to detect a range of biosignals. We reported the fabrication of a nanosheet composed of PEDOT:PSS coated on a poly(styrene-b-butadiene-b-styrene) (SBS) nanosheet with a total thickness of 339 ± 91 nm, as shown in Fig. 6.7a [15]. This conductive nanosheet was used as a skin-contact electrode to accurately record the surface

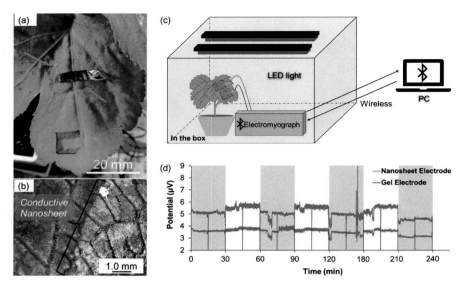

Fig. 6.8 (**a**) Photographs of nanosheet electrodes attached to a plant leaf after 14 days. (**b**) Microscopy image of the conductive nanosheet on the plant leaf. (**c**) Schematic diagram of the experimental setup for wireless measurement of the biopotential signal. (**d**) Results of biopotential measurement of plant leaves exposed to dark (gray field) and light (bright field) conditions (Adapted from ref. [16])

electromyogram (sEMG) signals on the thenar of a palm (Fig. 6.7c) during baseball pitching shown in Fig. 6.7b.

We used the conductive nanosheets to detect changes in the voltage generated in plant leaves anticipating that the observed conformability to the surface of human skin would translate to physical adhesion to uneven surfaces such as plant veins [16].

The conductive nanosheet followed and physically adhered to uneven structural surfaces such as plant leaf veins without chemical adhesives because it was ultrathin and lightweight (thickness: 300 nm, weight: 150 μg), as shown in Fig. 6.8a, b. Conventional bio-electrodes made of gels cause discoloration of the leaf surface during the measurement due to the acrylic adhesive. The combination of the conductive nanosheets and a Bluetooth system indicated in Fig. 6.8c enabled measuring potential changes on plant leaf surfaces for approximately 1500 h. In addition, it was found that the potential pattern changed when light-emitting diodes (LEDs) were used to periodically irradiate the leaf surface, as shown in Fig. 6.8d. The minimally invasive measurement using the conductive nanosheets is expected to have applications in agriculture and food science as a breakthrough method for analyzing the biological activity of plant.

6.2.2.2 Low Voltage-Driven Dielectric Elastomer Actuators (DEAs)

Dielectric elastomer actuators (DEAs), a type of soft polymer actuator, have high flexibility, low weight, large actuation strain, and high energy density compared to conventional actuators such as electromagnetic motors and hydraulic cylinders. The basic structure of a DEA is the same as that of a single-layer capacitor, consisting of a dielectric elastomer (DE) layer sandwiched between two compliant electrodes (Fig. 6.9). The electrostatic attraction generated by applying a kV-scale potential difference to the electrodes causes the DE layer to contract in the thickness direction and expand in the in-plane direction [17]. The relationship between the applied voltage and the deformation strain in the thickness direction (S_z) is expressed as

$$S_z = \varepsilon_0 \varepsilon_r \frac{V^2}{Yt^2} \tag{6.1}$$

where ε_0, ε_r, V, and t are the vacuum permittivity, the dielectric constant, the applied voltage, and the thickness of the DE layer, respectively, and Y is the elastic modulus of the DE layer. Reducing the thickness of a DE layer is essential to reduce the driving voltage of the DEA. We have developed low voltage-driven DEAs that can be operated below 72 V, making them harmless to skin and organs [18]. Using a 600 nm thick DE layer composed of PDMS and a pair of 200 nm thick nanosheet electrodes composed of PEDOT:PSS and SBS nanosheets, the DEA shown in Fig. 6.10a exhibited an oscillatory response with a 50 V sinusoidal signal wave in the 1–30 kHz range. The use of nanosheets (i.e., reducing the DE thickness) is a promising approach in terms of the electrical breakdown field strength (E_b) described by Eq. (6.2), which is the empirical formula to model the E_b of PDMS-based DEAs and the conformability to human skin.

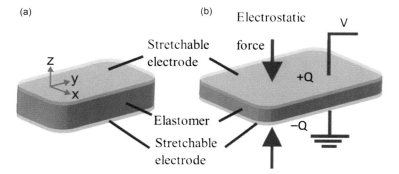

Fig. 6.9 Schematic diagram of the DEA fundamental structure and deformation principle in (**a**) the initial state and (**b**) the actuation state (Adapted from ref. [18])

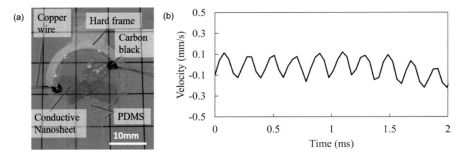

Fig. 6.10 (a) Photograph of the fabricated low voltage-driven DEA. (b) Velocity responses on the DEA surface at 5 kHz, 50 V (Adapted from ref. [18])

$$E_b = 147 t^{-0.23 \pm 0.02} \lambda^{1.77 \pm 0.03} \tag{6.2}$$

where t is the initial DE thickness and λ is the stretching ratio. Equation 6.2 implies that minimizing the elastomer thickness allows optimizing the E_b of the DE and reducing the driving voltage of DEAs [19].

6.2.2.3 Multilayered DEAs

Using the fibrous networks of single-walled carbon nanotubes (SWCNTs), we developed stretchable nanosheet electrodes (referred to as SWCNT-SBS nanosheets) composed of an SBS nanosheet coated with an ultrathin layer of SWCNTs [20]. A SWCNT aqueous dispersion was applied to a surface of an SBS nanosheet by using a gravure coating system, followed by drying with driers in the coating system. SWCNT-SBS nanosheets with a thickness of 101 nm were self-supporting, and nanosheets with a thickness of 491 nm had a Young's modulus of 80.9 MPa, 250% elongation at break, and a sheet resistance of 4.3 kΩ sq.$^{-1}$, while the nanosheets composed of SBS alone had a Young's modulus of 70.5 MPa and 340% elongation at break. We assumed that the sliding and buckling of the SWCNT fibers during the stretching [21] contributed to the mechanical properties of the SWCNT-SBS nanosheets compared to those of the SBS (Fig. 6.11). Using the self-supporting SWCNT-SBS nanosheets as electrodes and Ecoflex 00–30 sheet (Smooth-On, Inc.) as a dielectric elastomer (DE) layer, and staking the electrodes and the DE layers alternatively, a ten-layered DEA was fabricated without adhesive on three substrates with different stiffness: glass, Ecoflex 00–30, and a urethane elastomer model skin. Ecoflex 00–30 and the model skin were chosen as representations of flexible substrates with comparable Young's modulus to human epidermis. The low flexural rigidity (105 N m) of the ten-layered DEA ensures conformability to an index finger, as shown in Fig. 6.12a. Applying an actuation voltage of 2100 V caused the DEAs to exhibit a twofold larger displacement on the Ecoflex 00–30 substrate than on the

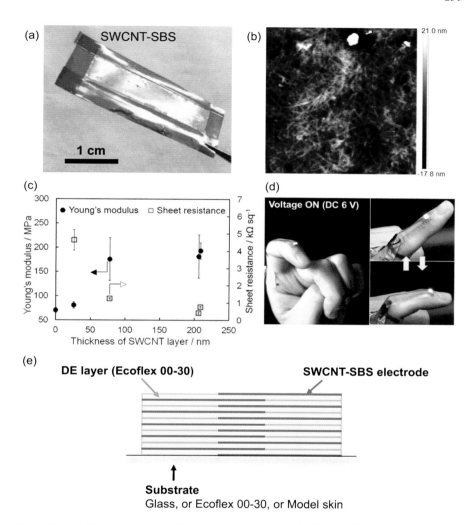

Fig. 6.11 (**a**) Photograph of an SBS nanosheet coated with single-walled carbon nanotubes (SWCNT-SBS nanosheet). (**b**) AFM height image of the SWCNT-SBS nanosheet. (**c**) Thickness dependency of the Young's modulus (filled circle) and sheet resistance (open square) of the SWCNT-SBS nanosheets. (**d**) Photographs of a blue LED connected to the SWCNT-SBS nanosheet on an index finger (applied DC voltage: 6 V) (Adapted from ref. [20]) (**e**) Cross-sectional scheme of a DEA

glass substrate. This ability to conform to the surface of the skin will enable its use in skin-contact haptic devices.

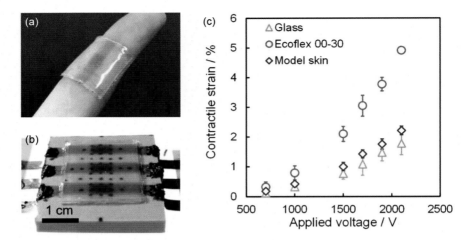

Fig. 6.12 (a) Photograph of a DEA attached to the surface of an index finger. (b) Photograph of the fabricated ten-layered DEA on urethane elastomer model skin substrates. (c) Voltage dependence of the contractile strain along the thickness direction (Adapted from ref. [20])

6.3 Development of Implantable Medical Devices

6.3.1 Electronic Devices That Use Bioadhesive Nanosheets for Cancer Therapy

Organ tissues are covered with ECM, and mucosa called mucopolysaccharides, which makes it difficult to attach devices without sutures. Therefore, we fabricated PDMS nanosheets with a thickness of less than 1 μm to attach the electronic devices encapsulated by the nanosheets to biological tissues [22]. The PDMS nanosheets prepared using a roll-to-roll gravure coating system showed a Young's modulus of 1.47 MPa. Modifying the surface of the PDMS nanosheets with polydopamine (PDA), which mimics the adhesive protein of a hard-shelled mussel, resulted in nanosheets with high stretchability on biological tissue such as muscle (Fig. 6.13a), 5 times higher adhesive force than the non-modified nanosheets, and 25 times higher adhesive strength than PDMS sheets with a thickness of 1 mm. Using the PDA-modified nanosheets to encapsulate electronic devices, the device could be attached to biological tissues without sutures. For example, an IC chip was attached to the peritoneal cavity of a rat for more than 1 month, and the information written in the IC could be read out from outside the body (Fig. 6.13b).

In addition, we fabricated implantable devices composed of bioadhesive nanosheets and an LED chip with a short-range wireless communication unit, as shown in Fig. 6.13c, to evaluate their effectiveness on photon cancer therapy (i.e., photodynamic therapy: PDT). The LED chip mounted under the skin of a tumor-bearing mouse was successfully turned on using a wireless power supply (13.56 MHz) generated from a power transmission board located under the mouse

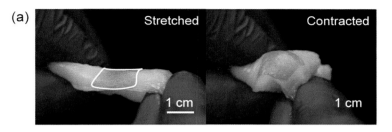

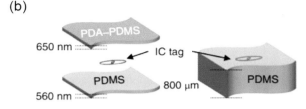

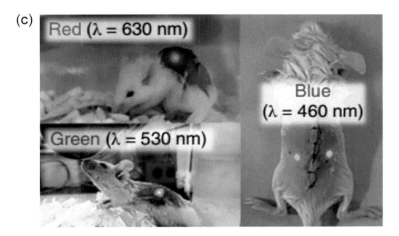

Fig. 6.13 (**a**) Photographs of a PDA-PDMS nanosheet on a chicken muscle taken while the muscle was stretched (left) and contracted (right). (**b**) Schematic diagram of the construction of the implantable IC device. Photograph showing the detection of the resonant frequency of embedded IC tags in the abdominal cavity of a rat. (**c**) Photographs of red, green, and blue light LEDs implanted subcutaneously (Adapted from ref. [22])

cage. In addition, we measured the changes in the tumor size of a rat administered a photosensitizer (i.e., 8 mg/kg of Photofrin) during continuous irradiation with LED light for 10 days. One tumor in ten under red light and six tumors in ten under green light were eradicated, leaving one ulcer. In other words, the implantable LED devices showed a significant antitumor effect after administration of Photofrin. This photon cancer therapy system showed that (1) the laser intensity of the LED (<100 μW cm^{-2}) was 1/1000 times lower than that of conventional PDT

(>100 mW cm^{-2}) and (2) green light, which has low tissue penetration and is not suitable for conventional PDT, showed an antitumor effect. These results suggest the potential for preventing heat damage to organs caused by excessive laser irradiation and for treating intractable deep-seated cancers.

6.3.2 Development of a Thin-Film Thermotherapy Device That Attaches to Biological Tissues

Heating has long been used for medical treatments, such as hemostasis of biological tissues and the direct excision of lesions by cauterization [23, 24]. Heating is a key technique in cancer therapies, such as thermotherapy, which exploits cellular sensitivity to high temperature, and therapies that combine anticancer drugs, radiation, and blood circulation stimulated by mild heating [25–27]. Since miniaturization of the heating device was expected to reduce the burden on patients, we developed a thin, compact heating device to ensure the local heating of biological tissues, as shown in Fig. 6.14 [28]. The heating device consisted of an induction heating (IH) circuit using Au ink patterned on a poly-DL-lactic acid (PDLLA) thin film. Annealing at 250 °C reduced the resistivity of the printed Au ink from 3.4×10^2 μΩ cm to 9.4 μΩ cm. However, the PDLLA has a low glass transition temperature (i.e., $T_g = 50 - 60$ °C). Therefore, we developed the following method: (1) The Au ink circuit was patterned on a polyimide film with superior heat

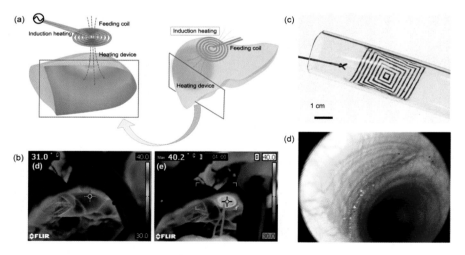

Fig. 6.14 (a) Schematic diagram of the IH device concept for local thermotherapy. (b) IR thermography of the IH device attached to the hepatic lobe before (right) and after (left) wireless powering. (c) Photograph of the device inserted into a transparent polyvinylchloride tube (inner diameter: 15 mm) using forceps. (d) Bronchoscopic camera image of the IH device on the endotracheal wall of a beagle dog (Adapted from ref. [28])

resistance up to 300 °C, and (2) the Au circuit annealed at 250 °C on the polyimide film was subsequently coated with PDLLA solution at room temperature. (3) After drying the PDLLA film, the Au circuit (i.e., the IH circuit) was transferred to the PDLLA thin film by peeling the PDLLA thin film from the polyimide film. With the above method, we obtained a low-resistance IH circuit on the flexible PDLLA thin film. This heater has a thin structure with a total thickness of 7 μm. The electrical and mechanical tolerance of the device to bending and folding allowed storage in a narrow tube with an inner diameter of 15 mm, which is required for endoscopic surgery. Applying an alternating magnetic field of ca. 30 MHz using a wireless power supply system to the thin IH heater increased the surface temperature by 28 °C in 1 min in air. On the surface of a beagle dog liver, the heating device successfully increased the local surface temperature by 7 °C in 1 min and 8 °C in 5 min, even in the presence of thermal diffusion caused by the abundant blood flow in the liver tissues. After the heating treatment, histopathological examination of the liver tissues did not reveal any burns. This technique, which enables the implantation of LEDs and heating devices inside the body, can transmit physical energy deep inside an organism with minimal invasion through combination with the bioadhesive nanosheets. The combination of thin-film-type devices, photo-responsive agents, luminescent pigments, and thermo-responsive polymers is expected to pave the way for application in novel medical treatments and diagnostic techniques.

6.4 Summary and Future Outlook

The patterned wiring and devices presented in this chapter—which combine polymer thin films and nanosheets with excellent biocompatibility and conventional printing technologies (e.g., gravure coating, inkjet printing)—are expected to contribute to medical, healthcare, and virtual reality-based information transfer technologies in the near future. Polymeric nanosheets can provide a tailoring layer to mediate the chemical and mechanical properties of commercial rigid devices and flexible and uneven biological tissues. In addition, they offer a new platform that can incorporate the functions of commercial devices into constructions with excellent biocompatibility. Functionalized thin-film electronics have small inertia and low constraints on the movement of a living body owing to their ultralightweight and low bending stiffness. Therefore, thin-film electronics allow us to regularly record and store information from movements and biological signals and control output (such as electrical and tactile stimulation) in response. It is also possible to develop technologies for early diagnosis and treatment of diseases through long-term biological information measurement. To take a final step toward practical application, we need to consider the wiring between the terminals of the device and the measurement and control devices, as well as external stimuli. It is necessary to achieve a flexible connection at the terminal/wiring interface and a design that shields against external stimuli such as wind, rain, rubbing against clothing, dirt, and electromagnetic noise.

Acknowledgment We thank Edanz (https://jp.edanz.com/ac) for editing a draft of this manuscript.

References

1. S. Takeoka, Y. Okamura, T. Fujie, Y. Fukui, Development of biodegradable nanosheets as nanoadhesive plaster. Pure Appl Chem **80**, 2259–2271 (2008). https://doi.org/10.1351/PAC200880112259
2. T. Fujie, Y. Okamura, S. Takeoka, Fabrication, properties, and biomedical applications of nanosheets. Funct Polym Film **2**, 907–931 (2011). https://doi.org/10.1002/9783527638482.CH29
3. T. Fujie, Y. Kawamoto, H. Haniuda, A. Saito, K. Kabata, Y. Honda, E. Ohmori, T. Asahi, S. Takeoka, Selective molecular permeability induced by glass transition dynamics of semi-crystalline polymer ultrathin films. Macromolecules **46**, 395–402 (2013). https://doi.org/10.1021/MA302081E
4. Y. Okamura, K. Kabata, M. Kinoshita, D. Saitoh, S. Takeoka, Free-standing biodegradable poly (lactic acid) nanosheet for sealing operations in surgery. Adv Mater **21**, 4388–4392 (2009). https://doi.org/10.1002/ADMA.200901035
5. Y. Kado, M. Mitsuishi, T. Miyashita, Fabrication of three-dimensional nanostructures using reactive polymer nanosheets. Adv Mater **17**, 1857–1861 (2005). https://doi.org/10.1002/ADMA.200500884
6. T. Fujie, Y. Okamura, S. Takeoka, Ubiquitous transference of a free-standing polysaccharide nanosheet with the development of a nano-adhesive plaster. Adv Mater **19**, 3549–3553 (2007). https://doi.org/10.1002/ADMA.200700661
7. J.A. Forrest, K. Dalnoki-Veress, J.R. Stevens, J.R. Dutcher, Effect of free surfaces on the glass transition temperature of thin polymer films. Phys Rev Lett **77**, 2002 (1996). https://doi.org/10.1103/PhysRevLett.77.2002
8. C. Jiang, V.V. Tsukruk, Freestanding nanostructures via layer-by-layer assembly. Adv Mater **18**, 829–840 (2006). https://doi.org/10.1002/ADMA.200502444
9. H. Endo, Y. Kado, M. Mitsuishi, T. Miyashita, Fabrication of free-standing hybrid nanosheets organized with polymer Langmuir–Blodgett films and gold nanoparticles. Macromolecules **39**, 5559–5563 (2006). https://doi.org/10.1021/MA052410J
10. R. Vendamme, S.Y. Onoue, A. Nakao, T. Kunitake, Robust free-standing nanomembranes of organic/inorganic interpenetrating networks. Nat Mater **5**, 494–501 (2006). https://doi.org/10.1038/nmat1655
11. T. Fujie, N. Matsutani, M. Kinoshita, Y. Okamura, A. Saito, S. Takeoka, Adhesive, flexible, and robust polysaccharide nanosheets integrated for tissue-defect repair. Adv Funct Mater **19**, 2560–2568 (2009). https://doi.org/10.1002/ADFM.200900103
12. T. Fujie, Development of free-standing polymer nanosheets for advanced medical and health-care applications. Polym J **48**, 773–780 (2016). https://doi.org/10.1038/pj.2016.38
13. T. Someya, M. Amagai, Toward a new generation of smart skins. Nat Biotechnol **37**, 382–388 (2019). https://doi.org/10.1038/s41587-019-0079-1
14. A. Zucca, K. Yamagishi, T. Fujie, S. Takeoka, V. Mattoli, F. Greco, Roll to roll processing of ultraconformable conducting polymer nanosheets. J Mater Chem C **3**, 6539–6548 (2015). https://doi.org/10.1039/c5tc00750j
15. K. Yamagishi, T. Nakanishi, S. Mihara, M. Azuma, S. Takeoka, K. Kanosue, T. Nagami, T. Fujie, Elastic kirigami patch for electromyographic analysis of the palm muscle during baseball pitching. NPG Asia Mater **11**, 80 (2019). https://doi.org/10.1038/s41427-019-0183-1
16. H. Taniguchi, K. Akiyama, T. Fujie, Biopotential measurement of plant leaves with ultra-light and flexible conductive polymer nanosheets. Bull Chem Soc Jpn **93**, 1007–1013 (2020). https://doi.org/10.1246/BCSJ.20200064

17. R. Pelrine, R. Kornbluh, Q. Pei, J. Joseph, High-speed electrically actuated elastomers with strain greater than 100%. Science **287**, 836–839 (2000). https://doi.org/10.1126/science.287.5454.836
18. A. Wiranata, M. Kanno, N. Chiya, H. Okabe, T. Horii, T. Fujie, N. Hosoya, S. Maeda, High-frequency, low-voltage oscillations of dielectric elastomer actuators. Appl Phys Express **15**, 011002 (2021). https://doi.org/10.35848/1882-0786/AC3D41
19. D. Gatti, H. Haus, M. Matysek, B. Frohnapfel, C. Tropea, H.F. Schlaak, The dielectric breakdown limit of silicone dielectric elastomer actuators. Appl Phys Lett **104**, 052905 (2014). https://doi.org/10.1063/1.4863816
20. T. Horii, K. Okada, T. Fujie, T. Horii, K. Okada, T. Fujie, Ultra-thin and conformable electrodes composed of single-walled carbon nanotube networks for skin-contact dielectric elastomer actuators. Adv Electron Mater, 2200165 (2022). https://doi.org/10.1002/AELM.202200165
21. D.J. Lipomi, M. Vosgueritchian, B.C.K. Tee, S.L. Hellstrom, J.A. Lee, C.H. Fox, Z. Bao, Skin-like pressure and strain sensors based on transparent elastic films of carbon nanotubes. Nat Nanotechnol **6**, 788–792 (2011). https://doi.org/10.1038/nnano.2011.184
22. K. Yamagishi, I. Kirino, I. Takahashi, H. Amano, S. Takeoka, Y. Morimoto, T. Fujie, Tissue-adhesive wirelessly powered optoelectronic device for metronomic photodynamic cancer therapy. Nat Biomed Eng **3**, 27–36 (2018). https://doi.org/10.1038/s41551-018-0261-7
23. F. Izzo, V. Granata, R. Grassi, R. Fusco, R. Palaia, P. Delrio, G. Carrafiello, D. Azoulay, A. Petrillo, S.A. Curley, Radiofrequency ablation and microwave ablation in liver tumors: an update. Oncologist **24**, e990–e1005 (2019). https://doi.org/10.1634/THEONCOLOGIST.2018-0337
24. C.J. Simon, D.E. Dupuy, W.W. Mayo-Smith, Microwave ablation: principles and applications. Radiographics **25**, S69–S83 (2005). https://doi.org/10.1148/RG.25SI055501
25. M. Mallory, E. Gogineni, G.C. Jones, L. Greer, C.B. Simone, Therapeutic hyperthermia: the old, the new, and the upcoming. Crit Rev Oncol Hematol **97**, 56–64 (2016). https://doi.org/10.1016/J.CRITREVONC.2015.08.003
26. A. Chicheł, J. Skowronek, M. Kubaszewska, M. Kanikowski, Hyperthermia – description of a method and a review of clinical applications. Reports Pract Oncol Radiother **12**, 267–275 (2007). https://doi.org/10.1016/S1507-1367(10)60065-X
27. E. De Bree, J. Romanos, D.D. Tsiftsis, Hyperthermia in anticancer treatment. Eur J Surg Oncol **28**, 95 (2002). https://doi.org/10.1053/EJSO.2001.1220
28. M. Saito, E. Kanai, H. Fujita, T. Aso, N. Matsutani, T. Fujie, Flexible induction heater based on the polymeric thin film for local thermotherapy. Adv Funct Mater **31**, 2102444 (2021). https://doi.org/10.1002/ADFM.202102444

Chapter 7
Solution-Processed Organic LEDs and Perovskite LEDs

Hinako Ebe, Takayuki Chiba, Yong-Jin Pu, and Junji Kido

Abstract Recent progresses on materials and device structures for solution-processed organic light-emitting devices (OLEDs) and perovskite light-emitting devices (perovskite LEDs) are discussed. Solution processable several materials such as fluorescent oligomer, phosphorescent dendrimer, lithium complex, zinc oxide, and polyvinylpyridine are designed and synthesized for achieving multilayer structure. The successful fabrication of solution-processed white phosphorescent OLEDs and tandem OLEDs will pave the way toward printable, low-cost, and large-area solid-state lighting application. Moreover, we discussed the recent research trends in perovskite LEDs. Herein, we proposed a strategy for fabricating efficient perovskite QD-LEDs and lead-free perovskite materials.

Keywords Solution-processed OLED · Solution processable material · Multilayer structure · Tandem OLED · Perovskite LEDs · Lead-free perovskite

7.1 Introduction

Solution processes such as spin coating, inkjet printing, slot-die coating or splay coating for organic light-emitting devices (OLEDs) are fascinating due to their potential advantages for a production of large-area devices at low cost, although vacuum evaporation (dry) processes are much ahead of the solution processes from

---

H. Ebe
Department of Organic Device Engineering, Research Center for Organic Electronics, Yamagata University, Johnan, Yonezawa, Japan

Faculty of Science, Yamagata University, Kojirakawa-machi, Yamagata, Japan

T. Chiba (✉) · J. Kido
Department of Organic Device Engineering, Research Center for Organic Electronics, Yamagata University, Johnan, Yonezawa, Japan
e-mail: T-chiba@yz.yamagata-u.ac.jp

Y.-J. Pu
RIKEN Center for Emergent Matter Science, Wako, Saitama, Japan

the mass production point of view. One of the key solutions to improve the performance of the devices is stacking of a number of successive layers of different functional materials. This multilayer structure allows for the separation of the charge-injecting, charge-transporting, and light-emitting functions to different layers, which leads to a dramatic increase in efficiency and lifetime. In the second section, we discuss our recent studies on fluorescent oligomers [1], phosphorescent dendrimers [2, 3], electron injection materials [4], and polymer binders [5] for solution-processed OLEDs. In the third section, we focus on solution-processed multilayer phosphorescent OLEDs using small molecules. On the basis of estimates from a solvent resistance test of small host molecules, we demonstrate that covalent dimerization or trimerization instead of polymer material can afford conventional small host molecules sufficient resistance to alcohols used for processing upper layers. This allows us to construct multilayer OLEDs through subsequent solution-processing steps, achieving record-high power efficiencies of 34 lm W^{-1} at 100 cd m^{-2} for solution-processed white phosphorescent OLEDs [6]. In the fourth section, we discuss the fabrication of a tandem OLED comprising two light-emitting units (LEUs) and a charge generation layer (CGL) between the indium tin oxide (ITO) anode and aluminum (Al) cathode using solution-based processes to simultaneously improve the luminance and device stability. A hybrid process of spin coating and thermal evaporation was utilized for the fabrication [7]. Each LEU with the configuration of first LEUs was fabricated using spin coating method. Ultrathin (1 nm) Al is deposited as the electron injection layer (EIL) in the first unit, and molybdenum oxide is subsequently deposited as the CGL by thermal evaporation. Tandem OLEDs using hybrid process showed almost twice the current efficiency of each light-emitting unit (LE). Additionally, fully solution-processed tandem OLEDs consisting of two LEUs and a CGL between the anode and cathode are fabricated. Zinc oxide (ZnO) and polyethyleneimine-ethoxylated (PEIE) nanoparticles bilayer is used as the EIL in the first LEU, and phosphomolybdic acid hydrate (PMA) is used as the electron acceptor of the CGL. Appropriate choice of solvents during spin coating of each layer ensures that a nine-layered structure is readily fabricated using only solution-based processes [8]. The determined driving voltage and efficiency of the fabricated tandem OLED are the sums of values of the individual LEUs.

7.2 Solution Processable Materials

7.2.1 Fluorescent Oligomer

Π-conjugated polymers have been extensively studied as solution processable emitting materials for the field of OLEDs since 1990 [9]. Precise control of molecular weight, end-group structure, and regioregular structure of the conjugated polymers for OLED has been established, but it is not possible to purify structural defects in a polymer chain itself thoroughly. However, monodisperse conjugated oligomers are

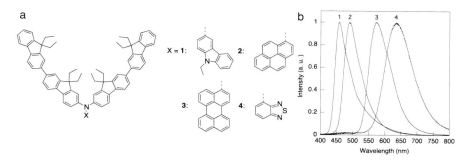

Fig. 7.1 (**a**) Solution-processable fluorescent compounds: bis(difluorenyl)amino-substituted carbazole *1*, pyrene *2*, perylene *3*, and benzothiadiazole *4*. (**b**) EL spectra of the devices with dye 1–4

able to have no structural defects and a better purity from conventional purification methods such as column chromatography, recrystallization, and sublimation. Four novel fluorescent dyes, bis(difluorenyl)amino-substituted carbazole *1*, pyrene *2*, perylene *3*, and benzothiadiazole *4* as solution processable light-emitting oligomer (Fig. 7.1a), are synthesized by palladium-catalyzed cross-coupling reaction. They are soluble in common organic solvents and can show a high glass transition temperature (T_g) and a good film-forming ability. The energy levels are related to the electronic properties of the central core; the electron-donating carbazole compound showed the lowest ionization potential and the electron—withdrawing benzothiadiazole compound showed the largest electron affinity. Emitting color can be easily controlled by a kind of central dyes, and outer fluorene oligomers can sterically prevent excimer formation between the emitting cores in a neat film. All compounds were purified from column chromatography and then thoroughly purified with a train sublimation for OLED application. These sublimable properties are one of advantages compared with the π-conjugated polymers in the purity point of view, because it is difficult to separate low molecular impurities having similar polarity to the target compounds by column chromatography. In practical, such impurities are regarded as detrimental for device stability. OLEDs with the configuration as ITO/poly(3,4-ethylenedioxythiophene): poly(styrenesulfonate) (PEDOT: PSS) (40 nm)/*1–4* (50 nm)/bis(2-methyl-8-quinolinolato) (biphenyl-4-olato)aluminum (BAlq) (50 nm)/LiF (0.5 nm)/Al (100 nm) were fabricated. PEDOT: PSS and the emitting layer were deposited by spin coating. BAlq and LiF/Al layer were deposited by evaporation under vacuum successively. Electroluminescence (EL) spectra of the compounds are well congruous with their photoluminescence (PL) spectrum of the film as shown in Fig. 7.1b, in which showed the emission color derived from the central dye ((1) sky blue, (2) blue green, (3) yellow, and (4) deep red). The outer oligofluorene groups did not affect the emission color because they have a wider energy gap than that of the central dye. The π-conjugations of the fluorene groups and the central dye do not seem to be fully delocalized. Photoluminescence quantum efficiency (PLQE) of the films is determined by using an integrating sphere system under nitrogen atmosphere. Polymer light-emitting device (PLED) of the compounds *1–3* exhibited higher than that of tris (8-quinolinolato)aluminum (Alq$_3$) film (22%) which was determined under the same conditions. Multicolor emissions from conjugated oligomer dyes having well-defined structures were achieved in their OLED fabricated from solution process.

7.2.2 Phosphorescence Dendrimer

The combination of the solution-process and the phosphorescent compounds can be an ideal choice to achieve low fabrication cost and high efficiency in OLEDs. For the solution process, substitution of functional dendrons on the complex is one of approaches to solubilize it, and P. L. Burn group has been done a lot of pioneering work on the dendrimer OLEDs [10, 11]. The dendron is bulky in volume, so that it can prevent intermolecular interaction between the emitting complexes, resulting in reduction of concentration quenching and high PLQE [12, 13]. From the OLED application point of view, those dendrons have to have enough high charge-transporting ability for low driving voltage [14] and have a larger triplet energy (T_1) level than that of the core complex not to quench the triplet exciton of the complex [15, 16]. In phosphorescent OLEDs, *m*-carbazolylbenzene (mCP) is one of well-known and widely used host materials, because its T_1 level is enough high (3.0 eV) to confine the phosphorescent emission of the iridium complex and has bipolar charge-transporting ability [17, 18]. 3,5-(*N,N*-di(4-(*n*-butyl)phenyl)amine) (DPA) is also used as hole transport substituent group. We designed and synthesized *(mCP)$_6$Ir* and *(DAP)$_6$Ir* (Fig. 7.2). The phosphorescent iridium complex, *(mCP)$_3$Ir*, attached three mCP dendrons having alkyl groups, and high efficiencies of the OLEDs using that complex. In *(mCP)$_3$Ir*, mCP dendrons are attached on each phenyl ring of tris(2-phenylpyridinato)iridium(III) (Ir(ppy)$_3$), and *(mCP)$_3$Ir* is a facial isomer, so that the three mCP dendrons are attached spatially the same side in the complex and surround only half side of Ir(ppy)$_3$ as shown in Fig. 7.3a. The fully surrounded Ir(ppy)$_3$ by six host dendrons *(mCP)$_6$Ir* shows Fig. 7.3b. Both of the complexes showed higher PLQE in a neat film than that of half-surrounded *(mCP)$_3$Ir* and *(AP)$_3$Ir*, well-supporting the results reported in the literature. PLQEs of the complexes in the neat film are important parameter to estimate the shielding effect of the surrounding dendrons to Ir(ppy)$_3$. PLQEs of the toluene solution and the films

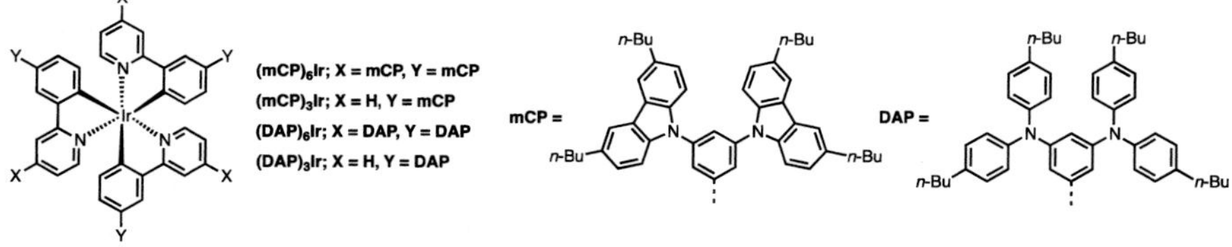

Fig. 7.2 Chemical structures of the dendronized iridium complex. Reprinted from ref. [3]. Copyright 2012, with permission from Elsevier

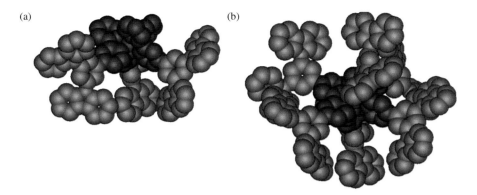

Fig. 7.3 The optimized structures of (**a**) half-surrounded (mCP)$_3$Ir and (**b**) fully surrounded (mCP)$_6$Ir by PM6 calculation. The butyl groups were replaced to hydrogen in calculation. Reprinted from ref. [3]. Copyright 2012, with permission from Elsevier

were measured by using an integrating sphere system under 331 nm excitation. In a diluted solution, all complexes showed higher PLQE than 70%, which are comparable to 85% of unsubstituted Ir(ppy)$_3$.

This result demonstrated that these surrounding dendrons are optically inert and do not affect the emission efficiency of Ir(ppy)$_3$ core. The fully surrounded complexes, *(mCP)$_6$Ir* and *(DAP)$_6$Ir*, showed high PLQE even in a neat film, which is comparable to PLQE in a dilute solution. On the other hand, the half-surrounded complexes, *(mCP)$_3$Ir* and *(DAP)$_3$Ir*, showed much lower PLQE in a neat film than that in a dilute solution. These complexes are facial isomers; therefore, in *(mCP)$_3$Ir* and *(DAP)$_3$Ir*, a some space around pyridyl groups of Ir(ppy)$_3$ core is opened, and their three-dimensional structure is like a hemisphere, resulting in only partial suppression of concentration quenching in a neat film of an iridium complex. However, in *(mCP)$_6$Ir* and *(DAP)$_6$Ir*, the bulky host dendrons fully surrounded Ir (ppy)$_3$ and effectively prevented the intermolecular interaction between Ir (ppy)$_3$s. There are still small amount of reduction of PLQEs from a solution to a neat film, due to the concentration quenching even in the fully substituted complexes. The substituted host dendrons are not enough large to completely suppress the interaction between the core complexes. Adachi et al. reported that an average distance between iridium complexes in a doped film critically influenced on PLQE [19]. Förster type energy transfer between Ir(ppy)$_3$ cores through an overlap of the emission and the absorption causes a decrease of neat film PLQE. If the average distance between iridium complexes is shorter than Förster radius, a strong quenching occurs. The stronger quenching of *(mCP)$_3$Ir* in the neat film than that of *(mCP)$_6$Ir* is due to the shorter average distance between the cores derived from a less number of bulky host dendrons of *(mCP)$_3$Ir* than that of *(mCP)$_6$Ir*. Substitution of more branched and larger dendrons to the core complexes is desirable to achieve the complete suppression of concentration quenching. Solution-processed OLED with *(mCP)$_6$Ir* exhibited high efficiencies, 19 lm W^{-1}, 32 cd A^{-1}, and 12% of external quantum efficiency (EQE) at 100 cd m^{-2} and 11 lm W^{-1}, 25 cd A^{-1}, 9.1% at 1000 cd m^{-2}. The energy levels of the surrounding dendrons intensely affected the charge injection into the emitting layer and the device performance.

7.2.3 Electron Injection Materials

PLEDs employ low work function metal, such as cesium, barium, or calcium, as an electron injection layer (EIL) and a cathode to enhance the electron injection to the emitting layer. However, these metals and the cathode are highly reactive with atmospheric oxygen and moisture, which results in degradation of the device. To avoid these problems, stable alkali metal fluorides, such as LiF or CsF, are commonly used in the EIL of dry-processed OLEDs. Cs_2CO_3 has been reported to be an effective EIL material in solution-processed OLEDs because it is soluble in alcohol solvents and can be coated from solution. The solution-processed Cs_2CO_3 EIL exhibits a high electron injection ability that is comparable to that of alkali metals [20, 21]. However, Cs_2CO_3 still has some disadvantages; it is hygroscopic and unstable in air, and it requires an ultrathin thickness because it is an insulating material. A strong chemical reduction is known to occur between Cs_2CO_3 and the thermally evaporated Al cathode.

Lithium phenolate complexes could be used to form an excellent EIL, and the device performance was much less sensitive to the thickness of the coating of these complexes because of their high electron-transporting ability compared with insulating Cs_2CO_3 [22, 23]. The lithium phenolate complexes also have stability against oxidation and are less hygroscopic. We reported the efficient solution processing of an EIL based on the lithium quinolate complex (Liq) that is dissolved into alcohol; in this EIL, a low driving voltage and improved stability of the PLEDs are achieved. Liq has high solubility in polar solvents, such as alcohols, and it has a smooth surface morphology. Therefore, Liq can be spin-coated onto the emitting polymer; the device prepared with spin-coated Liq as an EIL exhibited a lower turn-on voltage and had a higher efficiency than the devices prepared with spin-coated Cs_2CO_3 or with thermally evaporated calcium.

On the other hand, ZnO have recently been reported to be air-stable electron injection materials in PLEDs [24, 25]. To improve the electron injection ability of the solution-processed EIL that had a thickness of more than 10 nm, we utilized ZnO nanoparticles as a host for Liq or Cs_2CO_3 (Fig. 7.4a). ZnO can enhance the electron injection characteristics through the addition of alkali metal salts [26, 27]. The ZnO nanoparticles, which were synthesized from a zinc acetate precursor [28], were well dispersed into 2-ethoxyethanol at a concentration of 10 mg mL^{-1}.

The average diameter of the ZnO nanoparticles was estimated using transmission electron microscopy (TEM) and dynamic light scattering (DLS) and was approximately 10 nm, with the particles being monodispersed. The mixture of ZnO nanoparticles and Liq formed a thick film with a homogeneous and smooth surface, indicating that Liq is well dispersed around the ZnO nanoparticles. The combination

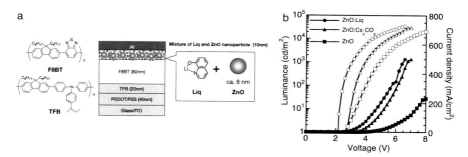

Fig. 7.4 (**a**) The chemical structure of Liq and the structure of the device. (**b**) Current density-voltage (solid symbol) and luminance-voltage (open symbol) characteristics. Reprinted with the permission from ref. [4]. Copyright 2011 American Chemical Society

of ZnO and Liq significantly reduced the driving voltage and improved the power efficiency compared to only ZnO or ZnO: Cs_2CO_3 (Fig. 7.4b). This inorganic–organic hybrid EIL is an effective approach for enhancing the efficiency and the stability of PLEDs, and the thickness can be sufficiently thick for reproducible large-scale devices.

7.2.4 Polymer Binder

The thickness of EILs comprised of compounds such as Liq and Cs_2CO_3 must be the ultrathin (<2 nm) to achieve efficient electron injection characteristics due to their poor electron transport properties. However, precise thickness control in the range of a few nanometers is practically impossible for large-scale device using solution processes such as spin coating and blade coating. In this context, only relatively thick EIL films (10–20 nm) can be mass-produced for large PLEDs using solution processing. Herein, we report the use of a mixture of poly(vinylphenylpyridine) and Liq for solution processable efficient, thick electron injection layers. Vinyl polymers with high solubilities in alcoholic solvents and good film-forming abilities, such as poly(4-vinylpyridine) (PV4Py) and poly[4-(4-vinylphenyl)pyridine] (PVPh4Py) (Fig. 7.5a), were used as a binders for Liq, and the effect of the π-conjugation of the polymers on the electron transport and injection characteristics was investigated. The influence of the position of the nitrogen in the pyridine rings was also investigated using poly[2-(4-vinylphenyl)pyridine] (PVPh2Py) and poly[3-(4-vinylphenyl)pyridine] (PVPh3Py).

In the UV–vis absorption spectra, the PVPhPys exhibited smaller energy gaps than that of PV4Py, because the additional phenyl group participates in extended π-conjugation compared with only the pyridine group (Fig. 7.5b). Among the PVPhPys, PVPh2Py exhibited a bathochromically shifted absorption peak compared to those of PVPh3Py and PVPh4Py. The greater π-conjugation of PVPh2Py is probably due to the greater planarity of the structure of 2-phenylpyridine, which

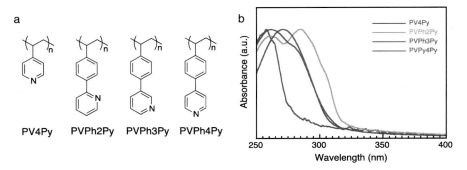

Fig. 7.5 (a) Chemical structures and (b) UV-vis absorption spectra of the poly(vinylpyridine) compounds

results because of the absence of a hydrogen at the ortho position and the consequent reduced steric hindrance. The influence of the concentration of Liq in a PVPh4Py: Liq mixture in the performance of the *ultrathin* layers of approximately 1.6 nm was investigated. In addition to devices prepared with EIL layers comprised of PVPh4Py with 10, 30, 50, or 70 wt%, two control devices were fabricated using *ultrathin* layers of only Liq and PVPh4Py.

The observed EL spectra of various devices are identical to the emission from F8BT, and no emission was observed from TFB or Liq. This result indicated that the holes and the electrons were confined within the F8BT and that the recombination of the charges occurred only in the F8BT. The device with 100 wt% PVPh4Py exhibited a high turn-on driving voltage of 3.0 V and driving voltages of 5.6 V and 8.4 V at 100 cd m^{-2} and 1000 cd m^{-2}, respectively. The EQE of 0.6% observed at 1000 cd m^{-2} for the devices with the 100 wt% PVPh4Py layer were lower than those of the device with PVPh4Py doped with Liq. This result suggested that PVPh4Py itself has poor electron injection property due to its shallow LUMO level of 1.9 eV. However, the device performance dramatically improved when Liq was added to the PVPh4Py. The driving voltage of the devices with PVPh4Py: Liq decreased with increasing Liq concentration from 10 wt% to 70 wt% due to increased electron injection into the F8BT from the Al cathode. The external quantum efficiencies were 4.9–6.9%. These driving voltages and efficiencies were nearly equivalent to those of the device with the *ultrathin* EIL layer comprised of 100 wt% Liq. In the device with 10 wt% Liq, balanced charge ratio resulted in the highest power efficiency of 23 lm W^{-1} and an EQE of 6.9% at 1000 cd m^{-2}. Notably, this power efficiency is the highest value reported in the literatures to date for devices with F8BT as the emissive layer (EML) [29, 30]. These results indicate that while PVPh4Py itself is not effective as an EIL, mixing it with Liq does not deteriorate the electron-injection properties of Liq and improves the driving voltages and efficiencies of the devices.

The performance of device with EILs of different thicknesses comprised of the mixtures of PVPh4Py and Liq was investigated. The three types of EILs with thicknesses of 1.6 nm, 8.6 nm, and 16 nm were deposited from solutions with

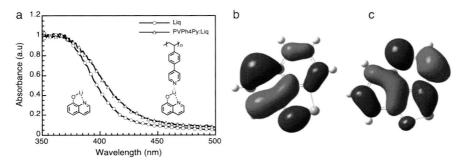

Fig. 7.6 (**a**) UV–vis absorption spectra of films of Liq alone and Liq with PVPh4Py film. (**b**) HOMO and (**c**) LUMO of Liq with structures optimized in the grand states by DFT calculation. Reproduced from ref. [5] by permission of Jon Wiley & Sons Ltd

different Liq concentrations using different spin-coating speeds. As the EILs thickness was increased from 1.6 nm to 16 nm, the driving voltages increased and the EQEs decreased. However, the increase in the voltage and the decrease in the EQE were suppressed in the EILs comprised of PVPh4Py and Liq compared to those for the EIL comprised 100 wt% Liq. The device with a 50 wt% mixed EIL exhibited the least dependence on the layer thickness, and the lowest driving voltage and the highest EQE for all of the devices was observed for an EIL with thickness of 16 nm. The high driving voltage and low efficiency of the device with the *thick* Liq layer are attributed to the poor electron transport properties of Liq itself. Conversely, mixing PVPh4Py with Liq could improve the electron transport properties of the EIL, and the driving voltage remained low, even for a *thick* EIL. Figure 7.6a shows that the UV absorption edge of the film prepared from the mixture of Liq and PVPh4Py with 50 wt% Liq was red shifted by 15 nm, corresponding to 0.11 eV, compared to that of the pure Liq film. Conversely, both the HOMO levels of Liq and the mixture of 50 wt% Liq and PVPh4Py were the same at 5.5 eV, as determined via by photoelectron yield spectroscopy. Consequently, the mixture of Liq and PVPh4Py had a smaller energy gap than that of Liq due to the lower LUMO level of the mixture than that of Liq. Therefore, to understand the distribution of the HOMO and LUMO level in Liq, DFT calculations were conducted. The HOMO is not located on the Li atom (Fig. 7.6b), but the LUMO is associated with the Li atom (Fig. 7.6c). These results suggest that the interactions between the Li atom of Liq and the pyridine ring of PVPh4Py affected the LUMO level. The reduced dependence of EIL performance on the layer thickness will be advantageous for the large-area coating processes, because it is difficult using solution processing to form uniform thin films with an accuracy of a few nanometers.

The position of the nitrogen in the pyridine rings had slight influence on the electron injection properties in the *ultrathin* layers (Fig. 7.7a). Conversely, in the devices with *thin* EILs (approximately 8.6 nm), the position of the nitrogen in the pyridine rings of the polymers had a greater influence on the driving voltage and efficiency. The driving voltage increased in the order of

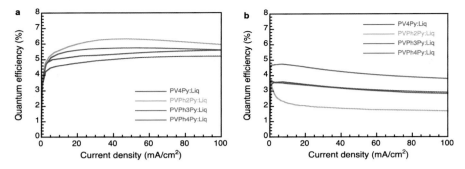

Fig. 7.7 External quantum efficiency–current density characteristics of PLEDs performance with (**a**) *ultrathin* EILs and (**b**) *thin* EILs using PV4Py, PVPh2Py, PVPh3Py, and PVPh4Py doped with 50 wt% Liq, respectively, and 100 wt% Liq

PVPh4Py < PVPh3Py < PVPh2Py. Alternatively, the EQE increased in the order PVPh2Py < PVPh3Py < PVPh4Py (Fig. 7.7b). These results suggest that the position of the nitrogen in the pyridine rings significantly affects the electron transport properties of the polymers rather than the electron injection properties. Sasabe et al. previously reported that the electron mobility of a series of oligo-phenylpyridine derivatives were strongly affected by the position of the nitrogen in the pyridine rings due to C-H⋯N hydrogen bonding interactions [31]. The glass transition temperatures (T_g) were determined via differential scanning calorimetry (DSC). The T_g of PVPh4Py was observed at 185 °C, which is higher than that of PV4Py (146 °C) due to the more rigid structure of the phenylpyridine. The T_g of PVPh4Py was also higher than those of PVPh2Py (162 °C) and PVPh3Ph (140 °C), suggesting that the location of the nitrogen at the 4-position of the PVPh4Py enables stronger intermolecular hydrogen bonding interactions than those in PVPh3Py and PVPh2Py. The denser packing of PVPh4Py that results from the stronger hydrogen bonding interactions probably leads to the enhanced electron transport properties observed for the *thick* films.

7.3 Solution-Processed Multilayer Small-Molecule OLEDs

7.3.1 Solubility of Small-Molecule Materials

Small molecular-based OLEDs typically consist of four or more multiple layers of different materials in precise opto-electrical design. Such multilayer structures allow for the separation of the charge-injecting, charge-transporting, and light-emitting functions to the different layers, thus leading to a marked increase in efficiency and lifetime [32–34]. Although stepwise vacuum evaporation easily achieves the required multilayer structures of small molecules at the expense of high manufacturing cost, it is more challenging in the case of solution processing, because depositing one layer would dissolve the layer beneath it. To achieve the multilayer structures by solution processing, research efforts have focused on π-conjugated polymers that afford a robust hydrophobic layer, on which a hydrophilic layer can be deposited from orthogonal solvents, such as water or water/alcohol mixture [35]. In situ cross-linking reactions have also been explored to afford covalently bound structures that are highly resistant to processing solvents [36–38]. Despite their high mechanical robustness and compatibility with subsequent solution processing, polymers are plagued by limited reproducibility in the device performance because of batch-to-batch variations with respect to molecular weight, polydispersity, regioregularity, and purity. Moreover, their efficiencies are still far below the fluorescent tubes [39]. The highest reported power efficiency of white polymer LEDs is 25 lm W^{-1} thus far [40, 41]. On the other hand, small molecules are very attractive because they have a well-defined molecular structure that offers more reproducibility of synthesis procedures and better understanding of molecular structure–device performance relationships. However, their thin-film assemblies, most of which are amorphous in nature, are easily broken up, even by the orthogonal solvents, because small molecules typically attach to each other only by weak intermolecular forces such as van der Waals, H-bonding, and π–π stacking interactions. Consequently, the highest reported efficiency of solution-processed small-molecule OLEDs still relies on a vacuum-evaporated electron-transporting layer (ETL), which is not practical for low-cost mass production of scalable devices [42]. Herein, we demonstrate highly efficient small-molecule OLEDs in which quadruple organic layers, including a molecular-emitting layer (EML) and ETL, are fully solution processed. The key feature of the devices is the use of newly developed small host molecules in the EML, which are sufficiently resistant to the orthogonal solvents such as alcohols, used for processing upper ETLs, thus allowing us to construct the multilayer structure through subsequent solution processing steps. While a robust host polymer is typically required to realize the multilayer structure, we simply modified conventional host molecules by covalent dimerization or trimerization to afford sufficient resistance to alcohols. With this approach, record-high efficiencies have been achieved for solution-processed blue, green, and white OLEDs.

Through experiments with 17 host molecules over a wide range of molecular weight from 243 to 1146 (Fig. 7.8), we found that their resistance to alcohols remarkably increasing molecular weight. Figure 7.8b shows the normalized remaining thickness of molecular thin films after rinsing with a variety of alcohols as a function of molecular weight, as measured by ultraviolet–visible absorption spectroscopy. From the best-fit cumulative distribution function, we determined that the threshold molecular weights for achieving 95% remaining thickness were 775, 811, 849, and 767 for methanol, ethanol, 1-propanol, and 2-propanol, respectively. This result demonstrates that even conventional host molecules can be compatible with the subsequent solution process simply by covalent dimerization or trimerization to exceed the threshold molecular weights, eliminating the need for polymeric counterparts. In addition, this figure covers a wide variety of building blocks, including arene, carbazole, triphenylamine, fluorene, benzothiophene, and even polar moieties such as benzophenone, pyridine, and triazine, making this

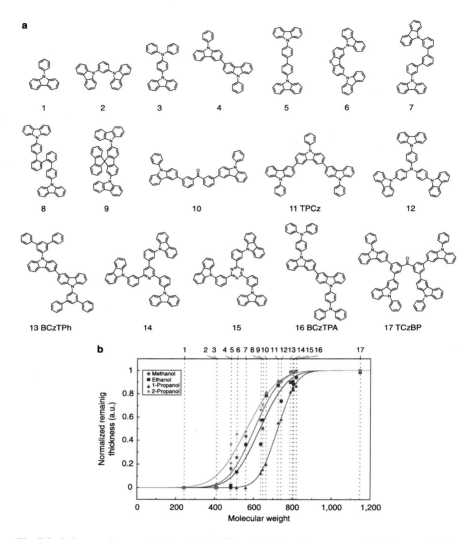

Fig. 7.8 Solvent resistance of molecular thin films. (**a**) Molecular structures of the host molecules arranged in order of increasing molecular weight. (**b**) Plots of normalized remaining thickness of the molecular thin films after rinsing with a variety of alcohols as a function of molecular weight. The solid lines represent the best fit to the cumulative distribution function. Reprinted by permission from Macmillan Publishers: ref. [6], copyright 1993

approach broadly applicable. We also note that our approach enables subsequent solution processing onto the EML without having to use water, which is detrimental to device efficiency and stability [43–45].

7.3.2 Green and Blue Phosphorescent OLEDs

On the basis of the above estimates, we elected to use two host molecules for green phosphorescent OLEDs: 3,3′:6′,3″-ter(9-phenyl-9 H-carbazole) (TPCz) and 3,3′,6,6′-tetrakis(9-phenyl-9H-carbazol-3-yl) benzophenone (TCzBP). The schematic energy diagram of the host molecules is shown in Fig. 7.9b. When mixing the two host molecules in an EML, holes should preferentially reside in the shallow HOMO of TPCz and electrons in the deep LUMO of TCzBP. Consequently, we can accurately optimize charge balance in the device by varying the ratio of the two host molecules to achieve high efficiency. By using the two molecules as hosts for Ir(ppy)$_3$, we fabricated green phosphorescent OLEDs, in which quadruple organic layers were fully solution processed. The device configuration was ITO (130 nm)/PEDOT:PSS (30 nm)/TFB (20 nm)/host:12 wt% Ir(ppy)3 (30 nm)/2,2′,2″-(1,3,5-benzinetriyl)tris(1-phenyl-1-H-benzimidazole) (TPBi) (50 nm)/Liq (1 nm)/Al (100 nm) (Fig. 7.9a). In these devices, TPBi was elected as an ETL because of its sufficient solubility in methanol, enabling the subsequent solution processing on the molecular EML. The most common host polymer poly(N-vinylcarbazole) (PVK) was also used for comparison. By mixing a 1:1 ratio of TPCz and TCzBP, we achieved a power efficiency of 52 lmW^{-1} at 100 cd m^{-2}. Indeed, the peak power efficiency reached an extremely high value of 96 lmW^{-1}. The corresponding external quantum efficiency (EQE) was 23%, which remained as high as 22% and 20% at 100 cd m^{-2} and 1000 cd m^{-2}, respectively. We also note that there is no perceivable change in luminance as a function of viewing angle (Lambertian factor: 1.03), eliminating the possibility of overestimating the efficiencies.

Despite the impressive efficiencies of the green phosphorescent OLEDs, our initial attempt at using standard blue phosphorescent emitter bis(2-(4,6-difluorophenyl)pyridine) (picolinate)iridium(III) (FIrpic) resulted in poor efficiencies. In the blue OLEDs, 4,4′-(3,3′-bi(9H-carbazole)-9,9′-diyl)bis(2,6-diphenyl) benzene (BCzTPh) and 4,4′-(3,3′-bi(9H-carbazole)-9,9′-diyl)bis(N,N-diphenyl)aniline (BCzTPA) were employed as host molecules and 2-propanol-soluble 1,3-bis (3-(diphenylphosphoryl)phenyl)benzene (BPOPB) was used as an ETL. The device configuration was ITO (130 nm)/PEDOT:PSS (30 nm)/TFB (20 nm)/host:12 wt% FIrpic (30 nm)/BPOPB (45 nm)/Liq (1 nm)/Al (100 nm) (Fig. 7.9a). While these host molecules have a sufficiently high T$_1$ level for efficient exothermic energy transfer to the phosphorescent blue emitter, the resulting device exhibited considerable emission from the host molecules at around 420 nm and with a low power efficiency of 6.5 lmW^{-1} at 100 cd m^{-2}. Alternatively, with three-coordinated tris (2-(4,6-difluorophenyl)pyridine)iridium(III) (Ir(Fppy)$_3$), the device efficiencies significantly increased to 36 lm W^{-1} for power efficiency and 20% for EQE at 100 cd m^{-2}. It is intriguing to note that Ir(Fppy)$_3$ performs five times as well as FIrpic in power efficiency, although these two blue emitters possess almost identical opto-electrical properties. In these devices, the majority of excitons would be generated near the EML/ETL interface because of the relatively large injection barrier between them (Fig. 7.9b) and would subsequently be harvested by the

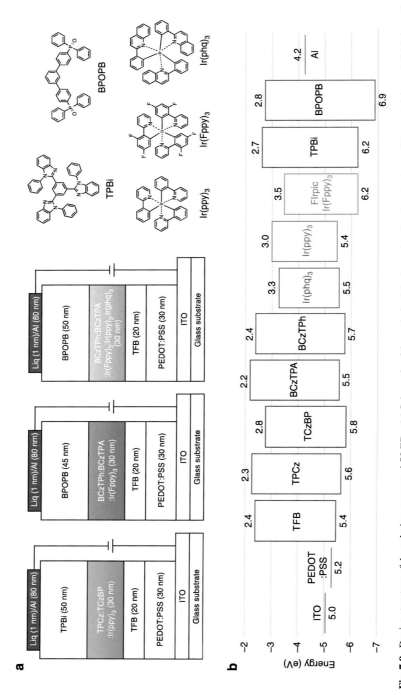

Fig. 7.9 Device structure of the solution-processed OLEDs. (**a**) Schematic of the optimized device and molecular structures of the materials used. (**b**) Schematic energy-level diagram of the materials. Reprinted by permission from Macmillan Publishers: ref. [6], copyright 1993

doped blue phosphorescent emitter. In addition, we have previously reported that the direct electron-trapping process of a blue phosphorescent emitter played a major rule in efficient electron injection at the EML/ETL interface in evaporated OLEDs [46]. We thus hypothesized that polar picolinate ligand containing FIrpic would dissolve away from the EML surface upon the subsequent solution processing of the ETL from the 2-propanol solution, resulting in the unwanted host emission and poor device efficiencies.

To verify the hypothesis, we performed depth-profiling measurements of the devices by time-of-flight secondary ion mass spectrometry (TOF-SIMS). These measurements involved Ar_{2500}^+ gas cluster ion beam etching starting [47] from the ETL surfaces to the underlying substrates. This direction of etching allows for collecting the composition of the EMLs without alternating their original position at the EML/ETL interfaces (peak fronts), while etching and primary ion beams cause peak tailing. Figure 7.10 displays the TOF-SIMS depth profiles of the solution-processed devices with the different blue phosphorescent emitters, FIrpic and Ir(Fppy)$_3$, in comparison with reference devices with a vacuum-evaporated ETL, for which a well-defined interface is expected to exist between the EML and ETL. We focused on $[C_{22}H_{12}F_4IrN_2]^+$ ions with $m/z = 573$ as signatures both of FIrpic and Ir(Fppy)$_3$ to obtain sufficient intensity in the dilute emitters embedded in the host matrix. We also monitored the corresponding molecular ions for the other molecules. The depth resolution was 11.5 nm under our experimental conditions. One observed form the TOF-SIMS depth profiles that the composition at the solution-processed EML/ETL interface was significantly varied between the FIrpic and Ir(Fppy)$_3$ system. Ir(Fppy)$_3$ showed a slight reduction in intensity in the interface region of roughly 10–20 nm for the solution-processed device, whereas a noticeable reduction in FIrpic intensity occurred almost over the entire EML, particularly at the EML/ETL interface. The corresponding reduction in the concentration of the blue emitters was also quantitatively observed by high-performance liquid chromatography analysis; the reduction in concentration of emitters was 48% and 8% for FIrpic and Ir(Fppy)$_3$, respectively, upon rinsing with pure 2-propanol. These results indicate that almost half of the FIrpic molecules dissolved from the EML while depositing the ETL, and the host molecules were compositionally rich in the resulting interface. On the other hand, the Ir(Fppy)$_3$ molecules existed over the entire EML, including the interface even after the deposition of the ETL. The improved device efficiencies upon introduction of Ir(Fppy)$_3$ therefore arise from the efficient electron injection at the EML/ETL interface through the direct electron-trapping process of Ir(Fppy)$_3$. We also note that a very small amount of the EML composition migrated and uniformly distributed into the solution-processed ETL.

7.3.3 White Phosphorescent OLEDs

We fabricated solution-processed-multilayer white OLEDs by incorporating green-emitting Ir(ppy)$_3$ (0.2 wt%) and red-emitting tris(2-phenyl-1-quinoline)iridium(III)

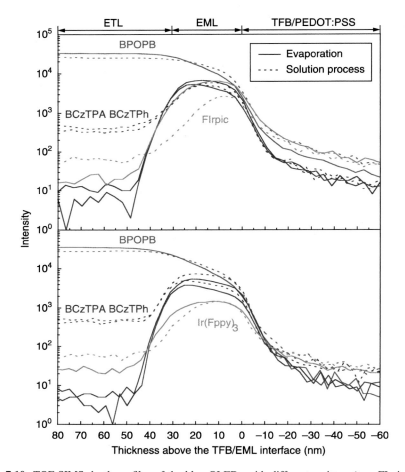

Fig. 7.10 TOF-SIMS depth profiles of the blue OLEDs with different emitters (top: FIrpic and bottom: Ir(Fppy)$_3$). The dashed lines represent depth profiles of the devices using solution-processed BPOPB. The solid lines represent data for reference devices using evaporated BPOPB. Reprinted by permission from Macmillan Publishers: ref. [6], copyright 1993

(Ir(phq)$_3$) (0.7 wt%) into the blue EML. Although the EL spectra of white OLEDs typically depend on current density (in other words, luminance) [20, 48, 49], the resulting solution-processed white OLED surprisingly showed no perceived change in the EL spectra under varying current density. The corresponding color shift in the Commission Internationale de l'éclairage (CIE) coordinates was as small as $\Delta x, y = 0.002, 0.002$ between 100 cd m^{-2} and 1000 cd m^{-2}. For comparison, the evaporated ETL exhibited relatively strong green and red emissions and a gradual blue shift in the CIE coordinates of $\Delta x, y = 0.054, 0.002$ between 100 cd m^{-2} and 1000 cd m^{-2}.

The stable EL spectra can be explained if the green and red emitters were partially washed away from the EML surface upon the solution processing of the ETL, causing electron trapping and recombination preferentially on the blue emitter at the EML/ETL interface. This would provide uniform exciton distribution among the three emitters, and thus the stable EL spectra, because the only exciton generation path left for the green and red emitters is energy transfer from the blue emitter staying at the EML/ETL interface. This hypothesis was confirmed with PL spectroscopy, showing that the red and green emissions of the EML decreased upon rinsing with pure 2-propanol. Similar stable EL spectra have also been observed in evaporated OLEDs, in which the blue emitter is placed at the recombination interface and spatially separated from other emitters as in this case.

Remarkably, we achieved a high power efficiency of 34 lm W^{-1} and an EQE of 21% at 100 cd m^{-2} for a white emission with a color rendering index (CRI) of 70 and CIE coordinates of 0.43, 0.43 without the use of any outcoupling enhancement. The peak power efficiency and EQE reached 45 lm W^{-1} and 22%, respectively. To the best of our knowledge, these efficiencies are considerably higher than the highest efficiencies ever reported for white polymer LEDs. In addition, when outcoupling all photons trapped in the glass substrate, these efficiencies increase by a factor of 1.96 as confirmed by using an index-matched hemisphere lens. As a result, the maximum achievable power efficiency and EQE are expected to be 88 lm W^{-1} and 41%, respectively. We also note that the solution-processed device showed lower driving voltages and higher efficiencies compared with the corresponding device with an evaporated ETL. The superior performance with a solution-processed ETL was also observed in the green phosphorescent OLEDs using TPBi as an ETL.

7.4 Solution-Processed Tandem OLEDs

7.4.1 Solution-Evaporation Hybrid Tandem OLEDs

Whereas the luminance of OLEDs increases with the current density, high currents promote the degradation of the organic materials [50]. To simultaneously improve the luminance and device stability, Kido et al. developed tandem OLEDs, comprising several stacked LEUs interconnected by CGLs [51]. In general, tandem OLED fabricated by evaporation can have more than ten layers between the anode and the cathode [52–58]. Whereas stepwise vacuum evaporation-based processes can generate multilayered structures, such structures are a challenge to solution-processing techniques because solution-based coating of one layer can dissolve the layer beneath it. Electron injection from CGL into the first LEU is a key factor impacting the characteristics of tandem OLEDs. In vacuum-processed devices, alkali metal [52, 54, 55] or a bilayer of alkali metal halide and Al [58] effectively enhances electron injection and is used in the EIL of the first LEU. The CGL is composed of electron-accepting materials, such as MoO_3 [55], V_2O_5 [54], and WO_3 [59], and electron-donating materials, such as arylamine derivatives. It is important to match the Fermi level of electron-accepting materials and the HOMO level of electron-donating materials. However, such metals, metal halides, and metal oxides are not readily solution-processable because of their poor solubility in organic solvents.

A hybrid process of spin coating and thermal evaporation was utilized for the tandem OLED fabrication. Each LEU with the configuration of PEDOT:PSS/LE-polymer/EIL was fabricated by spin coating. Ultrathin Al was deposited as the EIL in the first unit, and MoO_3 was subsequently deposited as the CGL by thermal evaporation. Low work function metals cannot be used as the EIL for solution-based processing of tandem devices because of their high reactivity with organic solvents, which results in severe degradation of the device. Cs_2CO_3-doped ZnO nanoparticles were used as an EIL on the LE polymer to improve the electron injection from the cathode.

The surface morphology of a spin-coated metal oxide nanoparticle layer appears to be rough, with many gaps due to agglutination of nanoparticles. Consequently, the thin layer of metal oxide nanoparticles cannot protect the first-LEU organic layer from the spin-coating solvent of the second-LEU organic layer. Thus, we chose PV4Py as a binder to improve the film morphology of the $ZnO:Cs_2CO_3$ mixture and facilitate the formation of a uniform and dense film to prevent the solvent from soaking into the first LEU. The thermally evaporated MoO_3 layer is insoluble in organic solvents, such as toluene, p-xylene, and dichlorobenzene. Thus, an electron-donating layer can be spin-coated on top of the MoO3 layer. Poly(4-butylphenyl-diphenyl-amine) (poly-TPD) was used as an electron-donating and hole-transporting layer and was spin-coated onto the MoO_3 layer using a dichlorobenzene solution. This combination of MoO_3 and poly-TPD as a prospective CGL may be successful because bilayers of MoO_3 and arylamine derivatives, such as NPD or TPD, can work as an efficient CGL in tandem OLEDs [55]. Poly-TPD is insoluble in toluene and p-xylene; therefore, an LE polymer such as F8BT can be spin-coated onto the poly-TPD layer using a p-xylene solution without dissolving the bottom layer. The efficient solution-based processing of EILs in the CGL containing MoO_3/poly-TPD bilayers was employed for the construction of a tandem device as shown in Fig. 7.11a.

At high luminance values of 1000 cd m^{-2}, first-LEU and second-LEU exhibited efficiencies of 6 cd A^{-1} and 4 cd A^{-1}, respectively (Fig. 7.11b). The efficiency of second LEU was lower than that of the first LEU due to the decrease in the charge balance and the increase in the driving voltage for MoO_3 as an HIL. Nevertheless, the current efficiency of the tandem device increased to 10 cd A^{-1}, which is the sum of the efficiency of the two single devices. The LUMO level of poly-TPD was 2.3 eV, which was just shallow enough to block electrons from the second LEU to the first LEU [60]. The conduction band (CB) of ZnO was 7.4 eV, which was deep enough to block holes from the first LEU to the second LEU. Thus, the emissions are attributed to the recombination of charges that were generated in the CGL without current leakage. These results demonstrate that MoO_3/poly-TPD can function as an effective CGL.

7 Solution-Processed Organic LEDs and Perovskite LEDs

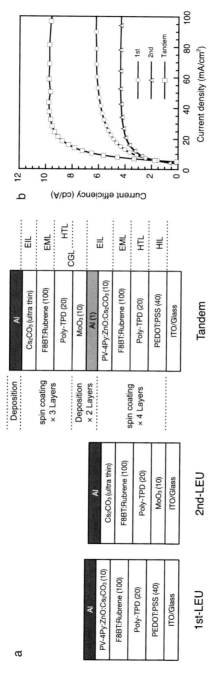

Fig. 7.11 (a) Device structure of the first LEU, second LEU, and tandem OLED. (b) Current efficiency–current density characteristics of the first LEU, second LEU, and tandem OLED. Reproduced from Ref. [7] by permission of The Royal Society of Chemistry

7.4.2 Fully Solution-Processed Tandem OLEDs

Recently, solution-processed tandem OLEDs were reported by Colsmann and co-workers [61] with an inverted structure of polymer OLEDs. To the best of our knowledge, there is no report on the tandem OLEDs having the regular configuration of ITO anode and Al cathode. Figure 7.12 shows tandem OLED structure comprising two LEUs (first LEU and second LEU) and a CGL between the anode and the cathode using only solution-based processes. The driving voltage and efficiency of the fabricated tandem OLED are the sums of corresponding values of the component LEUs. These results demonstrate that the solution-processed CGL successfully generated electrons and holes and that the generated electrons and holes were injected into first LEU and second LEU, respectively, when a voltage was applied, resulting in charge recombinations in each LEU. Recently, PEIE and ZnO nanoparticles have been reported as efficient electron-collecting layers between a semiconducting organic layer and a cathode in organic photovoltaics (OPV) [62, 63] and electron-injection layers between a cathode and the semiconducting organic layer in OLEDs [64–66]. These properties emerge from their ability to reduce the

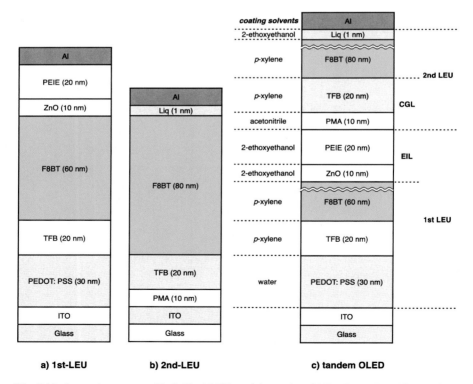

Fig. 7.12 Layered sequence of individual LEUs and the tandem OLED. Reproduced from ref. [8] by permission of John Wiley & Sons Ltd

work function of the cathode. PEIE along with ZnO nanoparticles in the EIL of first LEU and F8BT as the emitting polymer were used. The PEIE and ZnO nanoparticles can be coated as EILs onto the F8BT layer from 2-ethoxyethanol solution and dispersion, respectively. Furthermore, it is required that these EILs not be soluble in the solvent used for coating subsequent layers, which include layers of p-type electron-accepting and hole-transporting materials (HTL) in second LEU, consisting of the CGL. The solvent used for coating the electron acceptor cannot be water or alcohols, both of which dissolve PEIE.

Therefore, we chose phosphomolybdic acid hydrate $(MoO_3)_{12} \cdot H_3PO_4 \cdot (H_2O)_x$ (PMA) [67] as the electron acceptor of the CGL and acetonitrile as the solvent; PMA is soluble in acetonitrile, whereas PEIE, ZnO, and F8BT are not. Furthermore, TFB is chosen as the HTL of second LEU. For coating, a solution of TFB is prepared in p-xylene because p-xylene does not dissolve PMA, PEIE, or ZnO. However, F8BT is soluble in p-xylene; therefore, we carefully monitored the tendency of ZnO and/or PEIE layers to resist the dissolution of underlying F8BT in p-xylene using AFM surface images and the intensity of UV–Vis absorption after rinsing with the solvent (Fig. 7.13). Rinsing PEIE-coated F8BT with p-xylene reduces the thickness of the underlying F8BT layer, as indicated by the reduction in the intensity of UV–Vis absorption (Fig. 7.13a). The loss of F8BT is attributed to the nonuniform deposition of the PEIE layer (from its solution in 2-ethoxyethanol) on the F8BT layer due to the large difference in the surface energies of the compounds. Consequently, the roughness of the PEIE (0.67 nm) allows p-xylene to readily permeate into the F8BT layer. Similar dissolution of the F8BT underlayer on rinsing ZnO-coated F8BT is observed (Fig. 7.13b). It is likely that the roughness of the ZnO layer (2.19 nm) permits p-xylene to permeate the layer and dissolve F8BT. It must be noted here that both ZnO and PEIE are not soluble in p-xylene. To ensure the formation of a uniform layer of PEIE, we first deposited a layer of ZnO nanoparticles onto F8BT using a dispersion of the same in 2-ethoxyethanol. Once the ZnO layer is dried, it is not able to be re-dispersed into 2-ethoxyethanol. Subsequently, PEIE (in 2-ethoxyethanol) is uniformly coated onto the ZnO layer. The surface roughness of the F8BT/ZnO/PEIE layer (30 nm) is much smaller than that observed for F8BT/ZnO or F8BT/PEIE. The absence of any change in the intensity of UV–Vis absorption on rinsing F8BT/ZnO/PEIE with p-xylene (Fig. 7.13c) clearly demonstrates that a uniformly coated PEIE layer can prevent p-xylene. When compared with the device containing EIL composed only of ZnO, the driving voltage of first-LEU device is lower, and the efficiency is higher, demonstrating the superior electron injection property of the ZnO/PEIE bilayer. The second-LEU device (Fig. 7.12b), fabricated with the electron acceptor PMA as HIL, showed a similar low driving voltage, indicating that holes and electrons were generated at the interface of electron acceptor PMA and electron donor TFB, and these two layers worked properly as solution-processed CGL. These devices corresponding to the first-LEU and the second-LEU were combined to fabricate a solution-processed tandem OLED (Fig. 7.12c), which consists of nine layers except electrodes.

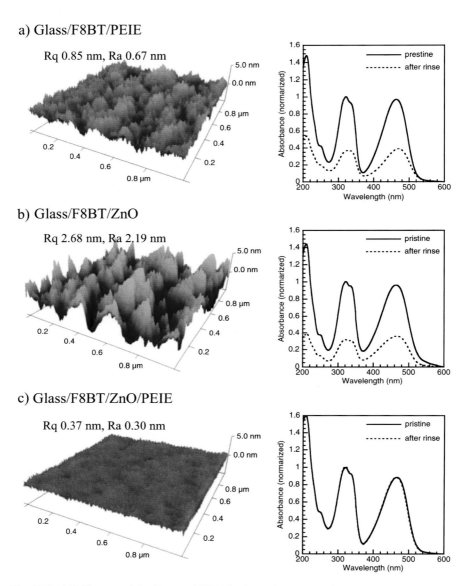

Fig. 7.13 AFM images of the films, and UV–Vis absorption spectra of the pristine and *p*-xylene-rinsed films: (**a**) Glass/F8BT (80 nm)/PEIE (20 nm), (**b**) Glass/F8BT (80 nm)/ZnO (10 nm), and (**c**) Glass/F8BT (80 nm)/ZnO (10 nm)/PEIE (20 nm). The Rq is square surface roughness, and the Ra is average surface roughness. Reproduced from ref. [8] by permission of John Wiley & Sons Ltd

The insolubility of each layer in the solvent used for coating the subsequent layer is carefully monitored, allowing for the successful stacking of the nine layers on the ITO substrate by a series of solution processes. Finally, Al is deposited in vacuum,

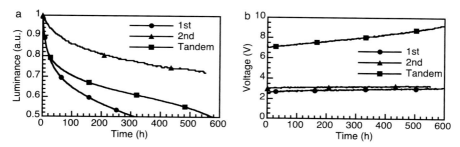

Fig. 7.14 Lifetime of the devices at the same current density (7.5 mA cm^{-2}). Initial luminances were 800 cd m^{-2} for the first LEU, 300 cd m^{-2} for the second LEU, and 1200 cd m^{-2} for the tandem OLED. Reproduced from ref. [8] by permission of John Wiley & Sons Ltd

and solution processing of the electrodes remains a formidable challenge. The driving voltage of the tandem OLED is nearly equal to the sum of the driving voltages of individual LEU devices at low current density. The driving voltage of the tandem OLED increases gradually. In tandem OLEDs, the two LEUs are connected in series; therefore, the current in each LEU must be same, whereas the voltage must be the sum of the voltages applied across each LEU. This additive driving voltage in the tandem OLED demonstrates that the interfaces in the device do not offer large resistance to increase voltage, and in particular, the electrons accepted by PMA from TFB in the CGL are smoothly injected into the F8BT of first LEU through the electron injecting of ZnO/PEIE bilayers. Investigation into the stability of the device at the same current density (Fig. 7.14) shows that the absolute value of the device lifetime is strongly dependent on the nature of the materials used. However, a distinct difference between the single LEU device and the tandem OLED is observed. The tandem OLED and first LEU show similar degradation tendencies in terms of the drop in luminance with increasing voltage, although the luminance of the tandem OLED is twofold higher than that of first-LEU. This observation shows that the solution-based process used for stacking the two LEUs does contribute to device instability; the advantage of the tandem OLED manifests in the form of high luminance for the lifetime of the device. The stability of second LEU extends to periods much longer than that observed in case of the tandem OLED or first LEU. The luminance of second LEU, however, is much lower than that of the tandem OLED due to lower efficiency. One clear reason for the longer lifetime of second LEU is the absence of the PEDOT:PSS layer, recognized widely as being unstable. Another possible reason can be the ZnO/PEIE bilayer, which is found in the tandem OLED and the other individual LEUs, in common, that show relatively short lifetimes.

7.5 Development of High-Efficiency and Stable Perovskite Quantum Dot LEDs

7.5.1 Overview and Issues of Perovskite Quantum Dots (QDs)

Lead halide perovskite semiconductors (APbX$_3$, A = Cs$^+$, CH$_3$NH$_3^+$, CH(NH$_2$)$_2^+$, X = Cl$^-$, Br$^-$, I$^-$) have been used in various optoelectronic applications, such as solar cells [68, 69], LED [70–72], and lasers [73], because of their bandgap (E_g) tunability, high defect tolerance, large absorption coefficient, and long carrier diffusion length [74–76]. Furthermore, perovskite QDs with a high quantum confinement effect exhibit higher PLQY than bulk perovskite owing to their large exciton binding energy [77–79]. The emission wavelength of perovskite QDs can be controlled by the chemical composition ratio and quantum size effects. However, perovskite QDs easily form surface defects during purification and deposition, significantly decreasing the PLQY [80]. To passivate the surface defects of perovskite QDs and to realize highly efficient perovskite QD-LEDs, several strategies, such as surface modification processes [81–85] and composition engineering [86–88], have been reported. Moreover, it has been reported that fast energy transfer between nanocrystals can reduce non-radiative recombination in perovskite films [89–94]. This energy transfer system between perovskite nanocrystals was first reported by Weerd et al. [90].

The energy transfer between perovskite QDs is strongly dependent on the inner distance between perovskite QDs and the spectral overlap between the emission of large-E_g QDs and the absorption of narrow-E_g QDs, which can be explained by Förster resonance energy transfer (FRET). Typically, the E_g of QDs can be controlled by halide anion exchange, such as the bromide anion exchange in CsPbI$_3$–CsPb(Br/I)$_3$. However, the ionic features of PQDs with different composition ratios easily cause ion exchange and E_g shifts following the mixing of PQDs [95]. Therefore, PQDs with identical chemical compositions and different particle sizes can be mixed to facilitate energy transfer without E_g shift. Enomoto et al. reported FRET from small CsPbBr$_3$ QDs with blue emission to large CsPbBr$_3$ QDs with green emission [96]. FRET can improve the PLQY of green-emitting CsPbBr$_3$ owing to the suppression of concentration quenching. However, studies on the effects of FRET on the performance of PQD-LEDs are lacking. To bridge this knowledge gap, the study of efficient energy transfer between PQDs with different E_g values is required, which can potentially improve the PLQY of the film and device performance.

In this study, we focused on the energy transfer between PQDs with different E_g values to improve the PLQY and device performance [97]. To observe the energy transfer, CsPbI$_3$ QDs with different E_g were prepared by controlling the size of large QDs (LQD) 10.7 nm and small QDs (SQD) 7.9 nm. The E_g of LQD and SQD were 1.81 eV and 1.85 eV, respectively, resulting in a high spectral overlap between SQD emission and LQD absorption. To confirm the energy transfer between the QDs, mixed QDs (MQD) were prepared by mixing SQDs and LQDs (SQD:LQD = 4:6 v/

v). The MQD film enhanced LQD emission and exhibited a higher PLQY (52%) with a longer PL decay time (7.4 ns) compared to those exhibited by the LQD film (38% and 6.2 ns). Energy transfer was determined to be FRET by photoluminescence excitation (PLE) and PL decay times. Furthermore, the EQE of MQD based-LED was higher than that of LQD based-LED (EQE of 15%) because of the efficient FRET.

7.5.2 Synthesis of Red Perovskite QDs

The size-controlled $CsPbI_3$ QDs were synthesized using the hot injection and reprecipitation purification methods (Fig. 7.15). Cesium carbonate and oleic acid were vacuum dried in 1-Octadecen (ODE) in a three-neck flask at room temperature (~25 °C) and then dissolved at 150 °C to prepare Cs-oleate. PbI_2, ZnI_2, HI, and $(CH_3COO)_2Mn \cdot 4H_2O$ were then added to ODE, OA, and OAm in a three-neck flask and dissolved at 120 °C under a vacuum. Subsequently, the temperature of the synthesized Cs-oleate was increased to 150 °C, followed by a quick injection of Cs-oleate solution into the precursor mixture at 190 °C in an N_2 atmosphere. After a few seconds, the reaction mixture was rapidly cooled in an ice water bath to room temperature, and synthesized $CsPbI_3$ QDs were collected and stored.

The synthesized $CsPbI_3$ QDs were centrifuged to separate LQD precipitates and SQD supernatant, as shown in Fig. 7.16. In the SQD supernatant, methyl acetate was added and centrifuged to collect the precipitate. The precipitate was then dispersed in a SQD toluene dispersion. In the LQD precipitates, the toluene was added and then centrifuged to collect LQD toluene dispersion. After each toluene dispersion was prepared, methyl acetate was added to the toluene dispersion and centrifuged to collect the precipitate. This process was repeated twice. Finally, the $CsPbI_3$ QDs were dispersed again in octane and centrifuged to remove the undispersed $CsPbI_3$ QDs. Thereafter, $CsPbI_3$ QDs were purified using the reprecipitation method, and methyl acetate was used to remove impurities such as insulating ligands and the reaction solvent octadecene.

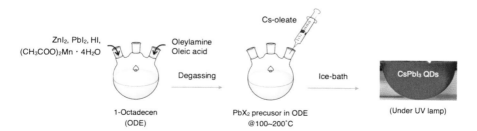

Fig. 7.15 Schematic of hot injection synthesis method

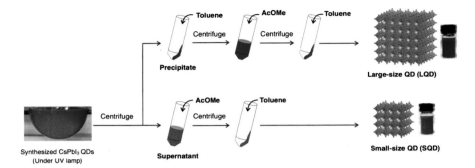

Fig. 7.16 Size separation and purification using reprecipitation method

7.5.3 Optical Properties of Perovskite QDs with Different Size

The size of each QDs was measured using transmission electron microscopy (TEM) (Fig. 7.17a–b). The average size was 10.7 nm for the LQD and 7.9 m for SQD, respectively. Figure 7.17c shows the X-ray diffraction (XRD) patterns of each QD. The LQD and SQD were attributed to the cubic phase of $CsPbI_3$ (ICSD 181288). Figure 7.17d shows the dependence of the PL and UV-vis absorption spectra on QD size (0.01 mg/mL, solvent, toluene). The PL peak wavelength of the LQD (PL peak at 665 nm) was red-shifted from that of the SQD (PL peak at 653 nm). The optical E_g was expanded by reducing the QD size owing to the quantum size effects. Furthermore, we analyzed the composition ratio of the $CsPbI_3$ QDs by X-ray photoelectron spectroscopy (XPS), as shown in Fig. 7.17e–f. The Pb:I ratios were 1:3.3 and 1:3.1 for the SQD and LQD, respectively, indicating almost no halogen defects.

7.5.4 Energy Transfer Between Perovskite QDs

The effect of energy transfer on the optical properties using MQD was investigated (SQD:LQD = 4:6 v/v). To confirm the dependence of the inner distance between QDs, dilute solutions, and films were prepared. Figure 7.18a–b show the PL spectra of the LQD, SQD, and MQD. The PL peak and FWHM of the MQD film were significantly similar to those of the neat LQD film (Table 7.1). On the other hand, the PL spectra of the MQD dilute solution show a blue shift as the FWHM broadens compared to that of the LQD because of the emission of both SQD and LQD. The PL decay was analyzed to further determine the energy transfer dynamics between the QDs, as shown in Fig. 7.18c–d. The PL decay of the MQD dilute solution was similar to that of the LQD and SQD. In contrast, the PL decay time of the MQD film was longer than those of the SQD and LQD films. Furthermore, the radiative recombination (k_r) and non-radiative recombination (k_{nr}) of the PQD from the PL

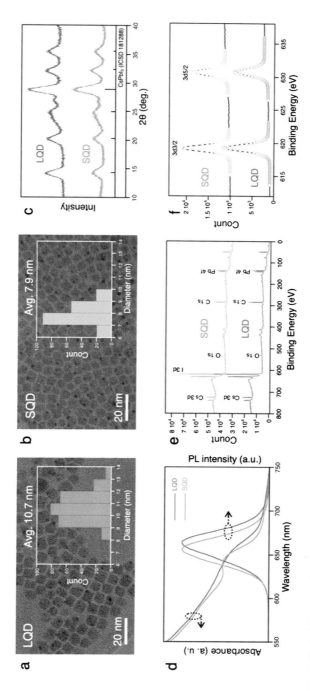

Fig. 7.17 TEM images of (**a**) LQD and (**b**) SQD (inset: size histogram of QDs). (**c**) XRD patterns of LQD and SQD films. (**d**) Normalized PL spectra (excitation wavelength: 400 nm) and UV-vis absorption spectra of LQD and SQD dilute solution. XPS spectra of LQD and SQD films: (**e**) wide ranges from 0 eV to 800 eV and (**f**) I 3*d*. Reproduced from ref. [97] by Copyright © 2022, American Chemical Society

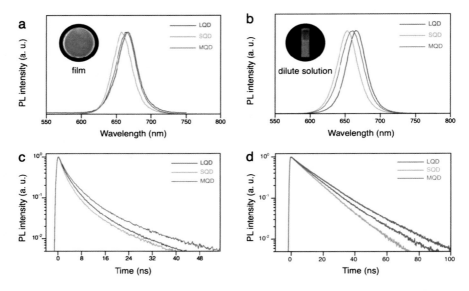

Fig. 7.18 PL spectra (excitation wavelength: 400 nm), PL decay (detection wavelength, 665 nm; excitation wavelength, 407 nm) of LQD, SQD, and MQD with SQD:LQD = 4:6 v/v (**a**, **c**) the films (**b**, **d**) diluted solution. Reproduced from ref. [97] by Copyright © 2022, American Chemical Society

Table 7.1 Summary of optical properties of LQD, SQD, and MQD (dilute solution/film). The solution concentration is 0.01 mg/mL

	PL peak (nm)	FWHM (nm)	PLQY (%)	$<\tau>$ (ns) at 665 nm	$k_r \times 10^7$ (/s)	$k_{nr} \times 10^7$ (/s)
LQD	665/667	33/33	73/38	18.0/6.1	4.0/6.2	1.5/10.2
SQD	653/658	34/32	70/29	13.0/5.7	5.6/5.1	2.4/12.5
MQD	661/665	38/34	72/52	15.6/7.4	4.6/7.0	1.8/6.5

decay time and PLQY were calculated (Table 7.1). The MQD film exhibited a lower k_{nr} than the LQD and SQD films owing to the energy transfer, achieving a higher PLQY of the MQD film.

UV-vis absorption and PLE measurements were performed to directly observe the energy transfer between the QDs, as shown in Fig. 7.19. The absorption peaks in both the film and dilute solution were observed at 660 nm for LQD and 645 nm for SQD. The absorption spectra of the MQD included peaks of both LQD and SQD. The PLE spectra of the film and the dilute solution were measured at a detection wavelength of 690 nm. In the PLE spectra of the MQD film, PLE peaks at 645 nm and 660 nm were observed, which correspond to the absorption of SQD and LQD. In contrast, the PLE peak of SQD at 645 nm was not observed in the dilute MQD solution because of the suppression of energy transfer from the SQDs to the LQDs. These results indicate that the energy transfer in the PQDs strongly depends on the inner distance between the QDs and confirm the presence of FRET in the MQD film.

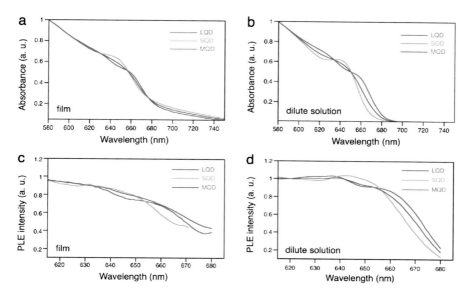

Fig. 7.19 UV-vis absorption spectra of LQD, SQD, and MQD in (**a**) the films and (**b**) dilute solution. PLE spectra of LQD, SQD, and MQD in (**c**) the films and (**d**) dilute solution (detection wavelength of 690 nm). Reproduced from ref. [97] by Copyright © 2022, American Chemical Society

7.5.5 Development of Red Perovskite QD-LEDs

The PQD-LEDs were fabricated using the following structure: ITO (130 nm)/ PEDOT:PSS with Nafion (40 nm)/poly-TPD (20 nm)/CsPbI$_3$ QDs/TPBi (50 nm)/ Liq (1 nm)/Al (100 nm). The corresponding energy diagram of the LED is shown in Fig. 7.20a. The energy diagram of CsPbI$_3$ QDs was estimated using ultraviolet photoelectron spectroscopy (UPS) and optical E_g. The electroluminescence (EL) peak was 673 nm for the LQD and 671 nm for the MQD-based LEDs, as shown in Fig. 7.20b and Table 7.2. The MQD-based LED shows high color purity with Commission Internationale de l'éclairage (CIE) coordinates of (0.72, 0.28), which can cover the BT. 2020 color gamut (Fig. 7.20c). The device performances of the LQD- and MQD-based LEDs are presented in Fig. 7.20d–f and Table 7.2. The MQD-LED exhibited a maximum luminance (L_{max}) of 703 cd m^{-2} and peak EQE of 15%, which were higher than those of the LQD-LED (L_{max} of 645 cd m^{-2} and peak EQE of 10%). Furthermore, the operational device stability of the LEDs was measured at a constant current density of 7.5 mA cm^{-2}, as shown in Fig. 7.21. The MQD-LEDs achieved an operational lifetime (LT$_{50}$) of 5 h at an initial luminance of 100 cd m^{-2}. These results suggest that FRET helps suppress non-radiative recombination, which is one of the key factors influencing the improvement in the device performance of CsPbI$_3$ QD-LEDs.

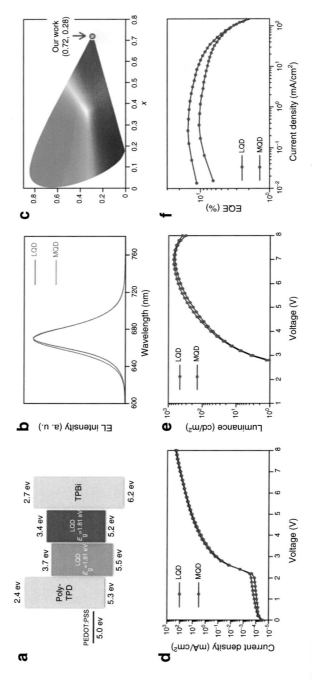

Fig. 7.20 Device characteristics of PQD-LEDs. (**a**) Energy diagram of the PQD-LED, (**b**) EL spectra at 25 mA cm^{-2}, (**c**) CIE color coordinates, (**d**) current density–voltage characteristics, (**e**) luminance–voltage characteristics, (**f**) EQE–current density characteristics. Reproduced from ref. [97] by Copyright © 2022, American Chemical Society

7 Solution-Processed Organic LEDs and Perovskite LEDs

Table 7.2 Summary of device characteristics of PQD-LEDs

	Von (V) 1 cd m^{-2}/ 100 cd m^{-2}	L_{max} (cd m^{-2})	P.E. (lm/W) 1 cd m^{-2}/Max EQE	C.E. (cd/A) 1 cd m^{-2}/Max EQE	EQE (%) 1 cd m^{-2}/Max EQE
LQD	2.8/4.5	645	1.80/1.89	1.60/1.93	8.67/10.44
MQD	2.7/4.3	703	2.82/2.86	2.46/2.91	12.75/15.05

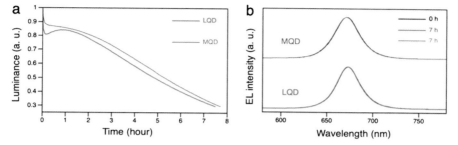

Fig. 7.21 (**a**) Operational lifetimes and of PQD-LEDs at a constant current density of 7.5 mA cm^{-2}. (**b**) Current density dependence of EL spectra. Reproduced from ref. [97] by Copyright © 2022, American Chemical Society

7.6 Development of Low-Oxidation Tin-Based Perovskite LEDs

7.6.1 Overview and Issues of Lead-Free Perovskite

Lead halide perovskite compounds of APbX$_3$ have excellent photoelectric properties and contain lead cations with severe toxicity [98–100]. Several lead-free perovskite compounds have been developed by controlling their chemical compositions and structures [101–108]. The compositional engineering has been used to replace Pb^{2+} with divalent cations of tin (Sn^{2+}) or germanium (Ge^{2+}). In structural control, low-dimensional perovskite nanocrystals, such as two-dimensional (2D) structures, can be formed by replacing the A-site cation with an organic molecule (phenylethylammonium cation (PEA)) [102, 109–113]. The 2D structure allows for a high PLQY and E_g control owing to the quantum confinement effect. In particular, the 2D tin-based perovskite PEA$_2$SnI$_4$ is a promising field-effect transistor material because of its superior carrier mobility and solution processability [114]. In addition, the PEA$_2$SnI$_4$ film showed red emission (PL peak of 620 nm) with high color purity (FWHM ~30 nm), which can meet the CIE color coordinates Rec. 2020 standard [101, 115, 116]. However, PEA$_2$SnI$_4$ emitters exhibit lower PL intensity than lead perovskite emitters because of crystal instability and high trap density. These issues are mainly caused by the oxidation of Sn^{2+} (Sn^{2+} to Sn^{4+}) owing to thermodynamic instability [117].

To suppress tin oxidation, previous studies have used antioxidant precursors such as Sn powder [118], SnF_2 [119, 120], and antioxidants such as sulfonates [121], valeric acid [122], and gallic acid [123]. In addition, it has been reported that the oxidation of Sn^{2+} is induced in precursor solvents. The common solvent DMSO shows high solubility for precursors such as SnX_2, while the oxygen atoms in DMSO promote the oxidation of Sn^{2+} [124]. Several solvents have been reported to achieve precursor solubility and inhibit oxidation [125, 126]. However, these candidate solvents have high boiling point, which causes compound decomposition and defect formation during high-temperature annealing.

On the other hand, crystallization of perovskite microcrystals via mechanochemistry achieves higher solubility than that of metal halide precursors [127–130]. Prochowicz et al. reported the pre-crystallization of $CH_3NH_3PbI_3$ powder to prepare precursors and found that the produced films achieved a low defect density [130]. Liao et al. prepared the precursor of nanocrystals by reprecipitation and enhanced the PLQY from 52% to 79% [127]. Thus, the pre-crystallization method for perovskite microcrystals is a promising strategy for achieving high solubility and low defect density.

Herein, we demonstrate two-step crystallization via microcrystal precursors to obtain PEA_2SnI_4 perovskite films with low Sn^{2+} oxidation and trap density [131]. PEA_2SnI_4 microcrystals were synthesized by conventional RP and were highly soluble in N,N-dimethylformamide (DMF). By spin coating the PEA_2SnI_4 precursor solution, we obtained a PEA_2SnI_4 film with a lower Sn^{4+} ratio and higher PLQY than those obtained by direct spin coating. Moreover, two-step PEA_2SnI_4-based LEDs improved device performance and proposed new strategies for realizing highly efficient lead-free perovskite LEDs.

7.6.2 Development of Tin-Based Perovskite by Two-Step Crystallization

PEA_2SnI_4 perovskites with lower oxide and trap densities were prepared by two-step crystallization, as shown in Fig. 7.22. Firstly, PEA_2SnI_4 microcrystals were synthesized by reprecipitation where γ-butyrolactone (GBL) solution containing PEAI and tin(II) iodide (SnI_2) was injected into dichloromethane (DCM), and the precipitate was collected by centrifugation. After vacuum-drying, the resulting powder exhibited red emission under UV irradiation, as shown in Fig. 7.22. Secondly, obtained PEA_2SnI_4 microcrystal was redissolved into the good solvent of DMF to prepare the precursor solution. The PEA_2SnI_4 precursor solution was then spin-coated to form PEA_2SnI_4 film (two-step film). To further investigate the effect of the two-step crystallization, the control PEA_2SnI_4 films were fabricated using a direct spin-coating method. The precursor solution of the control film was prepared by dissolving PEAI and SnI_2 in a DMSO mixture, as shown in Fig. 7.22. Moreover, each spin-coated film exhibited red emission, as well as PEA_2SnI_4 microcrystals.

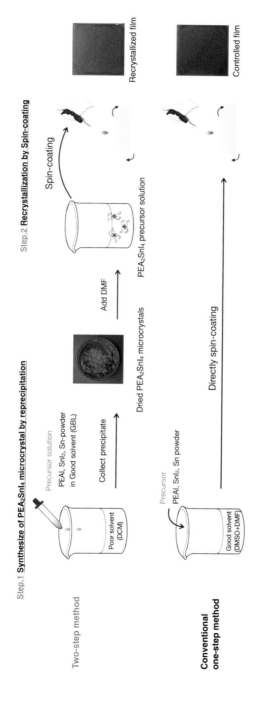

Fig. 7.22 Scheme of two-step and conventional approach preparation of PEA$_2$SnI$_4$ thin film

7.6.3 Crystal Structure and Chemical Composition of Tin-Based Perovskite

Figure 7.23a–c show X-ray diffraction (XRD) spectra and the grazing incidence wide-angle X-ray scattering (GIWAXS) diffraction patterns of each film. The strong (0 0 l) diffraction peaks, attributed to the 2D PEA$_2$PbI$_4$, were observed in the XRD spectra of two-step and control films. In addition, strong spots were obtained in the GAIWAXS patterns of both films, indicating a highly oriented 2D structure. The composition ratio of PEA$_2$SnI$_4$ films was analyzed by X-ray photoelectron spectroscopy (XPS) measurement, as shown in Fig. 7.23d–f. To confirm the Sn ratio, the Sn 3d core level was measured, which included the Sn^{2+} peaks at 495 eV and 486 eV and Sn^{4+} peaks at 496 eV and 488 eV (Fig. 7.23d). The Sn^{4+} of the two-step film (Sn^{4+}/(Sn^{4+}+Sn^{2+}) = 0.02) is significantly low as compared with control film (Sn^{4+}/(Sn^{4+}+Sn^{2+}) = 0.1). The suppression of the Sn^{4+} ratio by the two-step crystallization method can be attributed to the following factors: (1) The nonuse of oxidation agent, DMOS, as precursor solvent, and (2) the removal of Sn^{4+} impurities in the precursor by forming PEA$_2$SnI$_4$ microcrystals. Furthermore, the tin oxidation in PEA$_2$SnI$_4$ forms PEA$^+$ vacancies and increase I/(Sn^{4+}+Sn^{2+}) ratio, owing to the loss of two electrons. I/(Sn^{4+}+Sn^{2+}) ratio was also studied to further confirm the defects of PEA$_2$SnI$_4$ films (Fig. 7.23e). The determined I/(Sn^{4+}+Sn^{2+}) ratios of the two-step film (I/(Sn^{4+}+Sn^{2+}) = 4.22) were lower than that of the control film (I/(Sn^{4+}+Sn^{2+}) = 4.39). These results indicated that the two-step method can reduce the oxidation of Sn in PEA$_2$SnI$_4$ and suppress defect formation.

7.6.4 Optical Properties of Tin-Based Perovskite

To investigate the dependence of different precursors, UV–vis absorption and PL spectroscopy were performed for the two-step and control films, as shown in Fig. 7.24a–b. The UV–vis absorption peak attributed to 2D PEA$_2$SnI$_4$ was observed at ~610 nm. The PL spectrum of each film also exhibited nearly identical peaks (wavelength: 618–619 nm). The FWHM of the two-step film (26 nm) exhibited a narrower emission than that of the control film (30 nm). The PLQY exhibited 1.3% for the control films and 1.9% for the two-step films. Furthermore, PL decay for each PEA$_2$SnI$_4$ film was also studied. The PL decay curves were fitted with a multi-exponential function, as shown in Fig. 7.24c. The two-step film exhibited a longer PL decay lifetime than that of the control film, suggesting suppression of the non-radiative pathway.

To further investigate the non-radiative pathway, the temperature dependence of the PL spectrum from 300 K to 84 K in an N$_2$-filled environment was also studied, as shown in Fig. 7.25a–b. In a previous study, the dark state of the PEA$_2$SnI$_4$ film was observed below 100 K [132]. The PL peak of the two-step PEA$_2$SnI$_4$ film red-shifted with decreasing temperature. In addition, an additional peak at 650 nm appeared

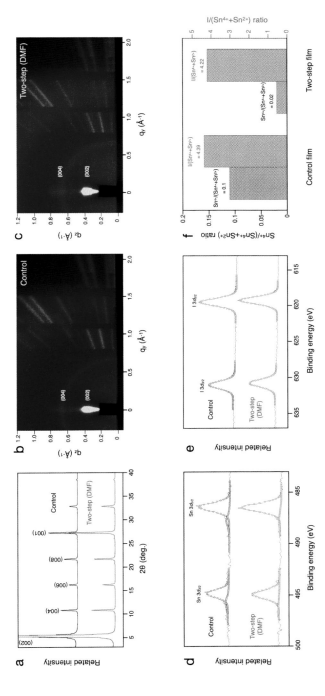

Fig. 7.23 Crystal structure analysis of PEA_2SnI_4 films: (**a**) XRD and GAIWAXS of (**b**) control- and (**c**) tow-step films. Chemical composition analysis: XPS measurement of (**d**) Sn 3*d* and (**e**) I 3*d*. (**f**) Composition ratio of $Sn^{4+}/(Sn^{4+}+Sn^{2+})$ and $I/(Sn^{4+}+Sn^{2+})$ for each film. Reproduced from ref. [131] by Copyright © 2022, American Chemical Society

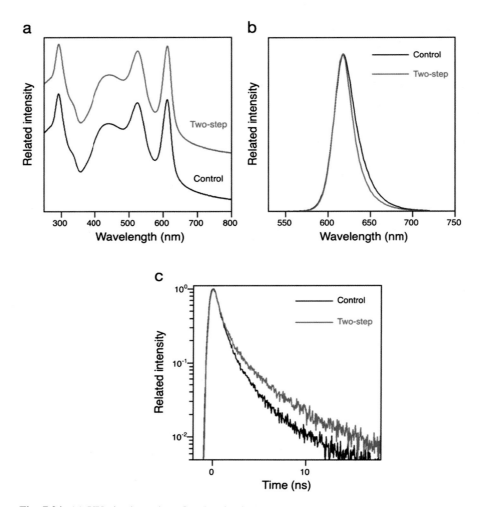

Fig. 7.24 (**a**) UV-vis absorption, (**b**) photoluminescence spectra, (**c**) photoluminescence decay spectra of films prepared from different precursors. Reproduced from ref. [131] by Copyright © 2022, American Chemical Society

within the lower-temperature region (less than 125 K) because of the formation of biexcitons. Dark excitons are more likely to be observed in lower-lying states below the emissive band edge, although they are blinded to the thermalized excitons. The two-step PEA_2SnI_4 film exhibited a higher PL intensity than the control film at 84 K, confirming the significant suppression of non-radiative recombination (Fig. 7.25b). Furthermore, the trap density of the PEA_2SnI_4 film by space charge limited current (SCLC) fitting using a hole-only device (HOD) was also estimated with the following structure: ITO (130 nm)/PEDOT:PSS (40 nm)/polyethylenimine ethoxylated (PEIE)/PEA_2SnI_4 films (~40 nm)/N,N'-diphenylbenzidine (NPD) (30 nm)/MoO_3

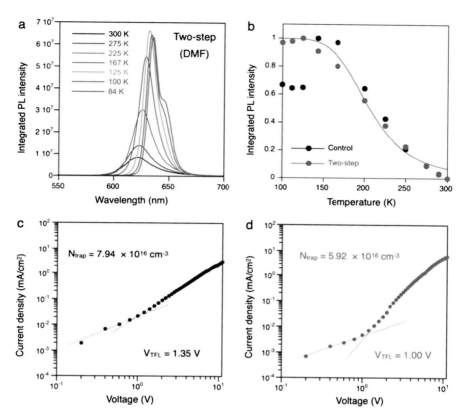

Fig. 7.25 (**a**) Temperature dependence of PL spectra of two-step PEA$_2$SnI$_4$ films. (**b**) Temperature to integrated PL spectra and SCLC of (**c**) control- and (**d**) two-step PEA$_2$SnI$_4$ films. Reproduced from ref. [131] by Copyright © 2022, American Chemical Society

(5 nm)/Al (100 nm). Figure 7.25c–d show the current density-voltage characteristics of the HOD and SCLC fitting. The trap-filling voltage (V_{TFL}) was 1.35 V for the control film and 1.00 V for the two-step film. The trap-state density (N_{trap}) was estimated using Eq. (7.1)

$$N_{trap} = \frac{2\varepsilon_0\varepsilon_r V_{TFL}}{ed^2} \qquad (7.1)$$

where ε_r is the relative permittivity, ε_0 is the vacuum permittivity, e is the elementary charge of the electron, and d is the thickness between the cathode and the anode. The estimated N_{trap} were 7.94×10^{16} cm^{-3} for the control film and 5.92×10^{16} cm^{-3} for the two-step film. These results suggest that the two-step crystallization method achieved a low defect density because of the reduction in the tin oxidation ratio (Sn^{4+}).

7.6.5 Development of Tin-Based Perovskite LED

PEA_2SnI_4-based LEDs were fabricated with the following structure: ITO (130 nm)/ PEDOT:PSS (40 nm)/PEIE/PEA_2SnI_4 perovskite (40 nm)/TPBi (50 nm)/Liq (1 nm)/ Al (100 nm). Figure 7.26a shows the energy diagram of the PEA_2SnI_4-based LEDs. The valence and conduction bands of the PEA_2SnI_4 film were obtained using UPS and optical E_g. The EL peak of each LEDs was 627 nm (FWHM of 30 nm) for the control LEDs and 625 nm (FWHM of 28 nm) for the two-step LEDs. Figure 7.26d–e show the current density-voltage and luminance-voltage characteristics of LEDs. The two-step LEDs achieved a lower turn-on voltage (2.9 V) and higher luminance (43 cd m^{-2}) than those of the control-based LEDs (3.5 V and 35 cd m^{-2}). Furthermore, the peak EQE of the two-step-LEDs achieved a higher EQE of 0.4% (Avg. EQE of 0.2%) than control-LEDs (peak EQE of 0.13%, Avg. EQE of 0.1%). The device performance was enhanced by suppressing the oxidation of tin and its defects. Furthermore, the operational device stability was measured at a constant current density of 1.25 mA cm^{-2}, as shown in Fig. 7.26f. The two-step LEDs achieved nearly twice the operational device lifetime (LT_{50} = 11 min) compared with that of the control LEDs (LT_{50} = 6 min).

7.7 Summary

In summary, solution-processed OLEDs have promising results of large area processing and low fabrication cost for lighting application. Several approaches have been studied to achieve efficiencies of solution-processed OLEDs. In this chapter, we discussed our recent works: (1) solution processable materials and (2) solution-processed multilayer structure. Fluorescent oligomers and phosphorescent dendrimers are synthesized as a solution processable light-emitting dye having well-defined structures. Liq, Cs_2CO_3, and ZnO nanoparticles are also studied as a solution-processed electron injection material. In addition, series of (vinylphenyl)pyridine-based polymer binders, PVPh2Py, PVPh3Py, and PVPh4Py, are synthesized for thick layer that can be mass-produced for large-area coating using solution processing. In the solution-processed multilayer white phosphorescent OLEDs using small molecules, we achieved high power efficiencies of 34 lm W^{-1} at 100 cd m^{-2} with stable electroluminescence spectra under varying current density. In addition, tandem OLED consisting of two LEUs and a CGL between the anode and the cathode is fabricated using only solution-based processes. Appropriate choice of solvents during spin coating of each layer ensures that a nine-layered structure is readily fabricated using only solution-based processes. The determined driving voltage and efficiency of the fabricated tandem OLED are the sums of values of the individual LEUs. These results indicate that the CGL formed by the solution-based process successfully generates electrons and holes under applied voltage. The formed electrons are efficiently injected into the first LEU

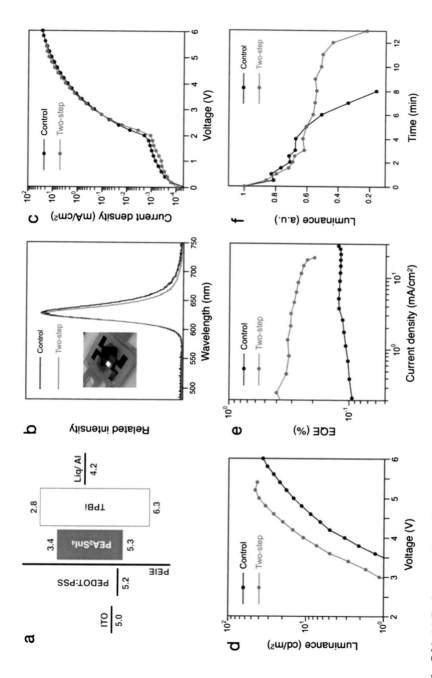

Fig. 7.26 (**a**) Device energy level diagram, (**b**) normalized EL spectra, (**c**) current density–voltage, (**d**) luminance–voltage, (**e**) EQE–current density characteristics, and (**f**) operational lifetimes at a constant current density of 1.25 mA cm^{-2}. Reproduced from ref. [131] by Copyright © 2022, American Chemical Society

through the ZnO/PEIE bilayer, and the holes are injected into the second LEU. The successful fabrication of solution-processed white phosphorescent OLEDs and tandem OLEDs will pave the way toward printable, low-cost, and large-area white light sources. In perovskite LEDs, we proposed a strategy to achieve efficient perovskite QD-LEDs and lead-free perovskite materials. To improve the PLQY and device performance, we demonstrated fast energy transfer from the larger-E_g QD to the narrower-E_g QD through FRET. The energy transfer between the QDs exhibited a higher PLQY and longer PL decay time than those of neat LQD films. The MQD−LED achieved a higher EQE of 15% than the LQD−LED. Moreover, we developed a lead-free PEA$_2$SnI$_4$ perovskite film with a low Sn^{2+} oxidation and trap density. To suppress tin oxidation (Sn^{2+} to Sn^{4+}), we demonstrated a two-step approach using the reprecipitation method. The obtained PEA$_2$SnI$_4$ film by the two-step recrystallization method exhibited a low Sn^{4+}/(Sn^{2+} + Sn^{4+}) ratio owing to the removal of Sn^{4+} impurities and nonuse of the oxidation agent, DMOS. These results indicate that the two-step PEA$_2$SnI$_4$-based LEDs improved the luminance and EQE and proposed new strategies for realizing efficient lead-free perovskite LEDs.

References

1. Y.-J. Pu, M. Higashidate, K. Nakayama, J. Kido, J. Mater. Chem. **18**, 4183 (2008)
2. N. Iguchi, Y.-J. Pu, K.I. Nakayama, M. Yokoyama, J. Kido, Org. Electron. **10**, 465 (2009)
3. Y.-J. Pu, N. Iguchi, N. Aizawa, H. Sasabe, K. Nakayama, J. Kido, Org. Electron. **12**, 2103 (2011)
4. T. Chiba, Y.-J. Pu, M. Hirasawa, A. Masuhara, H. Sasabe, J. Kido, ACS Appl. Mater. Interfaces **4**, 6104 (2012)
5. T. Chiba, Y.-J. Pu, S. Takahashi, H. Sasabe, J. Kido, Adv. Funct. Mater. **24**, 6038 (2014)
6. N. Aizawa, Y.-J. Pu, M. Watanabe, T. Chiba, K. Ideta, N. Toyota, M. Igarashi, Y. Suzuri, H. Sasabe, J. Kido, Nat. Commun. **5**, 5756 (2014)
7. T. Chiba, Y.-J. Pu, H. Sasabe, J. Kido, Y. Yang, J. Mater. Chem. **22**, 22769 (2012)
8. Y.J. Pu, T. Chiba, K. Ideta, S. Takahashi, N. Aizawa, T. Hikichi, J. Kido, Adv. Mater. **27**, 1327–1332 (2014)
9. J.H. Burroughes, D.D.C. Bradley, A.R. Brown, R.N. Marks, K. Mackay, R.H. Friend, P.L. Burn, A.B. Holmes, Nature **347**, 539 (1990)
10. P.L. Burn, S.C. Lo, I.D.W. Samuel, Adv. Mater. **19**, 1675 (2007)
11. S.C. Lo, P.L. Burn, Chem. Rev. **107**, 1097 (2007)
12. T. Sato, D.-L. Jiang, T. Aida, J. Am. Chem. Soc. **121**, 10658 (1999)
13. A.W. Freeman, S.C. Koene, P.R. Malenfant, M.E. Thompson, J.M. Fréchet, J. Am. Chem. Soc. **122**, 12385 (2000)
14. S. Gambino, S.G. Stevenson, K.A. Knights, P.L. Burn, I.D. Samuel, Adv. Funct. Mater. **19**, 317 (2009)
15. S.-C. Lo, R.E. Harding, C.P. Shipley, S.G. Stevenson, P.L. Burn, I.D. Samuel, J. Am. Chem. Soc. **131**, 16681 (2009)
16. J.J. Kim, Y. You, Y.-S. Park, J.-J. Kim, S.Y. Park, J. Mater. Chem. **19**, 8347 (2009)
17. V. Adamovich, J. Brooks, A. Tamayo, A.M. Alexander, P.I. Djurovich, B.W. D'Andrade, C. Adachi, S.R. Forrest, M.E. Thompson, New J. Chem. **26**, 1171 (2002)
18. B.W. DAndrade, S.R. Forrest, J. Appl. Phys. **94**, 3101 (2003)

19. Y. Kawamura, J. Brooks, J.J. Brown, H. Sasabe, C. Adachi, Phys. Rev. Lett. **96**, 017404 (2006)
20. J.S. Huang, G. Li, E. Wu, Q.F. Xu, Y. Yang, Adv. Mater. **18**, 114 (2006)
21. J. Huang, Z. Xu, Y. Yang, Adv. Funct. Mater. **17**, 1966 (2007)
22. J. Endo, T. Matsumoto, J. Kido, Jpn. J. Appl. Phys. **41**, L800 (2002)
23. Y.-J. Pu, M. Miyamoto, K.-i. Nakayama, T. Oyama, Y. Masaaki, J. Kido, Org. Electron. **10**, 228 (2009)
24. D. Kabra, M.H. Song, B. Wenger, R.H. Friend, H.J. Snaith, Adv. Mater. **20**, 3447 (2008)
25. H.J. Bolink, E. Coronado, J. Orozco, M. Sessolo, Adv. Mater. **21**, 79 (2009)
26. H.J. Bolink, H. Brine, E. Coronado, M. Sessolo, Adv. Mater. **22**, 2198 (2010)
27. D. Kabra, L.P. Lu, M.H. Song, H.J. Snaith, R.H. Friend, Adv. Mater. **22**, 3194 (2010)
28. B. Sun, H. Sirringhaus, Nano Lett. **5**, 2408 (2005)
29. L.P. Lu, D. Kabra, R.H. Friend, Adv. Funct. Mater. **22**, 4165 (2012)
30. B.R. Lee, E.D. Jung, J.S. Park, Y.S. Nam, S.H. Min, B.S. Kim, K.M. Lee, J.R. Jeong, R.H. Friend, J.S. Kim, S.O. Kim, M.H. Song, Nat. Commun. **5**, 4840 (2014)
31. H. Sasabe, D. Tanaka, D. Yokoyama, T. Chiba, Y.-J. Pu, K.-i. Nakayama, M. Yokoyama, J. Kido, Adv. Funct. Mater. **21**, 336 (2011)
32. C.W. Tang, S.A. Vanslyke, Appl. Phys. **51**, 913 (1987)
33. J. Kido, M. Kimura, K. Nagai, Science **267**, 1332 (1995)
34. S. Reineke, F. Lindner, G. Schwartz, N. Seidler, K. Walzer, B. Lussem, K. Leo, Nature **459**, 234 (2009)
35. P.K.H. Ho, J.S. Kim, J.H. Burroughes, H. Becker, S.F.Y. Li, T.M. Brown, F. Cacialli, R.H. Friend, Nature **404**, 481 (2000)
36. X.H. Yang, D.C. Muller, D. Neher, K. Meerholz, Adv. Mater. **18**, 948 (2006)
37. R.Q. Png, P.J. Chia, J.C. Tang, B. Liu, S. Sivaramakrishnan, M. Zhou, S.H. Khong, H.S.O. Chan, J.H. Burroughes, L.L. Chua, R.H. Friend, P.K.H. Ho, Nat. Mater. **9**, 152 (2010)
38. N. Aizawa, Y.-J. Pu, T. Chiba, S. Kawata, H. Sasabe, J. Kido, Adv. Mater. **26**, 7543 (2014)
39. S. Reineke, M. Thomschke, B. Lussem, K. Leo, Rev. Mod. Phys. **85**, 1245 (2013)
40. B.C. Krummacher, V.E. Choong, M.K. Mathai, S.A. Choulis, F. So, F. Jermann, T. Fiedler, M. Zachau, Appl. Phys. **88**, 113506 (2006)
41. F. Huang, P.I. Shih, C.F. Shu, Y. Chi, A.K.Y. Jen, Adv. Mater. **21**, 361 (2009)
42. B.H. Zhang, G.P. Tan, C.S. Lam, B. Yao, C.L. Ho, L.H. Liu, Z.Y. Xie, W.Y. Wong, J.Q. Ding, L.X. Wang, Adv. Mater. **24**, 1873 (2012)
43. T. Ikeda, H. Murata, Y. Kinoshita, J. Shike, Y. Ikeda, M. Kitano, Chem. Phys. Lett. **426**, 111 (2006)
44. H. Yamamoto, J. Brooks, M.S. Weaver, J.J. Brown, T. Murakami, H. Murata, Appl. Phys. **99**, 33301 (2011)
45. H.T. Nicolai, M. Kuik, G.A.H. Wetzelaer, B. de Boer, C. Campbell, C. Risko, J.L. Bredas, P.W.M. Blom, Nat. Mater. **11**, 882 (2012)
46. H. Sasabe, N. Toyota, H. Nakanishi, T. Ishizaka, Y.J. Pu, J. Kido, Adv. Mater. **24**, 3212 (2012)
47. I. Yamada, J. Matsuo, N. Toyoda, Nucl. Instrum. Methods Phys. Res. B **206**, 820 (2003)
48. J.H. Zou, H. Wu, C.S. Lam, C.D. Wang, J. Zhu, C.M. Zhong, S.J. Hu, C.L. Ho, G.J. Zhou, H.B. Wu, W.C.H. Choy, J.B. Peng, Y. Cao, W.Y. Wong, Adv. Mater. **23**, 2976 (2011)
49. S.Y. Shao, J.Q. Ding, L.X. Wang, X.B. Jing, F.S. Wang, J. Am. Chem. Soc. **134**, 20290 (2012)
50. R. Meerheim, K. Walzer, M. Pfeiffer, K. Leo, Appl. Phys. **89**, 061111 (2006)
51. T.N.T. Matsumoto, J. Endo, M. Mori, N. Kawamura, A. Yokoi, J. Kido, SID 03 Digest **27**, 979 (2003)
52. L.S. Liao, K.P. Klubek, C.W. Tang, Appl. Phys. **84**, 167 (2004)
53. T. Tsutsui, M. Terai, Appl. Phys. **84**, 440 (2004)
54. F.W. Guo, D.G. Ma, Appl. Phys. **87**, 173510 (2005)
55. H. Kanno, R.J. Holmes, Y. Sun, S. Kena-Cohen, S.R. Forrest, Adv. Mater. **18**, 339 (2006)

56. M.Y. Chan, S.L. Lai, K.M. Lau, M.K. Fung, C.S. Lee, S.T. Lee, Adv. Funct. Mater. **17**, 2509 (2007)
57. T.W. Lee, T. Noh, B.K. Choi, M.S. Kim, D.W. Shin, J. Kido, Appl. Phys. **92**, 043301 (2008)
58. T. Chiba, Y.-J. Pu, R. Miyazaki, K.-i. Nakayama, H. Sasabe, J. Kido, Org. Electron. **12**, 710 (2011)
59. M.H. Ho, T.M. Chen, P.C. Yeh, S.W. Hwang, C.H. Chen, Appl. Phys. **91**, 233507 (2007)
60. G. Sarasqueta, K.R. Choudhury, J. Subbiah, F. So, Adv. Funct. Mater. **21**, 167 (2011)
61. S. Hofle, A. Schienle, C. Bernhard, M. Bruns, U. Lemmer, A. Colsmann, Adv. Mater. **26**, 5155 (2014)
62. Y.H. Zhou, C. Fuentes-Hernandez, J. Shim, J. Meyer, A.J. Giordano, H. Li, P. Winget, T. Papadopoulos, H. Cheun, J. Kim, M. Fenoll, A. Dindar, W. Haske, E. Najafabadi, T.M. Khan, H. Sojoudi, S. Barlow, S. Graham, J.L. Bredas, S.R. Marder, A. Kahn, B. Kippelen, Science **336**, 327 (2012)
63. A.K. Kyaw, D.H. Wang, V. Gupta, J. Zhang, S. Chand, G.C. Bazan, A.J. Heeger, Adv. Mater. **25**, 2397 (2013)
64. T. Xiong, F. Wang, X. Qiao, D. Ma, Appl. Phys. **93**, 123310 (2008)
65. Y.H. Kim, T.H. Han, H. Cho, S.Y. Min, C.L. Lee, T.W. Lee, Adv. Funct. Mater. **24**, 3808 (2014)
66. S. Hofle, A. Schienle, M. Bruns, U. Lemmer, A. Colsmann, Adv. Mater. **26**, 2750 (2014)
67. Y. Zhu, Z. Yuan, W. Cui, Z. Wu, Q. Sun, S. Wang, Z. Kang, B. Sun, J. Mater. Chem. A **2**, 1436 (2014)
68. E.H. Jung, N.J. Jeon, E.Y. Park, C.S. Moon, T.J. Shin, T.Y. Yang, J.H. Noh, J. Seo, Nature **567**, 511 (2019)
69. Q. Jiang, Y. Zhao, X.W. Zhang, X.L. Yang, Y. Chen, Z.M. Chu, Q.F. Ye, X.X. Li, Z.G. Yin, J.B. You, Nat. Photonics **13**, 460 (2019)
70. H.C. Cho, S.H. Jeong, M.H. Park, Y.H. Kim, C. Wolf, C.L. Lee, J.H. Heo, A. Sadhanala, N. Myoung, S. Yoo, S.H. Im, R.H. Friend, T.W. Lee, Science **350**, 1222 (2015)
71. Y. Cao, N. Wang, H. Tian, J. Guo, Y. Wei, H. Chen, Y. Miao, W. Zou, K. Pan, Y. He, H. Cao, Y. Ke, M. Xu, Y. Wang, M. Yang, K. Du, Z. Fu, D. Kong, D. Dai, Y. Jin, G. Li, H. Li, Q. Peng, J. Wang, W. Huang, Nature **562**, 249 (2018)
72. W.D. Xu, Q. Hu, S. Bai, C.X. Bao, Y.F. Miao, Z.C. Yuan, T. Borzda, A.J. Barker, E. Tyukalova, Z.J. Hu, M. Kawecki, H.Y. Wang, Z.B. Yan, X.J. Liu, X.B. Shi, K. Uvdal, M. Fahlman, W.J. Zhang, M. Duchamp, J.M. Liu, A. Petrozza, J.P. Wang, L.M. Liu, W. Huang, F. Gao, Nat. Photonics **13**, 418 (2019)
73. J. Guo, T. Liu, M. Li, C. Liang, K. Wang, G. Hong, Y. Tang, G. Long, S.F. Yu, T.W. Lee, W. Huang, G. Xing, Nat. Commun. **11**, 3361 (2020)
74. L.N. Quan, F.P.G. de Arquer, R.P. Sabatini, E.H. Sargent, Adv. Mater. **30**, 180996 (2018)
75. N.-G. Park, Mater. Today **18**, 65 (2015)
76. V. D'Innocenzo, G. Grancini, M.J. Alcocer, A.R. Kandada, S.D. Stranks, M.M. Lee, G. Lanzani, H.J. Snaith, A. Petrozza, Nat. Commun. **5**, 3586 (2014)
77. L. Protesescu, S. Yakunin, M.I. Bodnarchuk, F. Krieg, R. Caputo, C.H. Hendon, A. Walsh, A. Walsh, M.V. Kovalenko, Nano Lett. **15**, 3692 (2015)
78. G. Nedelcu, L. Protesesc, S. Yakunin, M.I. Bodnarchuk, M.J. Grotevent, M.V. Kovalenko, Nano Lett. **15**, 5635 (2015)
79. M.V. Kovalenko, L. Protesescu, M.I. Bodnarchuk, Science **358**, 745 (2017)
80. K. Hoshi, T. Chiba, J. Sato, Y. Hayashi, Y. Takahashi, H. Ebe, S. Ohisa, J. Kido, ACS Appl. Mater. Interfaces **10**, 24607 (2018)
81. T. Chiba, K. Hoshi, Y.J. Pu, Y. Takeda, Y. Hayashi, S. Ohisa, S. Kawata, J. Kido, ACS Appl. Mater. Interfaces **9**, 18054 (2017)
82. T. Chiba, S. Ishikawa, J. Sato, Y. Takahashi, H. Ebe, S. Ohisa, J. Kido, Adv. Opt. Mater **8**, 2000289 (2020)
83. T. Chiba, Y. Takahashi, J. Sato, S. Ishikawa, H. Ebe, K. Tamura, S. Ohisa, J. Kido, ACS Appl. Mater. Interfaces **12**, 45574 (2020)

84. C.H. Bi, Z.W. Yao, X.J. Sun, X.C. Wei, J.X. Wang, J.J. Tian, Adv. Mater. **33**, 2006722 (2021)
85. H.M. Li, H. Lin, D. Ouyang, C.L. Yao, C. Li, J.Y. Sun, Y.L. Song, Y.F. Wang, Y.F. Yan, Y. Wang, Q.F. Dong, W.C.H. Choy, Adv. Mater. **33**, 2008820 (2021)
86. T. Chiba, Y. Hayashi, H. Ebe, K. Hoshi, J. Sato, S. Sato, Y.J. Pu, S. Ohisa, J. Kido, Nat. Photonics **12**, 681 (2018)
87. X.Y. Shen, Y. Zhang, S.V. Kershaw, T.S. Li, C.C. Wang, X.Y. Zhang, W.Y. Wang, D.G. Li, Y.H. Wang, M. Lu, L.J. Zhang, C. Sun, D. Zhao, G.S. Qin, X. Bai, W.W. Yu, A.L. Rogach, Nano Lett. **19**, 1552 (2019)
88. P.Z. Liu, W. Chen, W.G. Wang, B. Xu, D. Wu, J.J. Hao, W.Y. Cao, F. Fang, Y. Li, Y.Y. Zeng, R.K. Pan, S.M. Chen, W.Q. Cao, X.W. Sun, K. Wane, Chem. Mater. **29**, 5168 (2017)
89. L.N. Quan, Y.B.A. Zhao, F.P.G. de Arquer, R. Sabatini, G. Walters, O. Voznyy, R. Comin, Y.Y. Li, J.Z. Fan, H.R. Tan, J. Pan, M.J. Yuan, O.M. Bakr, Z.H. Lu, D.H. Kim, E.H. Sargent, Nano Lett. **17**, 3701 (2017)
90. C. de Weerd, L. Gomez, H. Zhang, W.J. Buma, G. Nedelcu, M.V. Kovalenko, T. Gregorkiewicz, J. Phys. Chem. C **120**, 13310 (2016)
91. X.Y. Chin, A. Perumal, A. Bruno, N. Yantara, S.A. Veldhuis, L. Martinez-Sarti, B. Chandran, V. Chirvony, A.S.Z. Lo, J. So, C. Soci, M. Gratzel, H.J. Bolink, N. Mathews, S.G. Mhaisalkar, Energy Environ. Sci. **11**, 1770 (2018)
92. S. Kumar, J. Jagielski, T. Tian, N. Kallikounis, W.C. Lee, C.J. Shih, ACS Energy Lett. **4**, 118 (2019)
93. A. Singldinger, M. Gramlich, C. Gruber, C. Lampe, A.S. Urban, ACS Energy Lett. **5**, 1380 (2020)
94. M.R. Yang, P. Moroz, E. Miller, D. Porotnikov, J. Cassidy, C. Ellison, X. Medvedeva, A. Klinkova, M. Zamkov, ACS Photonics **7**, 154 (2020)
95. V.K. Ravi, R.A. Scheidt, A. Nag, M. Kuno, P.V. Kamat, ACS Energy Lett. **3**, 1049 (2018)
96. K. Enomoto, R. Oizumi, N. Aizawa, T. Chiba, Y.J. Pu, J. Phys. Chem. C **125**, 19368 (2021)
97. H. Ebe, Y.K. Wang, N. Shinotsuka, Y.H. Cheng, M. Uwano, R. Suzuki, Y.T. Dong, D.X. Ma, S. Lee, T. Chiba, E.H. Sargent, J. Kido, ACS Appl. Mater. Interfaces **14**, 17691 (2022)
98. X.K. Liu, W. Xu, S. Bai, Y. Jin, J. Wang, R.H. Friend, F. Gao, Nat. Mater. **20**, 10 (2021)
99. L.N. Quan, B.P. Rand, R.H. Friend, S.G. Mhaisalkar, T.-W. Lee, E.H. Sargent, Chem. Rev. **119**, 7444 (2019)
100. A. Dey, J. Ye, A. De, E. Debroye, S.K. Ha, E. Bladt, A.S. Kshirsagar, Z. Wang, J. Yin, Y. Wang, L.N. Quan, F. Yan, M. Gao, X. Li, J. Shamsi, T. Debnath, M. Cao, M.A. Scheel, S. Kumar, J.A. Steele, M. Gerhard, L. Chouhan, K. Xu, X.G. Wu, Y. Li, Y. Zhang, A. Dutta, C. Han, I. Vincon, A.L. Rogach, A. Nag, A. Samanta, B.A. Korgel, C.J. Shih, D.R. Gamelin, D.H. Son, H. Zeng, H. Zhong, H. Sun, H.V. Demir, I.G. Scheblykin, I. Mora-Sero, J.K. Stolarczyk, J.Z. Zhang, J. Feldmann, J. Hofkens, J.M. Luther, J. Perez-Prieto, L. Li, L. Manna, M.I. Bodnarchuk, M.V. Kovalenko, M.B.J. Roeffaers, N. Pradhan, O.F. Mohammed, O.M. Bakr, P. Yang, P. Muller-Buschbaum, P.V. Kamat, Q. Bao, Q. Zhang, R. Krahne, R.E. Galian, S.D. Stranks, S. Bals, V. Biju, W.A. Tisdale, Y. Yan, R.L.Z. Hoye, L. Polavarapu, ACS Nano **15**, 10775 (2021)
101. M.Y. Chen, J.T. Lin, C.S. Hsu, C.K. Chang, C.W. Chiu, H.M. Chen, P.T. Chou, Adv. Mater. **30**, e1706592 (2018)
102. P. Cheng, T. Wu, J. Liu, W.Q. Deng, K. Han, J. Phys. Chem. C **9**, 2518 (2018)
103. T.C. Jellicoe, J.M. Richter, H.F. Glass, M. Tabachnyk, R. Brady, S.E. Dutton, A. Rao, R.H. Friend, D. Credgington, N.C. Greenham, M.L. Bohm, J. Am. Chem. Soc. **138**, 2941 (2016)
104. M.L. Lai, T.Y. Tay, A. Sadhanala, S.E. Dutton, G. Li, R.H. Friend, Z.K. Tan, J. Phys. Chem. Lett. **7**, 2653 (2016)
105. L. Lanzetta, J.M. Marin-Beloqui, I. Sanchez-Molina, D. Ding, S.A. Haque, ACS Energy Lett. **2**, 1662 (2017)
106. D. Yang, G. Zhang, R. Lai, Y. Cheng, Y. Lian, M. Rao, D. Huo, D. Lan, B. Zhao, D. Di, Nat. Commun. **12**, 4295 (2021)

107. M. Liu, H. Pasanen, H. Ali-Loytty, A. Hiltunen, K. Lahtonen, S. Qudsia, J.H. Smatt, M. Valden, N.V. Tkachenko, P. Vivo, Angew. Chem. Int. Ed. **59**, 22117 (2020)
108. Q. Wei, H. Li, Z. Ning, Trends Chem. **4**, 1 (2022)
109. Y. Gao, E. Shi, S. Deng, S.B. Shiring, J.M. Snaider, C. Liang, B. Yuan, R. Song, S.M. Janke, A. Liebman-Pelaez, P. Yoo, M. Zeller, B.W. Boudouris, P. Liao, C. Zhu, V. Blum, Y. Yu, B.M. Savoie, L. Huang, L. Dou, Nat. Chem. **11**, 1151 (2019)
110. Y. Gao, Z. Wei, P. Yoo, E. Shi, M. Zeller, C. Zhu, P. Liao, L. Dou, J. Am. Chem. Soc. **141**, 15577 (2019)
111. Y. Liao, H. Liu, W. Zhou, D. Yang, Y. Shang, Z. Shi, B. Li, X. Jiang, L. Zhang, L.N. Quan, J. Am. Chem. Soc. **139**, 6693 (2017)
112. X. Jiang, Z. Zang, Y. Zhou, H. Li, Q. Wei, Z. Ning, Acc Mater. Res. **2**, 210 (2021)
113. H. Shi, Z. Wang, H. Ma, H. Jia, F. Wang, C. Zou, S. Hu, H. Li, Z.A. Tan, J. Mater. Chem. C **9**, 12367 (2021)
114. C.R. Kagan, D.B. Mitzi, C.D. Dimitrakopoulos, Science **286**, 945 (1999)
115. H. Liang, F. Yuan, A. Johnston, C. Gao, H. Choubisa, Y. Gao, Y.K. Wang, L.K. Sagar, B. Sun, P. Li, G. Bappi, B. Chen, J. Li, Y. Wang, Y. Dong, D. Ma, Y. Gao, Y. Liu, M. Yuan, M.I. Saidaminov, S. Hoogland, Z.H. Lu, E.H. Sargent, Adv. Sci. **7**, 1903213 (2020)
116. Z. Wang, F. Wang, B. Zhao, S. Qu, T. Hayat, A. Alsaedi, L. Sui, K. Yuan, J. Zhang, Z. Wei, J. Phys. Chem. Lett. **11**, 1120 (2020)
117. M. Awais, R.L. Kirsch, V. Yeddu, M.I. Saidaminov, ACS Mater Lett. **3**, 299 (2021)
118. T. Nakamura, S. Yakumaru, M.A. Truong, K. Kim, J. Liu, S. Hu, K. Otsuka, R. Hashimoto, R. Murdey, T. Sasamori, H.D. Kim, H. Ohkita, T. Handa, Y. Kanemitsu, A. Wakamiya, Nat. Commun. **11**, 3008 (2020)
119. S. Gupta, D. Cahen, G. Hodes, J. Phys. Chem. C **122**, 13926 (2018)
120. S. Gupta, T. Bendikov, G. Hodes, D. Cahen, ACS Energy Lett. **1**, 1028 (2016)
121. C. Gao, Y. Jiang, C. Sun, J. Han, T. He, Y. Huang, K. Yao, M. Han, X. Wang, Y. Wang, Y. Gao, Y. Liu, M. Yuan, H. Liang, ACS Photonics **7**, 1915 (2020)
122. F. Yuan, X. Zheng, A. Johnston, Y.-K. Wang, C. Zhou, Y. Dong, B. Chen, H. Chen, J.Z. Fan, G. Sharma, Sci. Adv. **6**, eabb0253 (2020)
123. T. Wang, Q. Tai, X. Guo, J. Cao, C.-K. Liu, N. Wang, D. Shen, Y. Zhu, C.-S. Lee, F. Yan, ACS Energy Lett. **5**, 1741 (2020)
124. M.I. Saidaminov, I. Spanopoulos, J. Abed, W. Ke, J. Wicks, M.G. Kanatzidis, E.H. Sargent, ACS Energy Lett. **5**, 1153 (2020)
125. D. Di Girolamo, J. Pascual, M.H. Aldamasy, Z. Iqbal, G. Li, E. Radicchi, M. Li, S.-H. Turren-Cruz, G. Nasti, A. Dallmann, F. De Angelis, A. Abate, ACS Energy Lett. **6**, 959 (2021)
126. Y.J. Heo, H.J. Jang, J.H. Lee, S.B. Jo, S. Kim, D.H. Ho, S.J. Kwon, K. Kim, I. Jeon, J.M. Myoung, Adv. Funct. Mater. **31**, 2106974 (2021)
127. J.-F. Liao, Y.-X. Chen, J.-H. Wei, Y.-T. Cai, X.-D. Wang, Y.-F. Xu, D.-B. Kuang, Nano Res. **12**, 2640 (2019)
128. P. Ferdowsi, E. Ochoa-Martinez, U. Steiner, M. Saliba, Chem. Mater. **33**, 3971 (2021)
129. S. Sidhik, W. Li, M.H. Samani, H. Zhang, Y. Wang, J. Hoffman, A.K. Fehr, M.S. Wong, C. Katan, J. Even, Adv. Mater. **33**, 2007176 (2021)
130. D. Prochowicz, P. Yadav, M. Saliba, M. Saski, S.M. Zakeeruddin, J. Lewinski, M. Gratzel, ACS Appl. Mater. Interfaces **9**, 28418 (2017)
131. Y.H. Cheng, R. Moriyama, H. Ebe, K. Mizuguchi, R. Yamakado, T. Nishitsuji, T. Chiba, J. Kido, ACS Appl. Mater. Interfaces **14**, 22941 (2022)
132. G. Folpini, D. Cortecchia, A. Petrozza, A.R. Srimath Kandada, J. Mater. Chem. C **8**, 10889 (2020)

Chapter 8
Transient Properties and Analysis of Organic Photonic Devices

Hirotake Kajii

Abstract This chapter focuses on the transient properties of organic photonic devices such as OLEDs, OPDs, and emerging devices and the transient analysis using modulated signals. The applications of organic photonic devices generating optical and electrical modulated signals are described. Key advances in high-speed operation of 100 MHz are discussed. This chapter also shows how to generate high-speed optical and electrical modulated signals.

Keywords Transient properties · Transient analysis · Organic light-emitting diode · Organic photodetector · Organic light-emitting transistor

8.1 Introduction and Applications

Research and development of organic photonic devices have been rapidly performed on a global scale with the aim of creating new soft electronics based on organic molecules with π-bonds. Organic semiconductor materials have attracted considerable attention owing to their simple and low-cost processing and a high potential for optoelectronic applications. Organic light-emitting diodes (OLEDs) are capable of emission over a broad visible range, highly efficient, and require only a low driving voltage [1, 2]. OLEDs have been realized, which have a long lifetime and excellent durability for flat-panel display applications. There are some requirements for OLEDs when used not only in display applications but also as various solid-state lighting sources.

Organic photovoltaics, including organic solar cells and organic photodetector (OPD), are organic devices with photoelectric conversion characteristics. OPDs have a simple structure. The organic layers are sandwiched between the transparent ITO and metal electrode. Under illuminated light, the photocurrent was obtained. By choosing organic materials, it is easy to realize color-sensitive photodetectors.

H. Kajii (✉)
Graduate School of Engineering, Osaka University, Osaka, Japan
e-mail: kajii@eei.eng.osaka-u.ac.jp

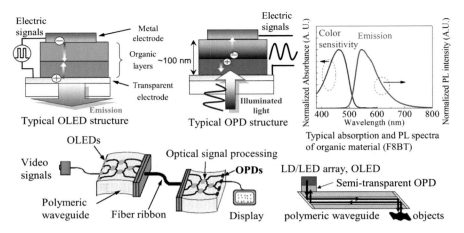

Fig. 8.1 Schematic images of the combination of polymeric waveguides and organic photonic devices

Therefore, OPDs have high potential, especially for signal processing and optical sensing systems. OPDs also have following advantages such as flexible, easy process, and semitransparent. For application, there are integrated optical devices using polymer waveguides, transparent scanner, and display scanner.

Various sensor and optical communication devices support the Internet of Things (IoT) and ICT society. At an early stage, the combination of polymeric waveguides and organic photonic devices has been proposed as flexible integrated photonic devices, as shown in Fig.8.1 [3]. Organic photonic devices such as OLEDs and OPDs also exhibit high potential for use in future information technology systems, especially for high-speed signal processing and optical sensing systems [3–6].

As an application of the optical link device using an OLED, the system, which transmits audio signals over a polymer optical fiber (POF), has been proposed [5]. Standard digital sound is stored as pulse code modulation (PCM) data. PCM is a digital scheme for transmitting analog data and can be used to digitize the signal amplitude of analog data such as voices and music. The yellow OLED, which is enabled to drive at lower voltages of below 5 V by using organic materials with high mobility, such as silole derivatives [7], can create a high-speed response in MHz order, as shown in Fig. 8.2. PCM optical output signals from the device lead to the optical digital input of the commercial audio capture through the POF. Then, the output PCM signal agrees with the input PCM signal. Transmission of clear audio signals reproduced by the OLED optical transmission system could be realized.

As an application of the optical sensor using a photoreflector, the semitransparent OPD, which consists of the heterostructure of copper phthalocyanine (CuPc) as a p-like material and N,N′-bis(2,5-di-tert-butylphenyl)3,4,9,10-perylenedicarboximide (BPPC) as an n-like material, was fabricated by using a-C: N/ITO cathode, as shown in Fig. 8.3a [8]. A semitransparent device has broad absorption and transparency ranges above 60% in visible light, as shown in

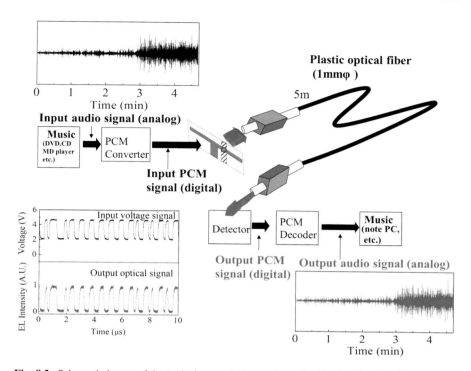

Fig. 8.2 Schematic image of the typical transmission system of audio signals using OLED

Fig. 8.3b, and a cutoff frequency of more than 2 MHz at −2 V under red light irradiation. Figure 8.3c shows the photoresponses to the different reflectances of the papers in front of the OPD. As samples, black paper, white paper, and an aluminum sheet with reflectances of 0.2%, 2.0%, and 77.3% at 640 nm, respectively, were used. The order of the reflection intensities of the samples is black < white < aluminum. That is, the intensity of the photoresponse increased with increasing sample reflectance. The integrated device with an OLED is expected to be applicable to fabricating photodetectors on a polymer substrate such as a polymeric waveguide.

Next, applications of solution-processable photonic devices based on fluorene-type polymers are described below. Employing solution-processable organic semiconductor materials is useful for organic photonic device fabrication. It allows for the possibility of low-cost mass production and energy conservation and for making large-area devices. Fluorene-type polymers have emerged as an essential class of conjugated polymers owing to their high stability and relatively high mobility.

A typical fluorene-type, poly(9,9-dioctylfluorene-co-benzothiadiazole), F8BT, a well-known conjugated polymer with a high quantum yield of fluorescence, is a widely used green emitting reference polymer for a variety of applications. The frequency response up to 100 MHz and the cutoff frequency of above 50 MHz have been obtained from the printed polymer light-emitting diode (PLED) based on F8BT

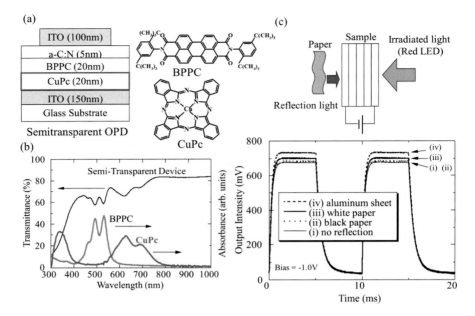

Fig. 8.3 (**a**) Device structure, (**b**) the transmission spectrum of a device, the absorption spectra of CuPc and BPPC, and (**c**) the photoresponses to the different reflectances of the papers in front of the OPD under 0.1 kHz modulation red light. A terminal resistance is 100 kΩ for measuring the output current

at high current densities. The F8BT-based PLED, which exhibits a short fluorescent lifetime of approximately 1 ns, has the potential for electro-optical conversion devices such as optical link device [9].

The video transmission system transmits high-quality National Television System Committee (NTSC) video signals over standard multimode optical fiber, as shown in Fig. 8.4. An F8BT device acts as the light source for generating optical pulses as electro-optical conversion processing. In this system, pulse frequency modulation (PFN) with the baseband frequency of MHz order, in which the pulse width is kept constant and the pulse generation cycle is varied, to ensure error-free transmission is employed. The output PFN and NTSC signals are in agreement with both the input signals, which shows the printed device can be applied as the electro-optical conversion device for transmitting the moving picture signals.

Fluorene-type polymers with various absorption wavelengths are one of the promising materials for large-area OPDs. The photoresponse properties of bilayer devices utilizing solution-processable starburst molecular, 1,3,5-tris [4-(diphenylamino)phenyl]benzene (TDAPB), as a donor-type layer were investigated, as shown in Fig. 8.5, because F8BT with an electron-withdrawing group was used as an acceptor-type material.

Phosphorescent materials such as iridium (Ir) complexes have heavy atom effects and induce intersystem crossing (ISC), which transforms excitons from singlet into

8 Transient Properties and Analysis of Organic Photonic Devices

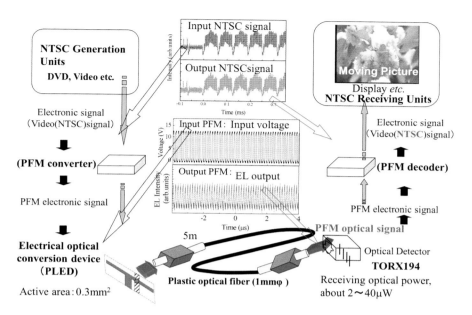

Fig. 8.4 Schematic diagram of a typical moving picture transmission system using printed OLED as the electro-optical conversion device

triplet. Triplet excitons have a longer lifetime than singlet excitons. This long lifetime can be expected to reduce exciton recombination and improve conversion efficiency [10, 11]. To improve the conversion efficiency of polymer photodetectors (PDs) fabricated by solution process, the properties of fluorene-type polymer photodetectors doped with Ir complexes have been investigated [12].

TDAPB layer was formed on the active layer without intermixing because tris (2-phenylpyridine)iridium(III) and Ir(ppy)$_3$ doped in F8BT films were insoluble in n-butyl acetate solution. Ir(ppy)$_3$ as a triplet material can enhance the incident-photon-to-current conversion efficiency (IPCE) of the bilayer device since its triplet level is lower than S_1 level of the host and higher than T_1 level of the host. These energy levels suggest that photogenerated excitons are easy to dissociate at the F8BT/TDAPB heterojunction.

The shape of IPCE directly reflects the F8BT absorption shape, as shown in Fig. 8.5b, which indicates that the relation between the spectra of IPCE from devices and the absorption spectrum of F8BT explicitly prove that photogeneration of charge carriers results from the dissociation of excitons excited in F8BT. F8BT devices had blue sensitivity. For bilayer devices, clear response pulses at 10 MHz were observed, as shown in Fig. 8.5c.

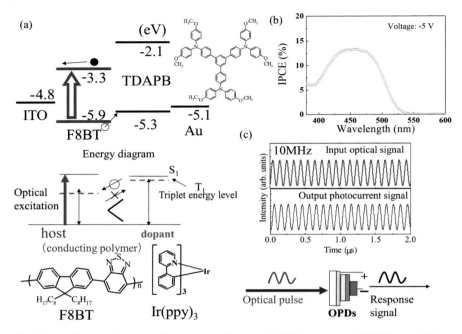

Fig. 8.5 (a) Schematic energy diagram and (b) typical IPCE spectrum of F8BT-based bilayer OPD, and (c) output signal at 10 MHz sinusoidal modulated laser light ($\lambda = 408$ nm) illumination under an applied bias of -5 V

For application, the transmission of moving pictures using an F8BT-based bilayer photodetector has been demonstrated, as shown in Fig. 8.6 [13]. Electrical moving picture signals are converted to NTSC signals and then to PFM signals by the electrical circuits. The system transmits PFN signals, which are converted from NTSC video signals, over a standard multimode polymeric optical fiber (POF). An ultraviolet (UV) LD ($\lambda = 408$ nm, 0.4 W/cm^2) was used to generate optical pulses as an example of optoelectrical processing. The optical signals are transmitted by the POF, and then they are received by the OPD. For a bilayer OPD of $(-)$ITO/F8BT:Ir(ppy)$_3$/TDAPD/Au $(+)$ under the applied bias of -6.5 V, the output PFM photocurrent signal was almost the same as the input PFM optical signal. The output PFM signals are converted to NTSC signals, and the transmitted pictures signals are displayed in the display device. The output NTSC signal agreed with the input NTSC signal, demonstrating that the polymer PDs fabricated by the solution process can be applied to short-range optical communication fields, such as optoelectrical conversion devices for optical links.

An additional advantage is that OLEDs are simple to fabricate on various kinds of substrates, including polymers. This means that mechanical flexibility is one of the key advantages of OLEDs. Polyimide film (PI) is resistant to thermal treatment and chemical solvents compared with other polymer films, such as polyethylene terephthalate (PET) and polyethylene naphthalate (PEN). In addition, since polyimide

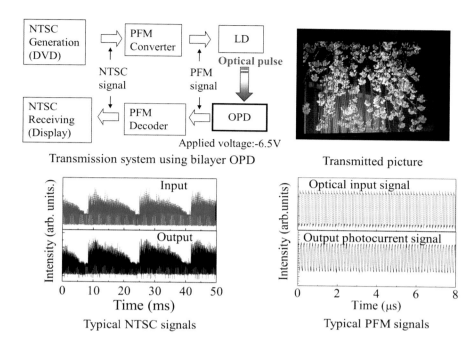

Fig. 8.6 Schematic diagram of a typical moving picture transmission system using printed OPD as the optoelectrical conversion device

film has high optical transparency and the refractive index can be controlled, it is a promising candidate as the substrate for optical devices. Transient EL properties and application of OLEDs fabricated on the 16 μm PI with silicon oxide and silicon nitride thin layers as passivation layers have been reported [5].

Flexibility in optoelectronic devices is essential for medical and bioengineering. Flexible organic photonic devices can be installed outside the human body as diagnostic and monitoring tools. Ultraflexible optical devices have been attracted extensively in next-generation wearable electronics. Photoplethysmography (PPG)-based monitoring devices which contain light sources and photodetectors have been widely developed for heart rate monitoring purposes. A pulse oximeter sensor composed of two types of green and red OLEDs and two OPDs has been proposed [14]. Ultraflexible organic sensors taped to the skin have been developed. By using the oximeter, which is interfaced with conventional electronics at 1 kHz, the acquired pulse rate and oxygenation can be calibrated. By fabricating an organic photonic device on adhesive tape, an ultraflexible reflective oximeter attached to a finger detects blood oxygen levels [15]. Ultraflexible organic sensors can be expected to lead to better, more stable monitoring than conventional devices. Wearable organic optoelectronic sensors, including a tissue-oxygenation sensor and a muscle-contraction sensor for medicine, have also been demonstrated [16].

From the viewpoints of interfacial engineering such as organic/organic, organic/inorganic, and organic/electrode interfaces, the research and development of the fabrication in organic photonic devices are essential for obtaining high-speed modulation.

This chapter focuses on the fundamental properties related to transient phenomena in organic photonic devices, such as OLEDs, OPDs, and emerging devices, which are driven by the modulated signals. This chapter also shows how to generate high-speed optical and electrical modulated signals.

8.2 Organic Light-Emitting Diodes

8.2.1 Fluorescent Organic Light-Emitting Diode

OLEDs based on fluorescent dyes and polymers, which exhibit short fluorescent lifetime, have the potential for high-speed electro-optical conversion devices. Tris (8-hydroxyquinoline)aluminum(III), Alq_3, is widely used in OLEDs as an emissive and electron-transport material. This section mainly focuses on the transient properties and transient analysis of OLEDs based on Alq_3. This section also describes how to generate high-speed optical modulated signals up to 100 MHz.

8.2.1.1 Transient EL in Alq_3-Based Devices

A schematic diagram of the measurement method of transient EL for OLEDs is shown in Fig. 8.7. To investigate the transient EL of OLEDs, the devices were driven at a 0.1 or 1 ms period and duty ratio of 1/10 (or 1/100) pulse voltage. The decay (rise) time is defined as the time required to change the optical response from 90% (10%) to 10% (90%) of its total intensity change. The delay time is dependent on the transient time of electrons in the Alq_3 layer because the hole mobility of the hole-transporting layer is higher than the electron mobility of Alq_3. Figure 8.8a shows applied voltage dependence of the rise, decay, and delay times in Alq_3-based devices of ITO/α-NPD(50 nm)/Alq_3 (30 nm)/Mg:Ag with various active areas. The rise, decay, and delay times decreased with increasing applied voltage owing to the decrease of resistivity and also decreasing active area owing to the decrease of capacitance.

The transient EL characteristics of the Alq_3-based device (0.01 mm^2 active area) driven at 10 ns period and duty ratio of 0.5 pulses are given in Fig. 8.8b. Clear light pulses by direct modulation of the device driven at the applied voltage pulses of 50 MHz are obtained. It is difficult to obtain optical pulses above 100 MHz using the Alq_3-based device because the fluorescent lifetime of Alq_3 is more than 10 ns [17].

The transient phenomenon is related to the charge and discharge of initial capacitance in the OLED. In order to enhance the response speed of the OLEDs, it is effective to apply a positive base voltage to the OLEDs in addition to the pulsed

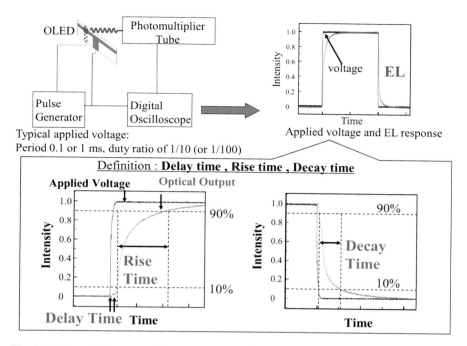

Fig. 8.7 Schematic diagram of the measurement method of transient EL for OLEDs

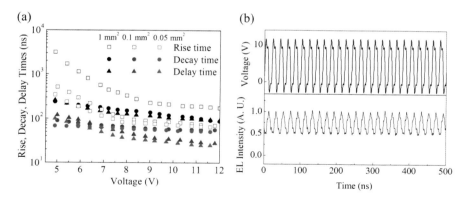

Fig. 8.8 (a) Applied voltage dependence of the rise, decay, and delay times in Alq$_3$-based device of ITO/α-NPD(50 nm)/Alq$_3$ (30 nm)/Mg:Ag with various active areas. (b) Transient EL characteristics of the Alq$_3$-based device (0.01 mm^2 active area) driven at 10 ns period and duty ratio of 0.5 pulses

voltage in many cases [6]. OLEDs are driven by lower pulsed voltage with a base voltage that is less than the turn-on voltage of the OLEDs.

Figure 8.9a shows the typical transient EL of Alq$_3$-based OLED with 0.03 mm^2 active area, using high-level voltage pulses of 5 V for low-level DC bias of 0 V and 2.4 V, for which 0.25-μs rectangular voltage pulses with a period of 2 MHz were

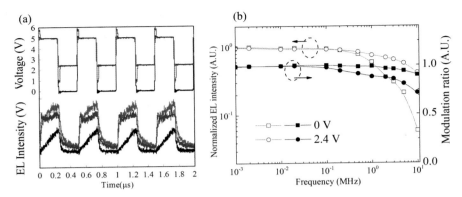

Fig. 8.9 (a) Voltage and optical output signals of a 0.03-mm² Alq$_3$-based OLED of ITO/α-NPD (50 nm)/Alq$_3$ (40 nm)/Mg:Ag measured with high-level voltage pulses of 5 V, with 0.25-μs rectangular voltage pulses applied at a repetition of 2 MHz for various bias values, and (**b**) the frequency dependence of EL intensity and modulation ratio of the device measured with high-level voltage pulses of 5 V for different low-level DC bias of 0 V and 2.4 V

applied. The EL intensity for positive DC bias of 2.4 V is approximately twofold bigger than that at 0 V, where EL intensity is defined as the total intensity change in optical response from the minimum value at a low-level bias to the maximum value at high-level voltage. The positive bias of 2.4 V leads to faster optical rise times. The frequency dependence of the EL intensity of the Alq$_3$-based device using high-level input voltage pulses of 5 V for different low-level DC biases of 0 V and 2.4 V, where EL intensities at various modulation frequencies are normalized by the EL intensity at 1 kHz, are given in Fig. 8.9b. Adding a low-level DC bias of 2.4 V in the high-frequency region of more than 1 MHz results in the improved frequency dependence of the device. However, the modulation ratio of the device on the application of 2.4 V is lower than that for a low bias of 0 V at low voltages since the fall time for the low-level bias of 2.4 V is longer than that for the low bias of 0 V. The disadvantage of longer fall time arises, as shown in Fig. 8.9a, suggesting that the slow dissociation of carrier occurs in the emissive layer owing to the reduced electric field.

To speed up rise and fall times, the overshoot or undershoot voltage pulses with a time duration of 25 ns are added to the rectangular input voltage pulse on the rising or falling edge, respectively. The improved transient EL was obtained by adding both the overshoot and undershoot voltage pulses to the periodic rectangular pulses. Adding the overshoot voltage to the rectangular voltage on the rising edge results in a faster rise time. In addition, adding the undershoot voltage to the rectangular voltage on the falling edge leads to improved falling time. The short response times of EL, which operate into the MHz regime at lower voltages, are achieved by improving the driving method.

8.2.1.2 Impedance Spectroscopy in Alq₃-Based Device Using Surface-Modified ITO

ITO electrode is often cleaned by using UV-ozone or O_2 plasma treatment. In addition, the surface treatment of electrodes using self-assembled monolayers (SAMs) is one of the crucial methods for improving the performance of organic devices [18–20]. The work function of ITO modified with fluoroalkyl phosphonic-based SAM, 1H,1H,2H,2H-perfluorooctanephosphonic acid (FOPA) increased to about −5.3 eV, compared with that of ITO (−4.8 eV).

For hole-only devices of ITO/α-NPD/MoO$_x$/Al without and with an ITO anode modified by FOPA, the current density increased by modifying the FOPA, as shown in Fig. 8.10, which results in the improvement of hole injection. The characteristics of the OLEDs of ITO/ α-NPD/Alq₃/LiF/Al without and with an ITO anode modified by FOPA were described. The device lifetime was taken from the initial luminance of 5000 cd/m² to 2500 cd/m². For ITO anode modified by FOPA, lifetime value is improving, and an Alq₃-based OLED with an estimated lifetime of over 180 h (time to reach 50% of initial intensity) at 5000 cd/m² was obtained, as shown in Fig. 8.11.

In order to improve the device lifetime of the OLEDs, it is necessary to figure out the degradation mechanism. The fundamental properties of devices are mostly investigated under DC operation. The current density–voltage (J–V) characteristics represent the overall result of the injection and transport of carriers through the organic layers and electrodes. FOPA treatment results in a lower driving voltage, as shown in Fig. 8.11b. After degradation, driving voltages increase for both devices. From the view of carrier transport dynamics, impedance spectroscopy is an effective method to study transport dynamics in devices [21–24].

This degradation mechanism of devices is analyzed using impedance spectroscopy measurement. In this measurement, impedance $[Z = V/I = V_1/I_1 \exp.(-j\varphi)]$ is determined from the current amplitude, and phase difference φ of the current

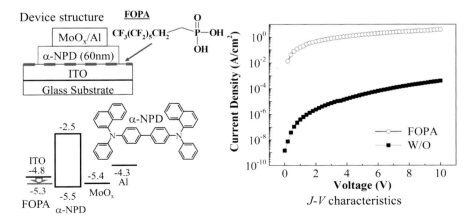

Fig. 8.10 Device structure and *J–V* characteristics of hole-only devices without and with FOPA

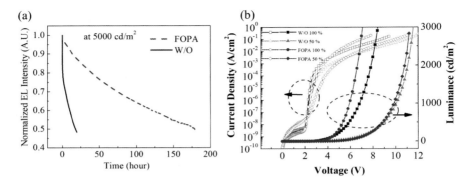

Fig. 8.11 Comparison of (**a**) typical device lifetimes and (**b**) *J-V-L* characteristics of OLEDs of ITO/α-NPD (60 nm)/Alq$_3$ (60 nm)/LiF/Al without and with FOPA

response signal $[I = I_0 + I_1 \exp.(j(\omega t + \varphi))]$ when a voltage signal $[V = V_0 + V_1 \exp.(j\omega t)]$ is applied to OLED device by adding small sine-wave voltage signals from μHz to MHz order to DC voltage, where ω is the angular frequency and t is time. The complex ac impedance (Z) is expressed as $Z = Z' - jZ'$, where Z' is the real part and Z' is the imaginary part.

It is well-known that holes are injected first owing to the negative interfacial charge at the α-NPD/Alq$_3$ interface. A small sinusoidal voltage with a frequency of 1 kHz was applied in OLEDs, and the *C-V* characteristics were investigated using the real components of $\varepsilon = 1/j\omega Z$. The capacitance value shows the geometric capacitance value when no charge is injected. As the applied voltage increases and the injection of holes into the α-NPD layer begins, the capacitance value increases. The voltage at which the capacitance value rises in the *C-V* characteristic can be estimated as the onset voltage related to hole injection. Figure 8.12 shows the *C-V* characteristics of OLEDs (active area: 4 mm^2) without and with FOPA before and after degradation. The onset voltages in OLEDs without and with FOPA were estimated to be 0.1 V and −0.8 V before degradation. FOPA treatment results in the smooth hole injection into the α-NPD layer. It is clear that after the degradation, the onset voltage of hole injection increased for both devices, which suggested that hole injection is worsened by degradation.

Resistance and capacitance can be estimated from the modulus plot ($M = j\omega Z$). Figure 8.13 shows schematic images of the typical Cole-Cole plot (M-plot) and Bode-plot of the device and the estimated equivalent circuit. In general, the equivalent circuit consists of parallel resistivity (*R*)-capacitance (*C*) circuits, which are mainly related to bulk resistance and capacitance of each layer, together with the contact resistance in series. Each diameter of the semicircle from the M-plot is estimated to be 1/*C*. Since the capacitance value *C* can be obtained from M-plot, the change in resistance value *R* can be estimated directly from the change in peak frequency of the Bode plot. The frequency shift of the peak is mainly dependent on the resistance (R_1, R_2).

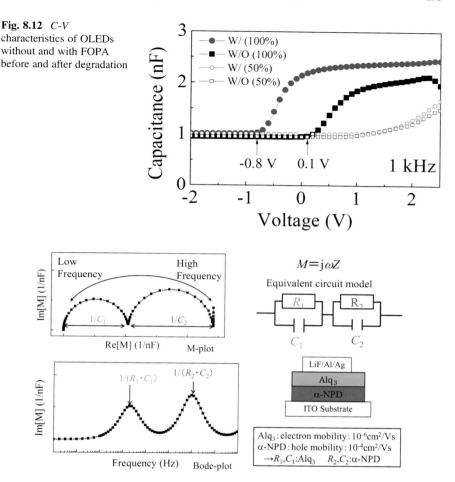

Fig. 8.12 *C-V* characteristics of OLEDs without and with FOPA before and after degradation

Fig. 8.13 Schematic images of typical Cole-Cole plot (M-plot) and Bode plot of the device and the estimated equivalent circuit

Figure 8.14a shows typical Bode plots of OLEDs without and with FOPA at various voltages. Since the frequency of each peak maximum in the Bode plot is inversely proportional to the time constant, the layers based on the mobility of the materials are determined since the hole mobility of α-NPD and the electron mobility of Alq$_3$ were the orders of 10^{-4} cm^2 V^{-1} s^{-1} and 10^{-6} cm^2 V^{-1} s^{-1}, respectively. The peak on the high-frequency side was determined to be the component derived from α-NPD.

The voltage dependences of the resistances of α-NPD and Alq$_3$ components obtained by fitting a modulus plot in OLEDs before degradation are shown in Fig. 8.14b. In the undegraded device, FOPA treatment reduced the resistance of the α-NPD layer. There was generally no significant difference in the resistance

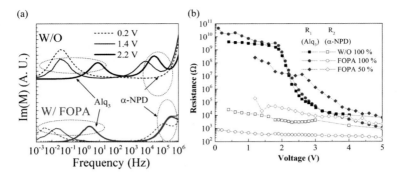

Fig. 8.14 (a) Typical Bode plot of OLEDs and (b) voltage dependences of the resistances of α-NPD and Alq₃ components

derived from the Alq$_3$ layer after threshold voltage. On the other hand, for OLED with FOPA, both the resistances of the α-NPD and Alq$_3$ layers increase after degradation and the longer operation time. These results suggested that the improved hole injection from ITO to a-NPD results in an improved lifetime.

8.2.1.3 High-Speed Modulation of OLED

The doping system is often used in OLEDs to prevent concentration quenching. The dye-doping method has practical importance because it is easy to improve device performance, such as color tunability, luminous efficiency, and device lifetime. This method is also effective in obtaining high-speed optical pulses from OLEDs with dye doped in Alq$_3$ as the emissive layer when the fluorescence lifetime of the dye is shorter than that of Alq$_3$. The fluorescence lifetime of the dye is one of the determining factors for generating high-speed optical pulses.

For high-speed modulation of OLED, it is considered that high hole and electron mobilities are required for the hole- and electron-transporting materials. In addition, high luminescence quantum efficiency and short fluorescence lifetime are also required for emitting materials. High-speed response from an OLED with a non-doping emissive layer is described [25]. Tris[4-(5-phenylthiophen- 2-yl)] amine (TPTPA), which exhibits the highest level of hole drift mobility (1.0×10^{-2} cm^2 V^{-1} s^{-1}) among those reported for organic disordered systems [26], is selected as an emissive material with high hole drift mobility.

Figure 8.15 shows the performances of OLEDs using an emissive material with high hole drift mobility, TPTPA, which exhibits a short fluorescence lifetime of 1.3 ns. MoO$_x$ was used to lower the hole injection barrier. KLET-03, purchased from Chemipro Kasei Kaisha, Ltd., of which the electron mobility is the order of 10^{-3} cm^2 V^{-1} s^{-1}, was used as a higher electron-transporting material. Clear blue emission was observed from TPTPA, which was used as the hole-transporting and emissive layer in the OLED. The luminance of approximately 17,000 cd/m^2 at 7.5 V

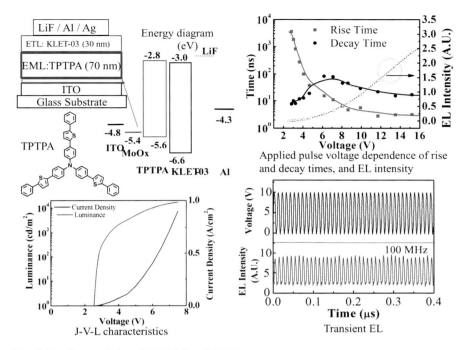

Fig. 8.15 Characteristics of TPTPA-based OLED

and the maximum current efficiency of 3.3 cd/A were achieved. The short response times of electroluminescence (ca. 10 ns) were achieved under the application of rectangular-shaped voltages. For OLEDs with smaller device sizes (below the active area of 0.03 mm^2), the RC time constant is approximately several ns order as the differential resistance of the device is below several hundred ohms at higher voltages (current densities). It is considered that the rise time is dominantly controlled by the RC time constant, and the diffusion of remanent carriers influences the decay time in organic layers, which can cause recombination in the emission layer. Optical pulses of 100 MHz were obtained from a TPTPA-based device, as shown in Fig. 8.15.

8.2.2 Phosphorescent Organic Light-Emitting Diodes

Employing phosphorescent materials yields high efficiencies because breaking the spin conservation rule allows both singlet and triplet excitons to contribute to emission. The phosphorescent organic light-emitting diodes (PHOLEDs) have demonstrated high external quantum efficiencies owing to the phosphorescent emission from a triplet state of phosphorescent dyes such as Ir and Pt complexes [27, 28]. External quantum efficiency (ηext) of OLEDs follows: $\eta_{ext} = \gamma \times \eta_r \times \phi_p \times \eta_p$, where γ is carrier balance factor, η_r is efficiency of exciton formation (~1 for phosphorescent materials), ϕ_p is the radiative quantum efficiency of the emitting material, and η_p is the light outcoupling efficiency.

8.2.2.1 Transient EL of Blue PHOLED

First, this section focuses on the transient properties of a blue PHOLED based on bis [(4,6-difluorophenyl)-pyridinato-N,C2'](picolinate) iridium(III) (FIrpic). A typical device had a blue phosphorescence emissive layer with FIrpic doped in 4,4'-bis (9-carbazolyl)-2,2'-dimethyl-biphenyl (CDBP). A poly(ethylenedioxythiophene): poly(styrenesulfonic acid) (PEDOT:PSS) as a hole injection layer. 1,3-Bis (N-carbazolyl)benzene (mCP) with a high triplet energy and a very deep highest occupied molecular orbital (HOMO) level is used as a buffer layer. A device consisted of an ITO-coated glass substrate, an α-NPD/mCP/CDBP:FIrpic (5 vol%) phosphorescent emissive layer, and a bis(2-methyl-8-quinolinato)-4-(phenylphenolato)aluminum (BAlq) electron transporting layer, as shown in Fig. 8.16. The PHOLED exhibited a blue emission based on FIrpic, approximately maximum efficiency of 14 cd/A and maximum luminance of 20,000 cd/m^2.

Figure 8.17a shows the current-density dependence of the rise and decay times and current efficiency of the FIrpic-based device with 1 mm^2 active area by applying the applied pulse voltage. There were two parts to the behavior of rise time. At a low current density, the rise time was longer than the decay time. The current density superlinearly increased with the voltage. Therefore, resistivity strongly depends on the applied voltage (current density). The rise time gradually decreased owing to the decrease of the resistivity of the PHOLED with increasing current density (applied voltage) and was approximately 1 μs at high current density.

On the other hand, the decay time slightly decreased with increasing current density. At higher current densities, the decay time was almost the same as the rise time and approximately 1 μs, which is almost the same as the phosphorescent lifetime of FIrpic.

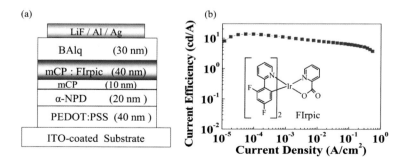

Fig. 8.16 (a) Device structure and (b) current efficiency characteristic of FIrpic-based device

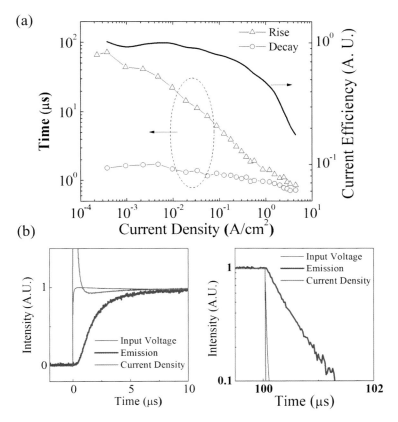

Fig. 8.17 (a) Current-density dependence of the rise and decay times and current efficiency of the FIrpic-based device with 1 mm² active area by applying the applied pulse voltage, and (b) the transient voltage, current, and EL signal of a FIpic-based OLED at 9 V for the rise and decay parts

Figure 8.17b shows the transient voltage, current, and EL signal of a FIpic-based OLED at 9 V for the rise and decay parts. At higher voltage of 9 V, the pulse current flowing through the device can be assumed to rise up instantaneously, and the steady state of current density is rapidly reached, compared with the rising part of blue emission from FIrpic. That is, the rise time of blue emission was longer than that of current, which suggests that an OLED does not immediately emit a blue light, although the carriers reach the recombination zone.

The rate equation [27] of PHOLEDs is given as follows:

$$\frac{d[^3M^*]}{dt} = -\frac{[^3M^*]}{\tau} - \frac{1}{2}k_{TT}[^3M^*]^2 + \frac{J}{qd} \quad (8.1)$$

where $[M^*]$ is the concentration of triplet excitons, τ is the phosphorescent recombination lifetime, k_{TT} is the triplet–triplet annihilation constant, J is the current density, q is the electron charge, and d is the length of the exciton formation zone. From this rate equation, the transient properties are strongly affected by phosphorescent recombination lifetime, τ.

The rate of τ is represented by the inverse of radiative recombination lifetime (τ_r) and nonradiative recombination lifetime (τ_{nr}): $1/\tau = 1/\tau_r + 1/\tau_{nr}$. At higher current densities, the increase of the nonradiative process results in the decrease in τ. This is because exciton interaction within an emissive layer and between an emissive layer and a hole or an electron transporting layer increases the probability of the nonradiative process. The reduced rise and decay times may also be due to high-density triplet excitons related to enhanced triplet-triplet annihilation [27–29]. Therefore, both the rise and decay times were observed to be below 1 µs. Therefore, the behavior of the FIrpic-based device is different from that of the Alq$_3$-based device. In addition, the transient properties of PHOLEDs were almost independent of the device sizes [30]. PHOLED based on FIrpic can be expected to be utilized as one of the light sources driven at less than 1 MHz.

8.2.2.2 Transient EL of White OLED with Phosphorescent and Fluorescent Dyes

White OLEDs with various emissive materials have been investigated, including blue, green, yellow, and red light emissions. White PHOLEDs are one of the higher efficient light-emitting devices but have the disadvantage of slow response time. That is, from the viewpoint of the frequency response of organic devices, white OLEDs can be expected to be applied to optical link devices, such as visible lighting transmission by fluorescence emission in white OLEDs with both the phosphorescent and fluorescent emissive layers. The typical transient characteristics of white OLEDs with blue phosphorescent and red fluorescent materials as emissive layers are described by applying the voltage pulses on the devices [31].

A typical device had a blue phosphorescence emissive layer with FIrpic doped in CDBP and a red fluorescence emissive layer with a 4-(dicyanomethylene)-2-i-propyl-6-(1,1,7,7-tetramethyljulolidyl-9-enyl)-*4H*-pyran (DCJTI) partial doping layer at the interface of α-NPD/Alq$_3$. Studies of the EL process of fluorescent OLEDs produced by the partial doping method have been reported [32–35]. A white device consisted of an ITO-coated glass substrate, an α-NPD/CDBP:FIrpic (5 vol%) phosphorescent emissive layer, a BAlq buffer layer, an α-NPD/DCJTI/Alq$_3$ fluorescent emissive layer, and a BAlq hole blocking layer, terminated with 0.5-nm-thick lithium fluoride (LiF) and Al/Ag electrodes, as shown in Fig. 8.18a. The typical average thickness of the partial doping layer of DCJTI was 0.03 nm. The device exhibited a maximum efficiency of approximately 7 cd/A and a maximum luminance of above 20,000 cd/m^2. The EL spectral shape of the device is less sensitive to changes in drive current, as shown in Fig. 8.18b. The white EL spectrum from blue phosphorescence FIrpic and red fluorescence DCJTI showed CIE (Commission Internationale de L'Eclairage) coordinates of (0.35, 0.43) with the addition of green emission of Alq$_3$.

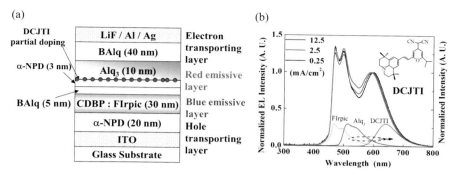

Fig. 8.18 (**a**) Device structure and (**b**) typical EL spectra of a device at various current densities, and FIrpic, Alq$_3$, and DCJTI spectra

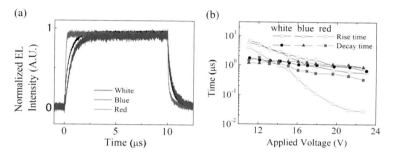

Fig. 8.19 (**a**) Typical transient EL signals of blue, red, and white emissions from a white OLED at 18 V, and (**b**) applied pulse voltage dependence of the rise and decay times of blue, red, and white EL emissions from a white OLED

The typical transient EL signals at 18 V and the applied pulse voltage dependence of the rise and decay times of blue, red, and white emissions from a white device are presented in Fig. 8.19. The transient characteristics of blue and red EL emissions from a white device based on FIrpic and DCJTI in the wavelength range from 425 nm to 500 nm and from 600 nm to 670 nm, using color filters and photomultiplier tubes, are explored, respectively. The device with an active area of 0.3 mm^2 was driven at a 1 ms period and a duty ratio of 1/100 pulses. The EL signal of white emission was almost the same as that of blue emission from FIrpic. The rise and decay times of blue emission from FIrpic were longer than those of red emission from DCJTI. At lower voltages, the rise time of red emission was longer than the decay time owing to the high resistivity and low mobility related to the device. At higher applied pulse voltages above 14 V, the rise time of red emission was shorter than the decay time and several ten times shorter than those of blue and white emissions. The decay time has fast and slow components [36], which are attributed to the photoluminescence (phosphorescent) lifetime of emissive materials or the RC time constant of devices and delayed EL from residual carriers in the emissive layer,

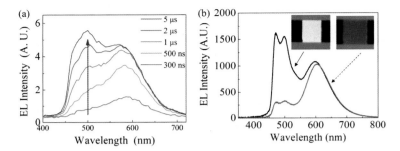

Fig. 8.20 (a) Time-resolved EL spectra of a white device, and (b) EL spectra of a white device driven at the pulse widths of 100 μs and 100 ns, frequencies of 1 kHz and 100 kHz, and applied voltages of 4.6 V and 15.8 V, respectively

respectively. The behaviors of rise and decay times of blue and white emission in a white device were almost the same as those in a blue device based on FIrpic.

Utilizing the difference of transient characteristics between the emitters, emission color can be tuned by controlling the applied pulse condition. Figure 8.20a shows typical time-resolved EL spectra. At times less than 1 μs, the red emission from the DCJTI is the main contribution of transient EL spectra. After 1 μs, the contribution of the blue light from FIrpic occurs. In other words, at response rates with pulse widths sufficiently faster than 1 μs, the intensity of blue emission from the FIrpic is reduced because of the longer rise time. The EL spectra of a white OLED were red-shifted with decreasing pulse width from 100 μs to 100 ns, as shown in Fig. 8.20b, although the spectrum was largely insensitive to dc drive current. EL spectra were changed from a white emission to an orange emission with the CIE coordinate of (0.5, 0.42) by varying applied pulse width and frequency. The color-tuning technique can be expected to be applied to color-tunable lighting devices, the color adjustment of illumination, etc.

8.2.2.3 Transient EL of White to Near-Infrared PHOLEDs with Ir and Pt Complexes

PHOLEDs, including both white and near-infrared emissions, can be expected to be used as illuminations with the sensing function. White emission can be obtained from single-layer polymer blends using blue and red emissive dyes. It is well-known that blue and red lights are absorbed by the chlorophylls, and near-infrared light markedly increases the rate of photosynthesis. From the viewpoint of large-area devices, solution-processed visible to near-infrared PHOLEDs can also be applied to effective light sources for use in plant-growing facilities because light sources for plant cultivation are needed to emit light in the blue, red, and near-infrared spectral ranges responsible for phototropism, photosynthesis, and photomorphogenesis, respectively. A carbazole derivative, poly(9-vinylcarbazole), PVCz is used as a host material for PHOLED. For a printed PHOLED, the emissive layer, which

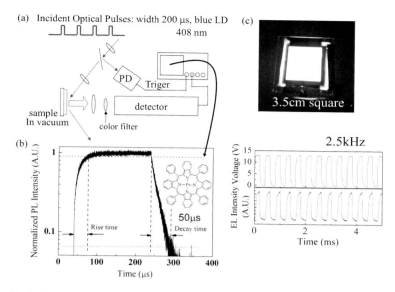

Fig. 8.21 (**a**) Typical transient PL measurement system, (**b**) typical transient PL of a PVCz:Pt (tpbp) film, and (**c**) typical transient EL of a white to near-infrared PHOLED (3.5 cm^2) with FIrpic, Ir(piq)$_3$ and Pt(tpbp) doped in PVCz as the emissive layer

consisted of the host polymer of PVCz doped with electron transport material of 1,3-bis[2-(4-tert-butylphenyl)-1,3,4-oxadiazol-5-yl]benzene (OXD-7) and various emissive dopants, FIrpic, tris(1-phenylisoquinoline)iridium(III) (Ir(piq)$_3$), and Pt-tetraphenyltetrabenzoporphyrin (Pt(tpbp)), was fabricated by spin-coating method. The doping ratios of red and near-infrared dopants were lower than that of a blue dopant for the host because the efficient cascade energy transfer occurred. White to near-infrared OLEDs covering the wavelength range from 450 nm to 850 nm, which consisted of ITO/PEDOT:PSS(40 nm)/PVCz:OXD-7:FIrpic:Ir (piq)$_3$:Pt(tpbp)(100:80:11.7:0.3:0.1 wt.%, 100 nm)/CsF/Mg:Ag/Ag, exhibited the maximum luminance of approximately 3000 cd/m^2 and the maximum current efficiency of 7 cd/A [30, 37].

A typical transient PL measurement system is shown in Fig. 8.21a. The transient response time in PL was measured using 200 μs-wide rectangular optical pulses from a 408 nm violet semiconducting laser diode (LD). Both the PL rise and decay profiles of a PVCz:Pt(tpbp) film were estimated to be approximately 50 μs, as shown in Fig. 8.21b. That is, the device can be expected to be utilized as pulse light sources driven at less than 10 kHz owing to the longer lifetime of Pt(tpbp) compared to other Ir complexes (~1 μs). 2.5 kHz modulated clear optical pulses were observed by applying 12 V pulse voltage to the device with an active area size of approximately 12 cm^2, as shown in Fig. 8.21c. White to near-infrared PHOLEDs can be suitable candidates for artificial lighting sources with kHz modulation for optical sensor and plant cultivation because of their advantages in large-area device fabrication.

8.2.2.4 Improvement of PEDOT:PSS for Blue PHOLEDs

A PEDOT:PSS layer on the transparent conducting oxide electrodes (TCOs) such as ITO, which acts as the hole injection layer, is the most widely used configuration to improve the interface between TCO and an organic layer. The hole injection layer of PEDOT:PSS was modified by blending Nafion® perfluorinated ionomer (PFI) [38] because exciton interaction between the hole injection layer and the EML increases the probability of the nonradiative process.

For the solution process, polymer materials are mainly used. In these materials, a carbazole derivative, PVCz, is generally used as a host material for PHOLEDs due to its high triplet level for some phosphorescent complexes. However, compared to polymer materials, small molecular materials have high purity and can be easily purified. A small molecular carbazole derivative, 3,6-di(9-carbazolyl)-9-(2-ethylhexyl)carbazole (TCz1) [39–41] which has side chains for solubility in organic solvent and high triplet level, in order to realize the highly efficient devices. Using PEDOT:PSS:PFI layer, the influences of charge balance and exciton distribution of solution-processed blue PHOLEDs with FIrpic doped in TCz1 are investigated.

The emissive layers (EMLs), which consisted of the host small molecular of TCz1 blended with hole transport material of 1,1-bis((di-4-tolylamino) phenyl) cyclohexane (TAPC) and electron transport material of 1,3-bis[(4-tert-butylphenyl)-1,3,4-oxadiazolyl]phenylene (OXD-7) and emissive dopant of FIrpic, were formed by spin-coating method into approximately 100 nm thick layers on PEDOT: PSS layer. In order to obtain highly efficient blue PHOLED, we optimized the doping concentrations of FIrpic, TAPC, and OXD-7 relative to the host of TCz1. The different compositions of TCz1: TAPC: OXD-7: FIrpic in weight ratio were prepared (100: x: 60: 22.5, $x = 0, 20$).

PLQY of an emissive layer is one of the important factors in achieving the high radiative quantum efficiency of the emitting material, ϕ_p. Figure 8.22 shows PL spectra of EML of TCz1:TAPC:OXD-7:FIrpic, PEDOT:PSS/EML, and PEDOT:PSS:PFI/EML films. The energy transfer between the TCz1 and FIrpic mainly occurred via the triplet state of TCz1 and the metal-ligand charge transfer (MLCT) state of FIrpic. The excited energy state is also transferred from the singlet state of TCz1 to the singlet MLCT state and then transferred to the triplet (MLCT) state via the intersystem crossing. For all films, the peak wavelength of approximately 470 nm with a slight shoulder and the blue light emission from FIrpic were observed. The PL intensity of PEDOT:PSS/EML was lower than that of EML. PLQYs of EML and PEDOT:PSS/EML films were estimated to be 58% and 29%, respectively. PL quenching was clearly observed owing to the exciton diffusion process between PEDOT:PSS and EML. On the other hand, PL intensity and PLQY of PEDOT:PSS:PFI/EML film were almost the same as that of EML film, as shown in Fig. 8.22 and Table 8.1. The composition of PEDOT:PSS:PFI of 1:6:50 in weight ratio was prepared. By adding the PFI, exciton quenching at the interface between PEDOT:

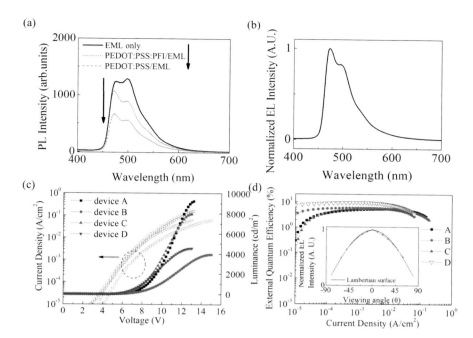

Fig. 8.22 (**a**) PL spectra of EML, PEDOT:PSS/EML, and PEDOT:PSS:PFI/EML films. (**b**) Typical EL spectrum of device D. (**c**) *J-V-L* characteristics and (**d**) current density dependence of EQE in the devices A, B, C, and D. The inset of Fig. 8.22d shows normalized EL intensity versus viewing angle characteristic of device D. The solid line corresponds to a Lambertian emitter

Table 8.1 Device performance of devices A, B, C, and D, including the voltage measured at 1000 cd/m^2, the maximum luminance (L_{max}), the maximum EQE (η_{ext_max}), and PLQY

Device	Voltage (V)	L_{max} (cd/m^2)	η_{ext_max} (%)	PLQY (%)
A:W/O	8.6	9200	5.8	29
B:PFI	9.8	3900	5.9	48
C:TAPC	8.2	7900	5.7	29
D:TAPC, PFI	8.0	4500	9.6	48

PSS and EML is suppressed. This result indicates that exciton diffusion is hard to occur between PEDOT:PSS:PFI and EML.

The device structures are as follows:

Device A: ITO/PEDOT:PSS(40 nm)/TCz1:OXD-7:FIrpic (110 nm)/CsF/Mg:Ag/Ag

Device B: ITO/PEDOT:PSS:PFI (60 nm)/TCz1:OXD-7:FIrpic (110 nm)/CsF/Mg:Ag/Ag

Device C: ITO/PEDOT:PSS (40 nm)/TCz1:OXD-7:TAPC:FIrpic (90 nm) /CsF/Mg:Ag/Ag

Device D: ITO/PEDOT:PSS:PFI (60 nm)/TCz1:OXD-7:TAPC: FIrpic (90 nm)/ CsF/Mg:Ag/Ag

Table 8.1 summarizes the device performance of the various PHOLEDs. Typical EL spectrum of device D is presented in Fig. 8.22b. For all devices, the peak wavelength and the shape of the EL spectrum agree with those of the PL spectrum, which indicates that the EL originates from the excited state of FIrpic.

Figure 8.22c, d shows the *J-V-L* characteristics and current density dependence of η_{ext} in devices A, B, C, and D. Typical normalized EL intensity versus viewing angle characteristic of device D is indicated in the inset of Fig. 8.22d. All devices show nearly Lambertian emissions. η_p is assumed to be almost 20%. All devices exhibited a gradual decrease in η_{ext} at the high current density owing to triplet-triplet annihilation. Device A without PFI showed a maximum luminance of 9200 cd/m^2 and a maximum η_{ext} of 5.8%. For device A, γ is estimated to be almost 1 as ϕ_p is 29% from PLQY of the PEDOT:PSS/EML film. For the device B with PFI, at lower current density, EQE characteristic was improved. The maximum η_{ext} of device B was almost the same as that in device A. This result indicates that the carrier balance factor is decreased by blending PFI as ϕ_p is 48% from PLQY of the PEDOT:PSS: PFI/EML film. The operation voltage at 1000 cd/m^2 increased to 9.8 V.

There is a high barrier for hole injection due to a wide bandgap between PEDOT: PSS, which has the work function of -5.1 eV, and TCz1, which has the ionized potential of -5.8 eV [6]. The performance of devices is often limited by hole injection. We investigated the blend effect of a hole transport material, TAPC, which has a high triplet level and lower HOMO level of -5.5 eV into an emissive layer. The operation voltages at 1000 cd/m^2 of devices C and D with TAPC decreased compared to that of device B. Therefore, by blending TAPC, hole injection ability was improved, and lower driving voltage was obtained. The maximum η_{ext} of device D with PFI was higher than that of device C without PFI and estimated to be approximately 10%. For device D, γ is estimated to be almost 1.

To further understand the factors affecting the electrical properties due to PFI, these behaviors of the electrical properties of the devices were analyzed using impedance spectroscopy measurements. Figure 8.23 show the modulus (*M*) plot of hole-only devices of ITO/PEDOT:PSS (40 nm) or PEDOT:PSS:PFI (60 nm)/TCz1: OXD-7:TAPC:FIrpic (90 nm)/MoO$_x$/Au/Ag without and with PFI, and PHOLEDs of ITO/PEDOT:PSS or PEDOT:PSS:PFI/TCz1:OXD-7:TAPC:FIrpic (90 nm)/CsF/ Mg:Ag/Ag without and with different thicknesses of PEDOT:PSS:PFI layers. For a hole-only device with PFI, the new semicircle was observed at the high-frequency region in the *M* plot, compared to a hole-only device without PFI. This shows that the equivalent circuit of the device can be expressed by units connected serially, each of which is composed of capacitance (*C*) and resistance (*R*) connected in parallel. The semicircle at the low frequency is assigned to EML by assuming the effective dielectric constant to be $\varepsilon = 3$. PEDOT:PSS is simply represented by resistance. This suggests that PEDOT:PSS acts as the hole injection layer. For the PHOLEDs with PFI, the diameter of the new semicircle at high-frequency region in the *M* plot increased with increasing the thickness of PEDOT:PSS:PFI. It is clear that the semicircle at high frequency is assigned to the PEDOT:PSS:PFI layer. This result

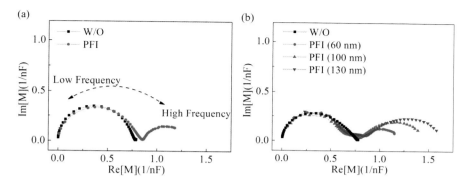

Fig. 8.23 M plots of (**a**) hole-only devices without and with PFI and (**b**) PHOLEDs without and with different thicknesses of PEDOT:PSS:PFI layers

indicates the accumulated carriers occurred, and PEDOT:PSS:PFI layer acts not only as a hole injection layer but also as a hole transport layer.

By blending PFI, the radiative quantum efficiency is improved owing to the suppression of exciton quenching. The devices in this study have a single EML, and the charge balance of holes and electrons injected in the EML might be influenced by the carrier transport characteristics of each material. Thus, good hole injection leads to the improvement of carrier valance factor, and this is the reason why the η_{ext} was improved.

8.3 AC-Driven Insulated Organic Electroluminescent Devices

Light-emitting device technology has been developed using two different driving modes, i.e., direct current (DC) and alternating current (AC). Field-induced EL of either inorganic or organic emissive materials under AC has been extensively developed [42–47]. AC-driven organic EL device (ACEL) exhibits the unique device architecture in which an emitting layer is separated with an insulator from the electrode, which offers novel design freedom.

Dielectric nanostructures for light management have attracted interest owing to their potential for improving OLED performance. Recently, current injection lasing from organic semiconducting lasers with distributed feedback gratings has been demonstrated [48]. Developing organic light-emitting devices with resonators is essential for enhancing optical coupling and suppressing optical losses. The concept of using printed inorganic/organic hybrid distributed Bragg reflectors (DBRs) utilizing inorganic semiconductors and insulating polymers included in ACEL with microcavity structure (μcACEL), as shown in Fig. 8.24, has been proposed to provide an approach to achieve spectral narrowing and strong forward directionality [49]. The large refractive index contrast of approximately 0.5 between inorganic

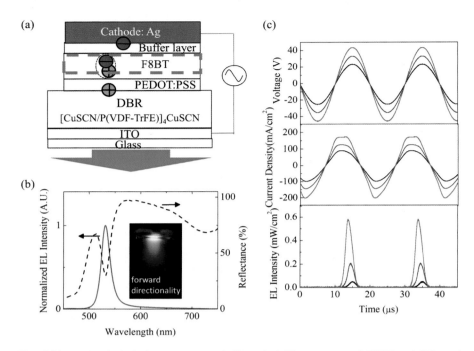

Fig. 8.24 (**a**) Typical device structure and (**b**) typical EL spectrum of F8BT μcACEL and reflectance spectrum of a film consisting of a μcACEL structure with Ag. (**c**) Typical transient voltage, current density, and EL signal of μcACEL

copper (I) thiocyanate, CuSCN, and insulating polymer of poly(vinylidene fluoride-trifluoroethylene), P(VDF-TrFE), results in the fabrication of solution-processed inorganic-organic hybrid dielectric DBRs with high reflectivity (approximately 90%) from nanostructures consisting of only four bilayers. A wide-bandgap inorganic semiconductor, copper (I) thiocyanate, CuSCN, exhibits high optical transparency in the visible light region and a high refractive index of around 1.9 [50]. P (VDF-TrFE), one of the ferroelectric and fluorine-based polymers, exhibits both a low refractive index at the optical frequency and high permittivity at the low frequency for ACEL operation.

The applied AC field assists in the formation of excitons and light generation. A μcACEL based on F8BT exhibited the spectral narrowing of the EL spectrum, the green emission with the CIE coordinates of (0.28, 0.68), and strong forward directionality. A film with Ag appeared with a sharp transmission dip with a peak transmissivity of approximately 70% centered at approximately 530 nm, which indicates that the resonant cavity mode represents a standing optical wave arising from the particular spacing between the semitransparent DBR and the Ag cathode. The peak wavelength of the EL spectrum in μcACEL is in good agreement with the dip of the peak of the reflectance spectrum.

The EL emission was not observed at low frequencies close to DC. EL intensity increased with AC voltage and frequency. This is because the current flowing through the device increases since the capacitive reactance of the device is inversely proportional to the frequency. The typical transient EL of μcACEL is shown in Fig. 8.24c. There exists a phase difference between the voltage and current waveforms owing to the capacitance of the dielectric mirror. Large sinusoidal displacement currents flow in the insulating layer during AC operation and linearly increase with applied AC voltages. The EL emission was observed only in the positive half of the AC cycle. PEDOT:PSS layer could allow the generation of holes in the subsequent negative voltage [51]. For the operation mechanism of this ACEL, holes are accumulated at the dielectric mirror side under a reverse AC electric field. Then the emission generation in the F8BT emissive layer occurs under a forward AC electric field owing to electron injection from the Ag electrode and hole injection from PEDOT:PSS layer.

This simple concept of using printed periodically alternating layers of inorganic CuSCN/insulating polymer could become attractive in the fabrication of large-area surface light-emitting devices with a strong directed emission pattern. Besides, ACELs can be applied to various interactive and sensing displays [52, 53].

8.4 Organic Light-Emitting Transistors

OLEDs operate in the viewpoint of device physics based on amorphous organic materials. The space-charge-limited current (SCLC) or trap-charge-limited current (TCLC) theory has been mainly employed to analyze the J–V characteristics of OLEDs. On the other hand, organic light-emitting transistors (OLETs) fabricated using crystalline organic materials are multifunctional devices that combine the light emission property of an OLED with the switching property of a field-effect transistor in a single-device architecture [54].

The field-effect transistor is a three-terminal active device that uses an electric field to control the current flow between the source/drain (S/D) electrodes. Single-layer ambipolar OLETs allow the efficient formation of excitons as provided by effective p/n junctions [55, 56]. In such devices, a saturated electron channel and a saturated hole channel are positioned in series to form a p/n junction within the OLET channel, resulting in charge recombination via excitons, which in turn leads to efficient light emission. For ambipolar OLETs, the equation for the electron and hole drain current, I_D, in the channel between S/D electrodes is as follows [57]:

$$I_D = \frac{WC_i}{2L}\left[\mu_e(V_G - V_{th,e})^2 + \mu_h\{V_D - (V_G - V_{th,h})\}^2\right] \quad (8.2)$$

where the gate dielectric capacitance, C_i, is equal to $\varepsilon\varepsilon_0 t^{-1}$ (ε_0, ε, and t are the permittivity, relative permittivity and thickness of the gate insulator), W is the channel width, L is the channel length, μ_e is the electron field-effect mobility, μ_h is

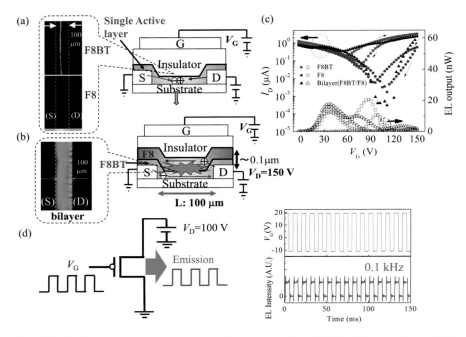

Fig. 8.25 Typical device structures and emission patterns of (**a**) F8BT or F8 single-layer OLET and (**b**) an F8BT/F8 heterostructure OLET. (**c**) Typical transfer and EL output characteristics of various OLETs (L: 100 μm, W: 2 mm) at V_D = 150 V. (**d**) Typical transient EL of an F8BT/F8 bilayer OLET (L: 20 μm, W: 2 mm) under applied V_G pulses of 0.1 kHz and V_D = 100 V

the hole field-effect mobility, $V_{th,e}$ is the electron threshold voltage, and $V_{th,h}$ is the hole threshold voltage.

Liquid-crystalline semiconducting polymers were self-organized, owing to the reorientation of molecules and the increase in the size of crystalline regions during thermal annealing. In particular, poly(alkylfluorene), a liquid–crystalline semiconducting polymer, exhibits blue emission and various morphological behaviors [58, 59]. Top-gate-type devices based on various crystalline fluorene-type polymers such as poly(9,9-dioctylfluorene), F8, have been reported to exhibit both ambipolar and light-emitting properties [60–63]. For the F8BT single-layer device with various S/D electrodes, such as ITO, Ag, Ag nanowire [64], and carbon nanotube [65], both the ambipolar characteristic and a line-shaped emission pattern are obtained, as shown in Fig. 8.25. The electron mobility in a device in which the F8BT film is in the mesophase is approximately ten times greater than that for F8BT films annealed at other temperatures [66]. The hole and electron field-effect mobilities of top-gated OLETs based on F8 and F8BT were estimated to be above 10^{-3} cm^2 V^{-1} s^{-1} by following the Eq. (8.2).

The multilayer active layer contributes to the improvement of OLET characteristics [67]. The concept of using an ambipolar bilayer heterostructure in polymer-based OLETs is introduced to provide a novel approach to achieving surface emission [68–71]. In-plane light emission pattern in the F8BT/F8 heterostructure device based on bilayer with F8 upper and F8BT lower layers has been achieved. Hole accumulation in the F8 upper layer affects electron injection from the electrode to the F8BT lower layer and the probability of recombining holes in the upper layer with electrons in the lower layer near the bilayer interface. Carrier balance between bilayers is a critical factor in achieving in-plane emission at the almost full-channel area when hole transport is dominant in the F8 upper layer, which acts as an electron-blocking layer, and electrons are injected into the F8BT lower layer. The optical pulses of above 0.1 kHz frequency are generated by directly modulating an F8BT/F8 bilayer OLET with an in-plane emission pattern with a channel length of 20 μm, as shown in Fig. 8.25d.

Since emission occurs in the narrow channel between the S/D electrodes, the OLET device design leads to the possibility of fabricating the micro-light source by patterning the S/D electrodes for an organic integrated optical sensor. The field-effect mobility of an OLET based on a crystallized film is higher than the bulk mobility measured in an OLED structure based on an amorphous film. OLETs can be driven at a high current density, which can be expected to achieve high-luminance light-emitting devices. However, the development of printed p-channel and n-channel materials with both high efficiency and mobility has been far more difficult than those with high mobility. Novel material synthesis and material techniques for printing processing for large-area flexible electronics become important regarding device integration on flexible substrates.

8.5 Organic Photodetectors

Many organic semiconductors have a high absorption coefficient in the order of 10^5 cm^{-1}. In general, a few hundreds nanometers of the active layer are thick enough to absorb an adequate amount of light. OPDs with thin active layer exhibit typical photovoltaic characteristics. The equation for IPCE of OPDs is as follows:

$$\text{IPCE} = \eta_A \times \eta_{ED} \times \eta_{CT} \times \eta_{CC} \qquad (8.3)$$

where η_A is the light absorption efficiency for exciton generation within the photoactive layer, η_{ED} is the exciton diffusion efficiency to the donor/acceptor (D/A) interface, η_{CT} is the charge transfer efficiency for exciton dissociation into a free electron and hole pair at the D/A interface, and η_{CC} is the charge collection efficiency at the electrodes. The D/A interface structure is essential to obtain improved characteristics.

From the viewpoints of organic photovoltaic cells, essential factors are absorption wavelength, filling factor, open circuit voltage, and short-circuit current. On the other hand, the electrical characteristics under reverse bias, frequency response, on/off ratio, and color sensitivity should also be evaluated for OPDs. One of the

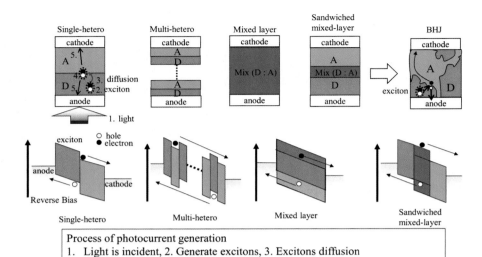

Fig. 8.26 Device structures and schematic images of energy diagram

Table 8.2 Comparison of four types of device structures

	Single-hetero	Multi-hetero	Mixed	Sandwiched mixed
Sensitivity	Low	High	Low	High
Speed	High	Low	Low	High

most important features of OPDs is the high-frequency response. Frequency response increases with applied reverse voltages because dissociation in an active layer is accelerated by applying the bias voltage.

CuPc has an absorption band at the blue and red regions with peaks at approximately 340 nm and 625 nm, respectively. BPPC shows an absorption band with two peaks at around 500 nm in the green region. Only the CuPc layer mainly absorbs red light. Under illuminated red light, frequency performances in four types of OPD with single-hetero, multi-hetero, mixed-layer, and sandwiched mixed layer based on CuPc and BPPC were described for an optoelectrical conversion device, as shown in Fig. 8.26 [6, 72, 73]. A comparison of four types of device structures is shown in Table 8.2.

The frequency responses of sandwiched mixed-layer OPDs with various composition ratios of CuPc are shown in Fig. 8.27a. The response speed improved with increased CuPc content owing to the increase of the carrier collection length, and at 75 vol.%, the response speed is higher. The mixed-layer device (75 vol.%) exhibited the worse frequency response, as shown in Fig. 8.27b. Sandwiched mixed-layer OPD was suitable for high-efficiency and high-speed device operations. The cutoff frequency was above 10 MHz for OPDs with the single-hetero and sandwiched mixed layer (active area of 0.03 mm^2). Furthermore, the frequency response of the

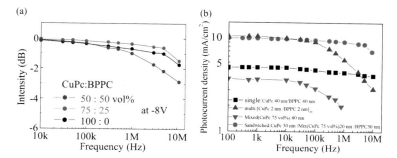

Fig. 8.27 (a) Frequency responses of the sandwiched mixed-layer OPDs of ITO/CuPc(30 nm)/CuPc:BPPC(20 nm)/BPPC(30 nm)/Ag with various composition ratios of CuPc, and (b) frequency dependence of photocurrent densities in four types of OPDs (active area: 0.03 mm^2) with single-hetero, multi-hetero, mixed-layer (75 vol.%), and sandwiched mixed layer (75 vol.%) at -8 V under illuminated red light (100 mW/cm^2)

mixed-layer device increased with the power of incident light owing to confined carriers in the hetero-barrier and trapped carriers in the heterointerface [72].

Fullerene derivatives doped in several conducting polymers act as effective quenchers and electron acceptors, and this photophysics is known as ultrafast photoinduced charge transfer. For the bulk heterojunction (BHJ) OPDs, η_{ED} and η_{CT} could be expected to be close to 100% for the BHJ devices as the photoactive region is close to the D/A interface. BHJ OPDs with large D/A interface areas have relatively high IPCE. The device performance of polymer photodetectors with BHJ is markedly influenced by the donor-acceptor interface.

Poly(3-hexyllthiophene) (P3HT) and [6,6]-phenyl C61-butyric acid methyl ester (PCBM) blends have been studied for organic bulk heterojunction photovoltaic cells because of high hole mobility and a broad absorption spectrum, and high photoelectric conversion efficiencies have consequently been achieved. However, for P3HT:PCBM device, the cutoff frequency of ~1 MHz has been reported, and the frequency response is not high [74].

For polymer photodetectors with BHJ, FOPA treatment results not only in the lowering of the injection barrier at the ITO/organic layer interface but also in the lowering of the contact resistance between ITO and the organic layer, which is estimated by impedance spectroscopy [19]. BHJ OPDs based on fluorene-type poly (9,9-dioctylfluorene-*co*-bithiophene), F8T2, and PCBM exhibited improved IPCE by using FOPA treatment and can operate at above 10 MHz at -4 V under the sinusoidally modulated violet light. For OPDs with smaller device sizes (below the active area of 0.03 mm^2), the RC time constant is approximately several ns order as the differential resistance of the device is below several hundred ohms under the illuminated red light of high intensity such as 100 mW/cm^2. Vacuum-deposited and printed OPDs with smaller active areas (0.01 mm^2) have the potential to detect 100 MHz optical pulses [73, 75].

The hemispherical focal plane organic detector arrays are integrated on a 1.0 cm radius plastic hemisphere by using direct transfer patterning technology [76]. The fabrication of large-area flexible scanners using an organic photoactive layer on a PEN substrate has been demonstrated, which indicates the possibility of flexible organic electronic devices [77]. A flexible organic image sensor array integrated into an active matrix backplane with a-Si:H thin-film transistors fabricated on PEN has been reported [78]. The coating method and new material development are essential for developing the OPD industry based on integrated devices for high-speed operation.

8.6 Conclusion

The fundamental properties of OLEDs and OPDs for optical link and sensor applications were discussed and reviewed. Printed OLEDs and OPDs can operate into MHz modulation. The transient EL response of PHOLED mainly depends on the phosphorescent recombination lifetime. OLEDs with smaller device sizes (below the active area of 0.03 mm^2), the RC time constant is approximately several ns order as the differential resistance of the device is below several hundred ohms at higher voltages (current densities). Transient EL of emerging devices such as ACEL and OLET for kHz modulation has been demonstrated. From the viewpoint of organic/organic semiconductor, oxide semiconductor/organic semiconductor, and electrode/organic semiconductor interfaces, it is essential to investigate fabrication methods further and driving methods for organic photonic devices that can be driven at high speed.

References

1. M.A. Baldo, S. Lamansky, P.E. Burrows, M.E. Thompson, S.R. Forrest, Appl. Phys. Lett. **75**, 4 (1999)
2. H. Uoyama, , K. Goushi, , K. Shizu, H. Nomura, C. Adachi, Nature, 492, (2012) 234–238
3. Y. Ohmori, H. Kajii, M. Kaneko, K. Yoshino, M. Ozaki, A. Fujii, M. Hikita, H. Takenaka, T. Taneda, IEEE J. Sel. Top. Quant. Electron. **10**, 70–78 (2004)
4. H. Kajii, T. Tsukagawa, T. Taneda, K. Yoshino, M. Ozaki, A. Fujii, M. Hikita, S. Tomaru, S. Imamura, H. Takenaka, J. Kobayashi, F. Yamamoto, Y. Ohmori, Jpn. J. Appl. Phys. **41**, 2746 (2002)
5. H. Kajii, T. Taneda, Y. Ohmori, Thin Solid Films **438–439**, 334–338 (2003)
6. T. Morimune, H. Kajii, Y. Ohmori, IEEE J. Disp. Technol. **2**, 170 (2006)
7. K. Tamao, M. Uchida, T. Izumizawa, K. Furukawa, S. Yamaguchi, J. Am. Chem. Soc. **118**(47), 11974–11975 (1996)
8. T. Morimune, H. Kajii, Y. Ohmori, Jpn. J. Appl. Phys. **44**, 2815–2817 (2005)
9. H. Kajii, T. Kojima, Y. Ohmori, IEICE Trans. Electron. **E94-C**, 190 (2011)
10. Y. Shao, Y. Yang, Adv. Mater. **17**, 2841 (2005)
11. Z. Xu, Y. Wu, B. Hu, Appl. Phys. Lett. **89**, 131116 (2006)
12. H. Kajii, A. Katsura, H. Ohmori, Y. Sato, T. Hamasaki, Y. Ohmori, J. Noncry Sol. **358**, 2504–2507 (2012)

13. H. Kajii, H. Ohmori, Y. Sato, A. Katsura, Y. Ohmori, Mater. Res. Soc. Symp. Proc. **1435**, 24–29 (2012)
14. C.M. Lochner, Y. Khan, A. Pierre, A.C. Arias, Nat. Commun. **5**, 5745 (2014)
15. T. Yokota, P. Zalar, M. Kaltenbrunner, H. Jinno, N. Matsuhisa, H. Kitanosako, Y. Tachibana, W. Yukita, M. Koizumi, T. Someya, Sci. Adv. **2**, e1501856 (2016)
16. A.K. Bansal et al., Wearable organic optoelectronic sensors for medicine. Adv. Mater. **27**, 7638 (2015)
17. H. Kajii, T. Tsukagawa, T. Taneda, Y. Ohmori, IEICE Trans. Electron. **E85-C**, 1245 (2002)
18. A. Sharma, B. Kippelen, P.J. Hotchkiss, S.R. Marder, Appl. Phys. Lett. **93**, 163308 (2008)
19. Y. Sato, H. Kajii, Y. Ohmori, Org. Electron. **15**, 1753 (2014)
20. H. Kajii, Y. Mohri, H. Okui, M. Kondow, Y. Ohmori, Jpn. J. Appl. Phys. **57**, 03DA03 (2018)
21. M. Meier, S. Karg, W. Riess, J. Appl. Phys. **82**, 1961 (1997)
22. J. Scherbel, P.H. Nguyen, G. Paasch, W. Brütting, M. Schwoerer, J. Appl. Phys. **83**, 5045 (1998)
23. T. Okachi, T. Nagase, T. Kobayashi, H. Naito, Thin Solid Films **517**, 1327 (2008)
24. S. Ishihara, H. Hase, T. Okachi, H. Naito, J. Appl. Phys. **110**(3), 036104 (2011)
25. T. Tamura, H. Kageyama, Y. Shirota, H. Kajii, Y. Ohmori, Mol. Crys. Liq. Crys. **538**, 98–102 (2011)
26. H. Ohishi, M. Tanaka, H. Kageyama, Y. Shirota, Chem. Lett. **33**, 1266 (2004)
27. M.A. Baldo, C. Adachi, S.R. Forrest, Phys. Rev. B **62**, 10967 (2000)
28. C. Adachi, M.A. Baldo, S.R. Forrest, Appl. Phys. Lett. **77**, 904 (2000)
29. H. Kajii, N. Takahota, Y. Wang, Y. Ohmori, Jpn. J. Appl. Phys. **50**, 04DK05 (2011)
30. H. Kajii, N. Takahota, Y. Wang, Y. Ohmori, Mater. Res. Soc. Symp. Proc. **1286**, 312 (2010)
31. H. Kajii, N. Takahota, Y. Sekimoto, Y. Ohmori, Jpn. J. Appl. Phys. **48**, 04C176 (2009)
32. T. Mori, K. Miyachi, T. Mizutani, J. Phys. D **28**, 1461 (1995)
33. H. Murata, C.D. Merritt, Z.H. Kafafi, IEEE J. Sel. Top. Quantum Electron. **4**, 119 (1998)
34. M. Matsumura, T. Furukawa, Jpn. J. Appl. Phys. **40**, 3211 (2001)
35. H. Kajii, K. Takahashi, J.-S. Kim, Y. Ohmori, Jpn. J. Appl. Phys. **45**, 3721 (2006)
36. C. Hosokawa, H. Tokailin, H. Higashi, T. Kusumoto, Appl. Phys. Lett. **60**, 1220 (1992)
37. H. Kajii, K. Kimpara, Y. Ohmori, Thin Solid Films **518**, 551–554 (2009)
38. T.W. Lee, Y. Chung, O. Kwon, J.-J. Park, Adv. Funct. Matter. **17**, 390–396 (2007)
39. M.-H. Tsai, Y.-H. Hong, C.-H. Chang, H.-C. Su, C.-C. Wu, A. Matoliukstyte, J. Simokaitiene, S. Grigalevicius, J.V. Grazulevicius, C.-P. Hsu, Adv. Mater. **19**, 862–866 (2007)
40. N. Seidler, S. Reineke, K. Walzer, B. Lüssem, A. Tomkeviciene, J.V. Grazulevicius, K. Leo, App. Phys. Lett. **96**, 093304 (2010)
41. N. Oda, H. Kajii, Y. Ohmori, Mol. Crys. Liq. Crys. **581**, 70–75 (2013)
42. P.D. Rack, P.H. Holloway, Mater. Sci. Eng **R21**, 171–219 (1998)
43. Y. Pan, Y. Xia, H. Zhang, J. Qiu, Y. Zheng, Y. Chen, W. Huang, Adv. Mater. **29**, 1701441 (2017)
44. T. Tsutsui, S.B. Lee, K. Fujita, Appl. Phys. Lett. **85**, 2382–2384 (2004)
45. A. Perumal, B. Lüssem, K. Leo, Appl. Phys. Lett. **100**, 103307 (2012)
46. A. Perumal, M. Fröbel, S. Gorantla, T. Gemming, B. Lüssem, J. Eckert, K. Leo, Adv. Funct. Mater. **22**, 210–217 (2012)
47. Y. Chen, Y. Xia, H. Sun, G.M. Smith, D. Yang, D. Ma, D.L. Carroll, Adv. Funct. Mater. **24**, 1501–1508 (2014)
48. A.S.D. Sandanayaka, T. Matsushima, F. Bencheikh, S. Terakawa, W.J. Potscavage Jr., C. Qin, T. Fujihara, K. Goushi, J.-C. Ribierre, C. Adachi, Appl. Phys. Express **12**, 061010 (2019)
49. H. Kajii, M. Yoshinaga, T. Karaki, M. Morifuji, M. Kondow, Org. Electron. **88**, 106011 (2021)
50. P. Pattanasattayavong, G.O.N. Ndjawa, K. Zhao, K.W. Chou, N. Yaacobi-Gross, B.C. O'Regan, A. Amassian, T.D. Anthopoulos, Chem. Commun. **49**, 4154–4156 (2013)
51. J.H. Lee, S.H. Cho, R.H. Kim, B. Jeong, S.K. Hwang, I. Hwang, K.L. Kim, E.H. Kim, T.-W. Lee, C.P. Ark, J. Mater. Chem. C **4**, 4434–4441 (2016)

52. E.H. Kim, S.H. Cho, J.H. Lee, B. Jeong, R.H. Kim, S. Yu, T.-W. Lee, W. Shim, C. Park, Nat. Commun. **8**, 14964 (2017)
53. S.W. Lee, S.H. Cho, H.S. Kang, G. Kim, J.S. Kim, B. Jeong, E.H. Kim, S. Yu, I. Hwang, H. Han, T.H. Park, S.-H. Jung, J.K. Lee, W. Shim, C. Park, ACS Appl. Mater. Interfaces **10**, 13757–13766 (2018)
54. A. Hepp, H. Heil, W. Weise, M. Ahles, R. Schmechel, H. von Seggern, Phys. Rev. Lett. **91**, 157406 (2003)
55. J. Zaumseil, C.R. McNeill, M. Bird, D.L. Smith, P.P. Ruden, M. Roberts, M.J. McKiernan, R.H. Friend, H. Sirringhaus, J. Appl. Phys. **103**, 064517 (2008)
56. M.C. Gwinner, D. Kabra, M. Roberts, T.J.K. Brenner, B.H. Wallikewitz, C.R. McNeill, R.H. Friend, H. Sirringhaus, Adv. Mater. **24**, 2728–2734 (2012)
57. R. Schmechel, M. Ahles, H. von Seggern, J. Appl. Phys. **98**, 084511 (2005)
58. Y. Ohmori, M. Uchida, K. Muro, K. Yoshino, Jpn. J. Appl. Phys. **30**, L1941–L1943 (1991)
59. M. Grell, D.D.C. Bradley, G. Ungar, J. Hill, K.S. Whitehead, Macromolecules **32**, 5810–5817 (1999)
60. J. Zaumseil, C.L. Donley, J.-S. Kim, R.H. Friend, H. Sirringhaus, Adv. Mater. **18**, 2708–2712 (2006)
61. H. Kajii, K. Koiwai, Y. Hirose, Y. Ohmori, Org. Electron. **11**, 509–513 (2010)
62. K. Koiwai, H. Kajii, Y. Ohmori, Synth. Met. **161**, 2107–2112 (2011)
63. I. Ikezoe, H. Tanaka, K. Hiraoka, H. Kajii, Y. Ohmori, Org. Electron. **15**, 105–110 (2014)
64. K. Hiraoka, Y. Kusumoto, I. Ikezoe, H. Kajii, Y. Ohmori, Thin Solid Films **554**, 184 (2014)
65. Y. Ohmori, H. Kajii, IEICE trans. Electron (Japanese Edition) **J99-C**, 659 (2016)
66. H. Tanaka, H. Kajii, Y. Ohmori, Synth. Met. **203**, 10–15 (2015)
67. S. Toffanin, R. Capelli, W. Koopman, G. Generali, S. Cavallini, A. Stefani, D. Saguatti, G. Ruani, M. Muccini, Laser Photonics Rev. **7**, 1011–1019 (2013)
68. H. Kajii, H. Tanaka, Y. Kusumoto, T. Ohtomo, Y. Ohmori, Org. Electron. **16**, 26–33 (2015)
69. H. Kajii, K. Hashimoto, M. Hara, T. Ohtomo, Y. Ohmori, Jpn. J. Appl. Phys. **55**, 02BB03 (2016)
70. T. Ohtomo, K. Hashimoto, H. Tanaka, Y. Ohmori, M. Ozaki, H. Kajii, Org. Electron. **32**, 213–219 (2016)
71. H. Kajii, Jpn. J. Appl. Phys. **57**, 05GA01 (2018)
72. T. Morimune, H. Kajii, Y. Ohmori, Jpn. J. Appl. Phys. **45**, 546–549 (2006)
73. T. Morimune, H. Kajii, Y. Ohmori, IEEE Photon. Technol. Lett. **18**, 2662 (2006)
74. L. Salamandra, L.L. Notte, C. Fazolo, M.D. Natali, S. Penna, L. Mattiello, R.D. Duca, A. Reale, Org. Electron. **81**, 105666 (2020)
75. T. Hamasaki, T. Morimune, H. Kajii, S. Minakata, R. Tsuruoka, T. Nagamachi, Y. Ohmori, Thin Solid Films **518**, 548–550 (2009)
76. X. Xu, M. Davanco, X. Qi, S.R. Forrest, Org. Electron. **9**, 1122–1127 (2008)
77. T. Someya, Y. Kato, S. Iba, Y. Noguchi, T. Sekitani, H. Kawaguchi, T. Sakurai, IEEE Trans. Electron Dev. **52**, 2502 (2005)
78. T.N. Ng, W.S. Wong, M.L. Chabinyc, S. Sambandan, R.A. Street, Appl. Phys. Lett. **92**, 213303 (2008)

Chapter 9
Microfluidic Self-Emissive Devices

Takashi Kasahara and Jun Mizuno

Abstract Recently, electrogenerated chemiluminescence (ECL) cells and liquid organic light-emitting diodes (OLEDs) have attracted attention for unique self-emissive display applications. A solution-based emitting layer is used in the ECL cells, while a liquid organic semiconductor-based emitting layer is used in the liquid OLEDs. Conventional ECL solutions have been typically prepared by dissolving a single luminescent material in an organic solvent. These devices have a simple structure and been fabricated by sandwiching an emitting layer between two glass substrates with transparent electrodes. Although their structure and easy fabrication process are advantageous for large-area displays, the luminescent performances have been still significantly poor in comparison with state-of-the-art solid-state OLEDs which are one of the most developed self-emissive devices. Our research group has attempted to improve the performances of both the ECL cells and liquid OLEDs and proposed novel device structures. In this chapter, we will introduce our progress on developing the host-guest ECL solutions, the highly efficient ECL solution doped with an emitting assist dopant, and microfluidic self-emissive devices having the ECL solutions and liquid organic semiconductors. Various ECL emissions can be realized by doping fluorescent guest materials such as perylene, quinacridone, and anthracene derivatives into the host solution. The ECL performances of 5,6,11,12-tetraphenyltetracene (rubrene) were found to be significantly enhanced by adding a styrylamine derivative to the rubrene solution. The microfluidic devices, which can exhibit both ECL and electroluminescence (EL) emissions, have been fabricated by microelectromechanical system processes and heterogeneous bonding techniques.

T. Kasahara (✉)
Department of Electrical and Electronic Engineering, Hosei University, Koganei-shi, Tokyo, Japan
e-mail: t.kasahara@hosei.ac.jp

J. Mizuno
Academy of Innovative Semiconductor and Sustainable Manufacturing, National Cheng Kung University, Tainan, Taiwan
e-mail: junmizuno@gs.ncku.edu.tw

Keywords Electrogenerated chemiluminescence · Electroluminescence · Liquid organic semiconductor · Host-guest system · Emitting assist dopant · Microfluidic devices

9.1 Introduction

Today, solid-state organic light-emitting diodes (OLEDs) have been used commercially in various displays such as mobile phones and high-resolution televisions. OLED is a self-emissive device consisting of functional organic thin films sandwiched between two electrodes. Electroluminescence (EL) is produced by the recombination of holes and electrons in an emitting layer (EML). In general, the EML has been composed of a wide-energy-gap host material and a guest material. Carbazole, anthracene, and fluorene derivatives have been used as the host [1–3]. Fluorescent, phosphorescent, and thermally activated delayed fluorescence (TADF) materials are doped into the host matrix as the guest at a concentration of only a few weight percent [4–6]. Hole injection/transport layers (HIL/HTL) and electron injection/transport layers (EIL/ETL) have been used to balance the number of holes and electrons in the EML. These layers are generally formed by a vacuum evaporation or a solution-coating method such as spin coating and inkjet printing. On the other hand, instead of organic thin films, self-emissive devices using a solution- or a liquid-based EML have been also investigated. An electrogenerated chemiluminescence (or electrochemiluminescence (ECL)) cell usually uses luminescent materials dissolved in an organic solvent, while a liquid OLED consists of a liquid organic semiconductor-based EML.

In general, an annihilation-type ECL is recognized to be a light emission phenomenon where the excited luminescent material is produced by electron transfer reactions between its reduced and oxidized species [7–10]. In recent years, much interest has been focused on the ECL cells because of their potential for future unique display applications. The basic structure of a conventional cell consists of only an ECL solution sandwiched by a pair of transparent electrodes [11–29]. Commercial indium tin oxide (ITO) and fluorine doped tin oxide (FTO) substrates have been used as the electrodes. The thickness of the EML is adjusted to be several μm to several tens of μm by using spacer materials such as glass bead, plastic film, and photoresist. Figure 9.1a shows a photograph of the ECL solutions under 365 nm UV irradiation. They are very simply prepared by dissolving luminescent materials in an organic solvent. 5,6,11,12-tetraphenyltetracene (rubrene), 9,10-diphenylanthracene (DPA), and ruthenium complex (tris(2,2′-bipyridine)ruthenium (II) hexafluorophosphate ($Ru(bpy)_3(PF_6)_2$)), whose chemical structures are shown in Fig. 9.1b, are representative luminescent materials, and their ECL characteristics are being investigated by researchers. Organic solvents such as acetonitrile (ACN), propylene carbonate (PC), and tetrahydrofuran (THF) have been used to prepare the solution. Figure 9.1c shows photographs of the fabrication process of a rubrene-based ECL cell. An epoxy-based negative photoresist (SU-8) can be used to form the spacer on a FTO anode substrate (Fig. 9.1c-1). First, the solution containing rubrene

9 Microfluidic Self-Emissive Devices

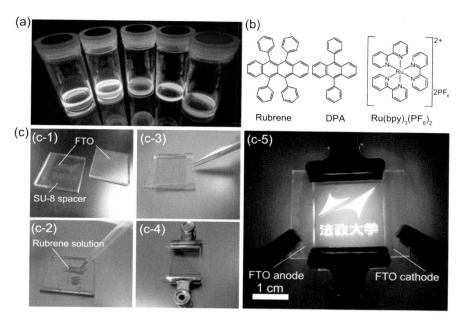

Fig. 9.1 (a) Photograph of the prepared ECL solutions under 365 nm UV irradiation. (b) Chemical structures of rubrene, DPA, and Ru(bpy)$_3$(PF$_6$)$_2$. (c) Photographs of the ECL cell fabrication process

is dropped on the anode substrate with the SU-8 spacer (Fig. 9.1c-2), and the FTO cathode substrate was then placed on top of the anode substrate (Fig. 9.1c-3). Finally, two substrates are clipped together (Fig. 9.1c-4). When a direct current (DC) voltage is applied to the cell, yellow ECL emission can be clearly observed (Fig. 9.1c-5). Besides, the ECL cell has been reported to be driven not only with the DC voltage but also with an alternating current (AC) voltage. The ECL intensity of the Ru(bpy)$_3^{2+}$-based cell having several tens of μm-thick spacer can be improved by the AC voltage because the generated oxidized and reduced species do not need to diffuse to the opposite electrode [20]. Electrochemical properties of the luminescent material are usually evaluated by cyclic voltammetry (CV) in a three-electrode cell. Figure 9.2a–c show the CV curves of rubrene, DPA, and Ru(bpy)$_3^{2+}$. Here, a glassy carbon disk was used as a working electrode, while a platinum wire coil and a silver wire were used as counter and reference electrodes, respectively. Tetrabutylammonium hexafluorophosphate (TBAPF$_6$) was added to a mixed solvent of ACN and 1,4-dioxane (1:2 (v/v)) at 100 mM as a supporting electrolyte. They are found to exhibit distinct oxidation and reduction waves. As shown in Fig. 9.2d, blue and orange-red ECL emissions are also obtained from the cell having the DPA and Ru(bpy)$_3^{2+}$ solutions, respectively. The key to improving the ECL performance is considered to balance the reduced and oxidized species.

In 2009, a first liquid OLED was proposed by Adachi's group at Kyushu University [30]. In their work, a solvent-free liquid carbazole derivative

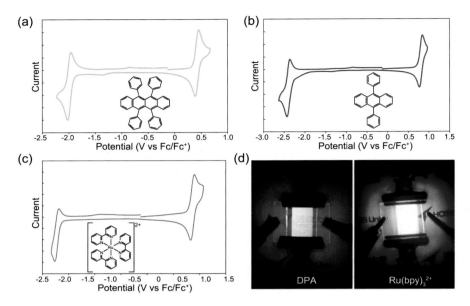

Fig. 9.2 CV characteristics of (**a**) rubrene, (**b**) DPA, and (**c**) Ru(bpy)$_3$(PF$_6$)$_2$ dissolved in a mixture of ACN and 1,4-dioxane (1:2 (v/v)) with 100 mM TBAPF$_6$ as a supporting electrolyte. (**d**) ECL emissions from the cells having DPA and Ru(bpy)$_3^{2+}$ solutions

(9-(2-ethylhexyl)carbazole (EHCz)) was used as the liquid host in the EML, and rubrene was doped into the EHCz host at 1 wt% as the guest. The device consisted of the rubrene-doped EHCz sandwiched between two ITO-coated glass substrates. Poly (3,4-ethylenedioxythiophene) polystyrene sulfonate (PEDOT:PSS) was coated on the ITO anode as the HIL, while a cesium carbonate (Cs$_2$CO$_3$)-based EIL was formed on the ITO cathode. The fabricated device [ITO anode/PEDOT:PSS (HIL)/ rubrene-doped EHCz (EML)/Cs$_2$CO$_3$ (EIL)/ITO cathode] exhibited the EL emission derived from rubrene. In 2011, Adachi's group improved the liquid OLED performance by using an electrolyte and a hole-blocking layer (HBL) [31]. In that work, TBAPF$_6$ was doped into the EHCz host at 0.1 wt% as the electrolyte, while titanium dioxide (TiO$_2$) layer was coated on the ITO cathode as the HBL by the sputtering method. A thiophene derivative (BAPTNCE) was used as a green guest. The maximum luminance of nearly 100 cd m^{-2} was obtained from the device [ITO anode/PEDOT:PSS (HIL)/EHCz doped with TBAPF$_6$ and BAPTNCE (EML)/TiO$_2$ (HBL)/ITO cathode]. When the voltage is applied to the device, anions (PF$_6^-$) and cations (TBA$^+$) of the electrolyte move toward the anode and cathode surfaces, respectively. Thereby, the electric dipole layers are formed at the interfaces between the EML and electrodes. This leads to the facilitation of the carrier injections from both the electrodes into the EML. Besides, because the TiO$_2$ layer has a deep valence band level (7.5 eV), a high energy barrier for holes is formed between the EML and HBL, which suppresses the hole leakage into the ITO cathode. As shown in Fig. 9.3a, solvent-free carbazole, naphthalene, and pyrene derivatives such as

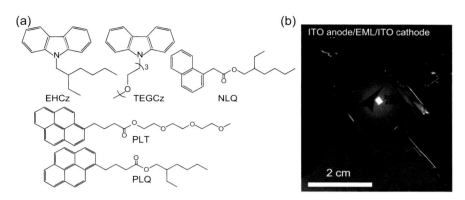

Fig. 9.3 (**a**) Chemical structure of liquid carbazole derivatives (EHCz and TEGCz), liquid naphthalene (NLQ), and liquid pyrene derivatives (PLT and PLQ). (**b**) Photograph of the liquid OLED having the PLQ host doped with DPT and the electrolyte

9-{2-[2-(2-Methoxyethoxy)ethoxy]ethyl}-9H-carbazole (TEGCz), 1-naphthaleneacetic acid 2-ethylhexyl ester (NLQ), 2-[2-(2-methoxyethoxy)ethoxy]ethyl 4-(pyren-1-yl)butanoate (PLT), and 1-pyrenebutyricacid 2-ethylhexyl ester (PLQ) have been reported to be used as the EMLs in the liquid OLEDs [32–40]. TEGCz and NLQ have been used as the liquid host, while a greenish blue EL emission has been obtained from the liquid OLEDs having PLT and PLQ as the emitter. Besides, PLQ has been also used as the liquid host for fluorescent guests. Figure 9.3b shows a photo of the EL emission from the liquid OLED having the PLQ host doped with 5,12-diphenyltetracene (DPT) and the electrolyte. This device was fabricated by sandwiching the liquid EML between the ITO anode and cathode.

Both the ECL cells and the liquid OLEDs have a simple structure and can be easily fabricated and driven in ambient air, which is advantageous for large-area display applications. In addition, as shown in Fig. 9.4a, the SU-8-based spacer is also easily fabricated. After the substrate is cleaned by dry cleaning method such O_2 plasma, the SU-8 resist is spin-coated on it to obtain a desired thickness. Then, the SU-8 layer is exposed to 365 nm UV lamp through the photomask. Finally, the substrate is developed in propylene glycol monomethyl ether acetate (PGMEA). As shown in Fig. 9.4b, only one light-emitting device that meets people's expectations is expected to be fabricated. These cells have the rubrene-based ECL solution as the EML. However, although ECL cells and liquid OLEDs have abovementioned unique properties, their luminescent characteristics such luminance and current efficiency have been still significantly low in comparison with those of solid-state OLEDs. Highly efficient OLEDs have been reported to have a maximum luminance of over 10,000 cd m^{-2} and a maximum current efficiency of over 40 cd A^{-1} one after another [41–43]. In addition, the OLED materials have been continually developed by many researchers and companies. By contrast, recently, novel ECL materials have been rarely reported.

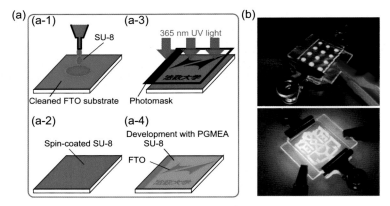

Fig. 9.4 (**a**) Fabrication process of the SU-8 spacer on an FTO substrate. (**b**) Rubrene-based ECL cells having various SU-8 spacer patterns

Our research group has focused on developing the highly luminescent solution- and liquid-based self-emissive devices from the point of view of both the EML material and the device structure [35–38, 44–55]. In 2013, we have proposed a microfluidic self-emissive device in order to integrate multiple ECL cells or liquid OLEDs on one substrate, and the device fabrication process has been developed by using a microelectromechanical system (MEMS) process and a heterogenous bonding technique [35]. This chapter introduces our recent studies on the host-guest ECL solutions, the highly luminescent rubrene-based ECL solution, and the functional microfluidic devices.

9.2 Fabrication of Electro-microfluidic Device on a Glass

Although conventional ECL cells and liquid OLEDs can be simply fabricated by sandwiching the EML between two glass substrates, multiple EMLs are difficult to be formed on one cell due to the use of the solutions or the liquids as the emitters. In the last few decades, microfluidic devices have been developed for a wide range of chemical and biological applications such as point-of-care diagnostics [56, 57], cell culturing [58], drug discovery [59, 60], and organic synthesis [61] because of their potential platforms for delivering and mixing small amounts of samples on a chip [62]. In general, the miniaturized components such as microchannels, microchambers, and micropumps are fabricated via MEMS technologies, including deposition, photolithography, and etching processes. Furthermore, several studies have been reported on incorporation of electrodes in the microchannels toward next-generation microfluidic devices [63, 64].

Figure 9.5a, b show the photograph of our developed electro-microfluidic device and its fabrication process. The device consists of several μm-thick microchannels

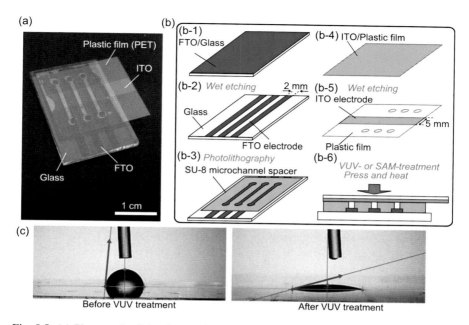

Fig. 9.5 (a) Photograph of the electro-microfluidic device and (b) its fabrication process. The device has the microchannels sandwiched between two transparent electrodes. (c) Water contact angles on the SU-8 before and after the VUV treatment

sandwiched between two electrodes. Transparent conductive oxides-coated glass and plastic film are used as the substrates. Polyethylene terephthalate (PET) and polyethylene naphthalate (PEN) have been used as the film substrate. The electrodes are patterned by photolithography and wet etching. The FTO layer is etched by hydrochloric acid with zinc powder, while the ITO layer is etched by aqua regia which is a mixture of nitric and hydrochloric acid. The SU-8-based microchannel spacer is formed on a glass substrate. Both the substrates can be bonded by using amine- and epoxy-terminated self-assembled monolayers (SAMs) [35–38, 44–47] or by a vacuum-ultraviolet (VUV)-assisted direct bonding technique [48–50]. The epoxy-amine boding process is well known and widely used for fabricating microfluidic devices [65]. The SAM-modified surfaces can be readily formed by immersing the O_2 plasma- or UV/O_3-treated substrates in the SAM solutions. We have used 3-aminopropyltriethoxysilane (APTES) and glycidyloxypropyltrimethoxysilane (GOPTS) for fabricating the microfluidic devices. On the other hand, in the case of the VUV-assisted direct bonding technique, a 172 nm Xe excimer lamp has been used to improve the surface hydrophilicity. As shown in Fig. 9.5c, the water contact angle on the SU-8 surface is drastically decreased after the VUV treatment. Polar functional groups such as OH, C=O, and O-C=O are expected to be formed on the surfaces of the SU-8 microchannel spacer and the ITO-patterned plastic film [66]. Two VUV-treated substrates can be bonded by applying appropriate physical pressure and heat.

Finally, multi-color EMLs are readily formed in one device by injecting emitters (ECL solutions or liquid organic semiconductors) into the microchannels.

9.3 Host-Guest ECL Solutions

In 1987, Tang and co-worker at Eastman Kodak fabricated an OLED device having a double-layer structure, and a maximum luminance of over 1000 cd m^{-2} was achieved at a DC voltage below 10 V [67]. Furthermore, in 1989, this group also proposed host-guest OLEDs [68]. In that work, tris(8-hydroxyquinoline)aluminum (III) (Alq$_3$) was used as the host, while coumarin and dicyanomethylene derivatives were used as guests. They found that the EL efficiency of the host-guest OLED was significantly increased in comparison with the non-doped OLED. Nowadays, the host-guest systems are indispensable for fabricating efficient OLED devices. In order to occur energy transfer from host to guest, the host materials have been required to have a wide energy gap between highest occupied and lowest unoccupied molecular orbitals (HOMO and LUMO) and a photoluminescence (PL) spectrum overlapped with an absorption spectrum of the guest.

On the other hand, the EML of the ECL cells is typically composed of a single luminescent material dissolved in an organic solvent. Thus, luminescent materials are required to be both oxidized and reduced in a well-balanced manner. However, few materials having such bipolar electrochemical property have been reported. Our research group has been investigating the solutions containing multiple luminescent materials, and recently, red, green, sky-blue, and white ECL solutions have been developed by referring to the EMLs of host-guest OLEDs [44–47, 52, 54, 55]. Compared with a conventional ECL solution containing a single species, the host-guest ECL system was found to be advantageous for tuning emission colors and achieving efficient ECL performances. The luminescent characteristics of the proposed solutions have been evaluated with the abovementioned electro-microfluidic device (see Fig. 9.5) or the simple structured cell. A mixture of ACN and 1,2-dichlorobenzene has been used as a solvent for preparing the host-guest ECL solutions, and 1,2-diphenoxyethane is added to the solution as an ion conductive assist dopant (ICAD). This ICAD system was proposed by Nishimura et al. in 2001 [12]. They demonstrated that the luminance of the rubrene-based cell is enhanced by using the 1,2-diphenoxyethane-doped solution. The radical cation of the luminescent materials is expected to be stabilized by forming complex with 1,2-diphenoxyethane and to collide with the radical anion efficiently.

9.3.1 Red Host-Guest Solution

Tetraphenyldibenzoperiflanthene (DBP), whose chemical structure is shown in the inset of Fig. 9.6a, is a well-studied fluorescent guest for highly efficient red OLEDs.

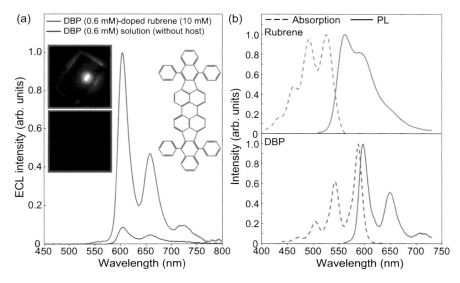

Fig. 9.6 (**a**) ECL spectra of the 5-μm-thick microfluidic device with DBP-doped rubrene and DBP solutions. The inset is a chemical structure of DBP. (**b**) Absorption and PL spectra of the rubrene host and the DBP guest. Reprinted from ref. [52] by permission of The Japan Institute of Electronics Packaging

In 2006, Okumoto et al. proposed the DBP-doped rubrene EML in OLEDs, and a high current efficiency of 5.4 cd A^{-1} was obtained [69]. This value was the highest among the red fluorescent OLEDs at that time. The EL characteristics of this EML have been investigated by many researchers [70–73]. In 2014, Adachi's group at Kyushu University enhanced the EL performance of the DBP-based OLEDs significantly by using the TADF material, which has a small energy gap between singlet and triplet energy states, as an assist dopant [74]. In that work, a maximum current efficiency of 25 cd A^{-1} was obtained. This system has been known as hyperfluorescence.

In ECL research, $Ru(bpy)_3(PF_6)_2$ is the most studied orange-red material, and several methods to improve its ECL performance have been reported [20–28]. However, the red ECL materials other than this ruthenium complex are rarely reported. In 2014, we have proposed a novel red solution containing both DBP and rubrene [44]. As shown in Fig. 9.6a, although ECL emission of the DBP solution is significantly week, the microfluidic device with the DBP-doped rubrene solution exhibits bright emission at 5.0 V [52]. The concentrations of the DBP guest and the rubrene host in the solution are 0.6 mM and 10 mM, respectively. Figure 9.6b shows the PL and absorption spectra of DBP and rubrene. It can be seen that their spectra are well overlapped with each other and the obtained ECL spectrum is identical to the PL spectrum of DBP. When the excited rubrene molecule is generated by collision of its radical anion and cation, this energy is expected to be transferred to the DBP guest. A maximum luminance of 63 cd m^{-2} is measured from the microfluidic device, whose EML is about 5 μm, at the DC voltage of 5.5 V [52].

9.3.2 Green Host-Guest Solution

Although many green guest materials for OLEDs such as coumarin, quinacridone, and anthracene derivatives and iridium complexes have been reported, there have been few green molecules that can be used for the ECL cells. Our research group has developed a novel green ECL solution by using DPA as the host [54]. In the solution, a quinacridone derivative, 5,12-dibutyl-1,3,8,10-tetramethylquinacridone (TMDBQA) was used as the guest, while a styrylamine derivative, 4,4′-bis [4-(diphenylamino)styryl]biphenyl (BDAVBi) was used as the assist dopant. Figure 9.7a illustrates their chemical structures. As shown in Fig. 9.7b, bright green ECL emission is produced at 5.5 V when both TMDBQA and BDAVBi are doped into the TBADN host solution. The obtained spectrum is almost the same as the PL spectrum of TMDBQA displayed in Fig. 9.7c. On the other hand, when both the host and the assist dopant are absent from the solution, the ECL emission was not observed from the device. Instead of DPA, an anthracene derivative, 2-*tert*-butyl-9,10-di(naphth-2-yl)anthracene (TBADN), can be also used as the host solution for TMDBQA [47]. In OLED research, TBADN, whose chemical structure is shown in the inset of Fig. 9.8a, has been widely used as both the host and the blue emitter

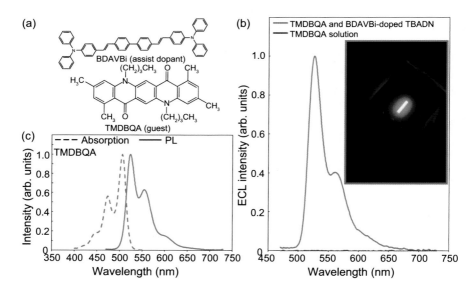

Fig. 9.7 (**a**) Chemical structures of TMDBQA and BDAVBi. (**b**) ECL spectra of the microfluidic ECL device with host-guest and guest solutions at 5.5 V. (**c**) Absorption and PL spectra of TMDBQA. Reprinted from ref. [54] by permission of The Japan Institute of Electronics Packaging

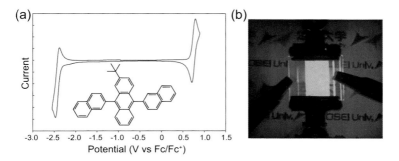

Fig. 9.8 (**a**) CV of TBADN in a mixture of ACN and 1,4-dioxane (1:2 (v/v)) with 100 mM TBAPF$_6$. (**b**) ECL emission from the cell having the TBADN solution at 7.0 V

[75, 76]. As shown in Fig. 9.8a, b, TBADN also exhibits distinct oxidation and reduction waves in the CV measurement, and its ECL emission can be obtained in the simple structured cell. The maximum luminance and the maximum current efficiency of the microfluidic device with the solution containing TMDBQA (guest), BDAVBi (assist dopant), and TBADN (host) have been measured to be 40.9 cd m^{-2} and 2.25 cd A^{-1}, respectively [47]. These values are the highest for a solution-based green ECL cell.

9.3.3 Sky-Blue Host-Guest Solution

Sky-blue ECL solutions can also be prepared by using the TBADN host and the guest dopant [50]. When an anthracene derivative, 9,9′-10,10′-tetraphenyl-2-2′--bianthracene (TPBA) (Fig. 9.9a), is doped into the host solution, sky-blue emission is clearly observed from the microfluidic device. As shown in Fig. 9.9b, the ECL intensity of the host-guest solution is higher than that of the guest solution. Because both the ECL spectra are identical with each other, the excited TPBA molecules are found to be produced efficiently by combining TPBA with the TBADN host. Figure 9.9c shows current density–voltage–luminance (*J–V–L*) characteristics of the microfluidic device with the host-guest solution. A luminance of 7.7 cd m^{-2} has been obtained at 6.0 V, which is still significantly lower than that of state-of-the-art sky-blue OLEDs. Thus, the methods for increasing the concentration of well-balanced radical ions have to be considered to further improve the ECL characteristics.

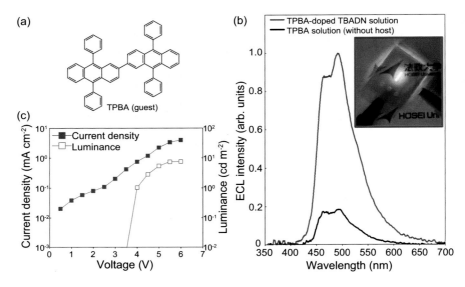

Fig. 9.9 (**a**) Chemical structure of TPBA. (**b**) ECL spectra of the device with host-guest and guest solutions at 6.0 V. (**c**) *J–V–L* characteristics of the device with the host-guest solution

9.3.4 White Host-Guest Solution

White OLEDs have attracted much attention for next-generation lighting source, and the EML containing both blue and yellow dopants has been widely used to produce the EL emission that covers the whole visible spectral range [77, 78]. In ECL research, Kobayashi's group at Chiba University has proposed AC-driven white ECL cell [14]. In that work, both the rubrene and DPA molecules are dissolved in *N*-methyl-2-pyrrolidone (NMP) with tetra-n-butylammonium perchlorate (TBAP) as an electrolyte. When a 300 Hz sine-wave AC voltage was applied to the cell, the ECL emissions of both rubrene and DBA were produced, which resulted in a white ECL emission. In 2019, we demonstrated the DC-driven white ECL cell employing the TBADN host solution doped with sky-blue and yellow guests [46]. The solution can be prepared by dissolving 2 mM BDAVBi and 0.34 mM tetra(*t*-butyl)rubrene (TBRb) in the solvent. As shown in Fig. 9.10a, white ECL emission is observed when the DC voltage of 6.5 V was applied to the cell consisting of two FTO-coated glass substrates with 5-μm-thick SU-8 spacer. Figure 9.10b shows the PL and absorption spectra of two guests. It was found that the ECL emissions of BDAVBi and TBRb are simultaneously produced. We also observed that the cell exhibited yellowish ECL emission at 4.0 V [46]. Besides, the emission color changed gradually from yellowish to whitish-yellow ECL emission when the voltage increased from 4.0 V to 6.0 V. Thus, the ECL emission may be produced not only by the energy transfer from the excited TBADN molecules to two guests but also collision between radical ions of the employed materials. We believe that a chromatic-stable

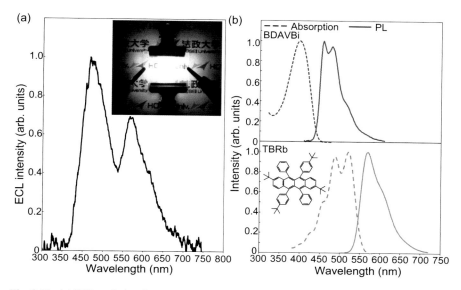

Fig. 9.10 (**a**) ECL emission from the cell having TBRb and BDAVBi-doped TBADN solution at 6.5 V. (**b**) Absorption and PL spectra of two guest molecules. Reprinted from ref. [46], Copyright 2020, with permission from Elsevier

white ECL device in a wide driving voltage can be developed when the behavior of radical ions of all the materials in the solution becomes clear.

9.4 Yellow ECL Solution Doped with an Emitting Assist Dopant

Rubrene is one of the most studied organic semiconducting materials and widely used in organic electronic devices such as OLEDs, organic field-effect transistors (OFETs), and organic light-emitting transistors (OLETs). In addition, its ECL phenomenon has been also studied since 1960s [7]. As described above, in 2001, Nishimura et al. proposed the ICAD system and prepared the rubrene solution doped with 1,2-diphenoxyethane as the ICAD in a mixed solvent of ACN and 1,2-dichlorobenzene in the ratio of 1:2 (v/v) [12]. In that work, the cell with the proposed solution exhibited a maximum luminance of 183 cd m^{-2} at the DC voltage of 8.0 V, which is 600 times higher than that of the cell without the ICAD. In 2017 and 2021, Kim et al. fabricated the AC-driven ECL cell with the rubrene solution, and the maximum luminance of over 30 cd m^{-2} was obtained [15, 17].

In 2023, we proposed a novel rubrene-based ECL cell having an emitting assist dopant and demonstrated that the ECL performances such as luminance, current efficiency, and device lifetime were drastically improved in comparison with conventional cell [55]. A styrylamine derivative (4-(di-p-tolylamino)-4′-[(di-p-

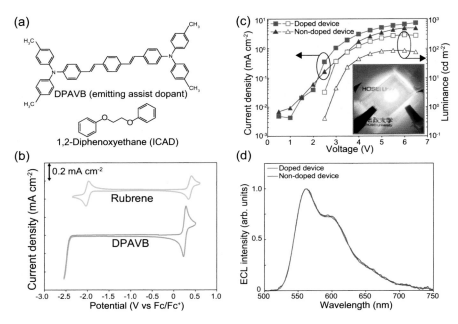

Fig. 9.11 (**a**) Chemical structures of DPAVB and 1,2-diphenoxyethane. (**b**) CVs of rubrene and DPAVB in the mixed solvent of ACN and 1,2-dichlorobenzene in the ratio of 1:2 (v/v) with 100 mM TBAPF$_6$ as the electrolyte. (**c**) J–V–L characteristics and (**d**) ECL spectra of the doped and non-doped devices

tolylamino)styryl]stilbene (DPAVB)), whose chemical structure is shown in Fig. 9.11a, was used as the emitting assist dopant. Figure 9.11b shows the CVs of rubrene and DVAVB in the mixed solvent of ACN and 1,2-dichlorobenzene in the ratio of 1:2 (v/v) with 100 mM TBAPF$_6$ as the electrolyte. It can be seen that DPAVB is more readily oxidized and more difficult to be reduced than rubrene. In other words, both HOMO and LUMO levels of DPAVB are shallower than those of rubrene. Figure 9.11c shows the J–V–L characteristics of the microfluidic device having the rubrene solution with and without DPAVB. The concentrations of rubrene and DPAVB in the mixed solvent are 10 mM and 4 mM, respectively, and 180 mM 1,2-diphenoxyethane (Fig. 9.11a) was added to the solution as the ICAD. It can be seen that the current density and luminance are drastically increased by doping DPAVB into the rubrene solution. The maximum luminance of the doped device was 292 cd m^{-2} at the DC voltage of 6.0 V, which is 3.4 times higher than that of the non-doped device (86.2 cd m^{-2}). The inset of Fig. 9.11c shows the photograph of the doped device at 6.0 V. Bright yellow ECL emission was readily observed by the naked eye in a lighted room. Furthermore, the maximum current efficiency of the doped device was measured to be 4.50 cd A^{-1} at 5.5 V, which is the highest efficiency reported to date. As shown in Fig. 9.11d, the obtained ECL spectra of the doped and non-doped devices at 5.0 V are identical to each other. It is likely that because DPAVB is more easily oxidized than rubrene, the excited rubrene molecules are generated by the electron transfer reaction not only between the radical anion and cation of rubrene (RUB$^{•-}$ and RUB$^{•+}$) but also between RUB$^{•-}$ and the radical cation of DPAVB (DPAVB$^{•+}$). Thus, a well-balanced number of radical anions (RUB$^{•-}$) and radical cations (RUB$^{•+}$ and DPAVB$^{•+}$) were produced in the doped device. Besides, the doped device showed the half-life time of 144 s at 4.0 V, which is 16 times longer than that of the non-doped device [55]. In future work, we will investigate the generation mechanism of the excited rubrene molecules in detail and prepare novel highly efficient ECL solutions containing the emitting assist dopant.

9.5 Microfluidic ECL Device Having an Electron Injection Layer

A ruthenium complex, Ru(bpy)$_3$(PF$_6$)$_2$, is a well-known orange-red ECL material, and in recent years, many studies have focused on improving ECL performances of the Ru(bpy)$_3^{2+}$-based cell [20–29]. In particular, metal oxide semiconductor nanoparticles (MOS NPs), which are coated on one side of the electrode, have been frequently used to enhance the ECL intensity. The surface area of the electrode can be increased by coating MOS NPs such as TiO$_2$ and zinc oxide (ZnO). The carriers are considered to be injected efficiently from the MOS NPs-coated electrode into Ru(bpy)$_3^{2+}$.

Recently, we have developed a fabrication process of the microfluidic ECL device having aluminum-doped ZnO (AZO) NPs as an EIL and improved device performances [49]. As shown in Fig. 9.12a, the AZO paste, which was prepared by mixing AZO NPs, polyethylene glycol, and deionized water, was coated manually on the FTO cathode surfaces in the SU-8 microchannel spacer using a pipette tip and then annealed at 250 °C on a hot plate to remove an organic substance. Figure 9.12b shows a scanning electron microscope (SEM) image of the AZO NPs. The NPs layer, whose thickness is approximately 1 µm, is found to be uniformly coated on FTO. Finally, the glass substrate with the SU-8 layer and the ITO anode-patterned PET film was bonded by the VUV-assisted direct technique as described in Sect. 9.2. Figure 9.12c shows the J–V–L characteristics of the device with and without the AZO NPs-based EIL. The solution containing 125 mM Ru(bpy)$_3$(PF$_6$)$_2$ in the PC solvent was injected into the 5-µm-thick microchannel as the emitter. It was found that both current density and luminance are increased by utilizing the AZO NPs-coated cathode. The luminance of the AZO NPs-based device was measured to be 175 cd m^{-2} at the DC voltage of 3.6 V, which is 3.5 times higher than that of the reference device (50 cd m^{-2} at 3.6 V). Besides, the current efficiency of 0.58 cd A^{-1} was obtained at 3.0 V, which is comparable to the AC-driven Ru(bpy)$_3^{2+}$-based cell [20, 21]. As shown in Fig. 9.12c, the obtained ECL spectra of both devices were identical to each other, indicating that the AZO-based device can enhance the performances without changing emission color. The proposed device

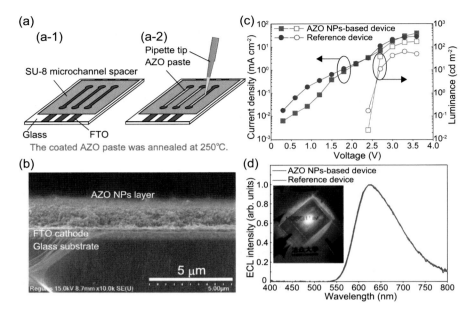

Fig. 9.12 (a) Patterning process of the AZO NPs layer on the FTO cathode. (b) SEM image of the AZO NPs layer on the FTO cathode. (c) *J–V–L* characteristics and (d) ECL spectra of the microfluidic device with and without AZO NPs-based electron injection layer at 3.0 V. Reprinted from ref. [49], Copyright 2022, with permission from Elsevier

also exhibited the half-life time of 30 s at a constant voltage of 3.0 V, which is 4.3 times longer than that of the reference device (7 s at 3.0 V). This suggests that many ruthenium complexes are well adsorbed on the AZO NP surfaces and an electron injection into them was facilitated. Consequently, the concentration of the reduced species (Ru(bpy)$_3^+$) was increased, which leads to the improvement of the balance of oxidized and reduced species of Ru(bpy)$_3^{2+}$ in the solution.

We also observed that the performance of the microfluidic device was enhanced by using the TiO$_2$ NPs-based EIL [48]. The TiO$_2$ paste coated in the microchannels was also annealed at 250 °C. The maximum luminance of the TiO$_2$-based device was 93.5 cd m^{-2} at 3.3 V. The TiO$_2$ NPs have been widely used as an ETL in dye-sensitized solar cells (DSSCs) and generally annealed at around 450–500 °C. However, both AZO and TiO$_2$ NPs in the abovementioned devices had to be annealed at 250 °C to prevent the degradation of the SU-8 microchannel spacer [48, 49]. Thus, an effect of the annealing temperature of the EIL paste on the performances of the Ru(bpy)$_3^{2+}$-based ECL cell has not been clearly understood. As shown in Fig. 9.13a, we fabricated the simple structured cell consisting of two FTO-coated glass substrates [53]. The TiO$_2$ paste, which was prepared by mixing polyethylene glycol, TiO$_2$ NPs (Aeroxide P25, Sigma-Aldrich), and deionized water, was spin-coated on the cathode substrate and then annealed at two different temperatures of 250 °C and 450 °C. Figure 9.13b shows the *J–V–L* characteristics of

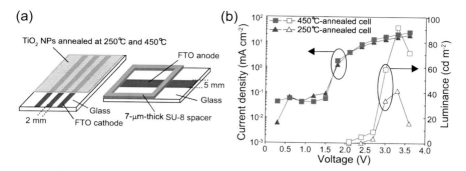

Fig. 9.13 (**a**) Schematic illustration of the simple structured cell having the TiO_2 NPs-based EIL. (**b**) J–V–L characteristics of the 250 °C- and 450 °C-annealed cells

two ECL cells. The 450 °C-annealed cell showed the improved luminance. The maximum luminance of the 450 °C-annealed cell was 91.9 cd m^{-2}, which is about 2.2 times higher than that of the 250 °C-annealed cell. This indicates that electron injection was facilitated by using the 450 °C-annealed TiO_2 NPs. In future work, several experiments such as SEM and X-ray diffraction (XRD) analyses have to be carried out to understand the mechanism of the improved performance. In addition, we will investigate the fabrication process of the microfluidic device having the TiO_2 NPs-based EIL treated by high-annealing temperature.

9.6 Microfluidic OLEDs

9.6.1 Evaluation of Microfluidic Device Having Liquid Organic Semiconductors

Recently, solvent-free functional molecular liquids have attracted attention for applications in novel organic electronic devices having stretchable and bendable features [79–82]. In particular, liquid pyrene derivatives are one of the most studied molecular liquids. As shown in Fig. 9.14a, we have studied the EL characteristics of the liquid pyrene derivative using an electro-microfluidic device [35–38]. The device has a 3 × 3 matrix of light-emitting pixels. The widths of the microchannels are designed to be 1000 μm, 1250 μm, and 1500 μm. The thickness of the EML can be adjusted by varying the SU-8 series (MicroChem Co.). Here, the 2.5-μm-thick and 6-μm-thick microchannels were fabricated by using SU-8 2002 and SU-8 3005, respectively. As shown in Fig. 9.14b, the SU-8 microchannels-patterned glass (anode substrate) and the ITO electrodes-patterned PEN film (cathode substrate) were bonded by using APTES- and GOPTS-SAMs. For the formation of the GOPTS-SAM only on the SU-8 layer, the ITO anodes in the microchannels were covered with a sacrificial layer (positive resist). This process plays an important role

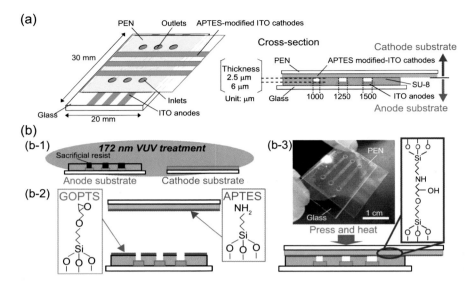

Fig. 9.14 (a) Design of a microfluidic OLED having a 3 × 3 matrix of light-emitting pixels. (b) Bonding process of anode and cathode substrates using APTTS- and GOPTS-SAMs. Reprinted from ref. [37], Copyright 2015, with permission from Elsevier

to fabricate shallow microchannels because the chemical reaction between APTES- and GOPTS-SAMs is expected to be prevented in the microchannels during the bonding process [35]. For the bonding process, both the anode and cathode substrates were treated by 172 nm VUV light for enhancing hydrophilicity of the SU-8, PEN, and ITO cathode surfaces. The VUV-treated anode and cathode substrates were subsequently immersed in 1% (v/v) GOPTS and 5% (v/v) APTES solutions prepared in deionized water, respectively. The anode substrate was subsequently rinsed with acetone, followed by isopropyl alcohol and deionized water to remove both the sacrificial resist and any unbound GOPTS-SAM, while the cathode substrate was rinsed with ethanol and deionized water. Finally, two substrates were bonded under contact pressure of 1.5 MPa at 140 °C for 5 min to form amine-epoxy bonds. The work functions of the ITO anode and APTES-modified ITO cathode, which were measured by a photoemission yield spectroscopy in air (Riken Keiki, AC-2), were 4.77 eV and 4.55 eV, respectively [37]. A reduced work function of the APTES-modified ITO may be attributed to the lone pair of electrons in the nitrogen atoms.

The fabricated device was clamped between two acrylic plates to ensure stable operation, as shown in Fig. 9.15a. The inlet nozzles are connected to one side of the acrylic plates. The liquid EMLs are formed on-demand by injecting liquid emitters from the inlet nozzles into the microchannels with syringes. The used emitters are individually collected from outlet nozzles. The spring-loaded probes embedded in the plates are utilized for the electrical connection between the device and source meter. As shown in Fig. 9.15b, PLQ (Nissan Chemical Co., Ltd.) was used as a

9 Microfluidic Self-Emissive Devices

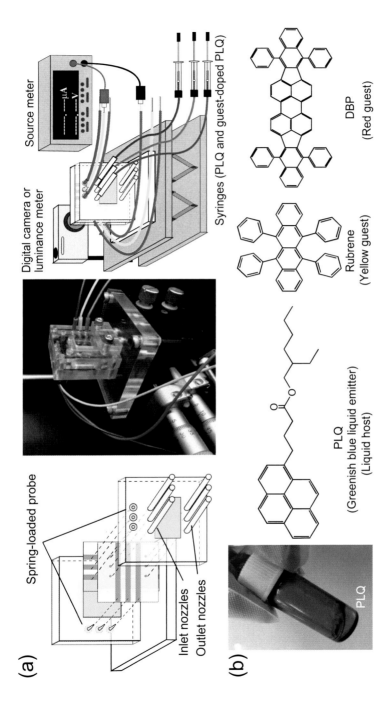

Fig. 9.15 (a) Experimental setup of the microfluidic device. (b) Chemical structures of PLQ (greenish blue emitter and liquid host), rubrene (yellow guest), and DBP (red guest). Reprinted from refs. [35, 37], Copyright 2013 and 2015, with permission from Elsevier

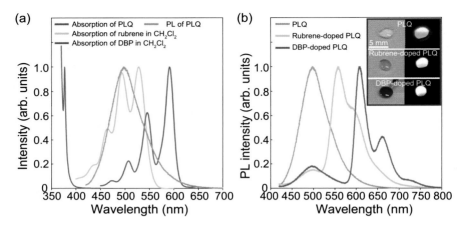

Fig. 9.16 (a) PL spectrum of the PLQ host and the absorption spectra of the host and guests. (b) PL spectra of the guest-doped PLQ under 365 nm UV irradiation. Reprinted from ref. [37], Copyright 2015, with permission from Elsevier

greenish blue emitter and a liquid host, while rubrene and DBP were doped into the PLQ host as the yellow and red guests, respectively [37]. Tributylmethylphosphonium bis(trifluoromethanesulfonyl)imide (TMP-TFSI) was also added to the EML at 0.25 wt% as the electrolyte. Dichloromethane (CH_2Cl_2) was used as the solvent for mixing the host, guest, and electrolyte, and then only the solvent was removed in the vacuum oven. The doped concentrations of rubrene and DBP in the EML were 2 wt% and 0.4 wt%, respectively.

As shown in Fig. 9.16a, the PLQ host has a strong UV absorption feature and exhibits the greenish blue PL emission with a maximum wavelength of 500 nm under the 365 nm UV light. Furthermore, it can be seen that the PL spectrum of PLQ has spectral overlaps with the absorption spectra of rubrene and DBP dissolved in the CH_2Cl_2 solvent. The photographs of the prepared liquid emitters are displayed in the inset of Fig. 9.16b. Similar to the neat PLQ, the guest-doped PLQ was found to be maintained in the liquid state. Figure 9.16b shows the PL spectra of the guest-doped PLQ. The 365 nm UV light was used for the selective excitation of only the PLQ host. Although small contributions from PLQ (around 500 nm) were observed, the maximum PL wavelengths of the rubrene-PLQ and DBP-doped PLQ were confirmed to be at 557 nm and 609 nm, respectively. This result indicates that Förster energy transfer occurred from the PLQ host to two guests.

Figure 9.17a shows photographs of the 6-μm-thick microfluidic device with the PLQ-based EMLs under 365 nm UV light and under the DC voltage of 70 V. It was found that PLQ is filled in the microchannels without leakage at the bonded interface, and the device exhibits greenish blue EL emissions at the light-emitting pixels. Furthermore, as shown in Fig. 9.17b, the multicolor EMLs are simply formed on one device. The EL spectra of the 6-μm-thick microfluidic device with PLQ, rubrene-doped PLQ, and DBP-doped PLQ are shown in Fig. 9.18a. It was found that

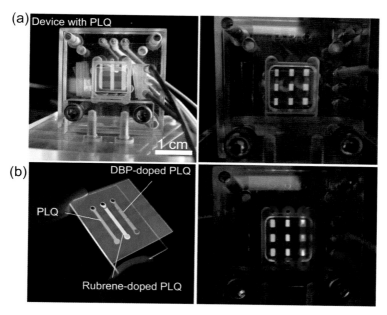

Fig. 9.17 (a) Device with the PLQ EML under 365 nm UV light and DC voltage of 70 V. (b) Device with PLQ, rubrene-doped PLQ, and DBP-doped PLQ EMLs. Reprinted from refs. [35, 37], Copyright 2013 and 2015, with permission from Elsevier

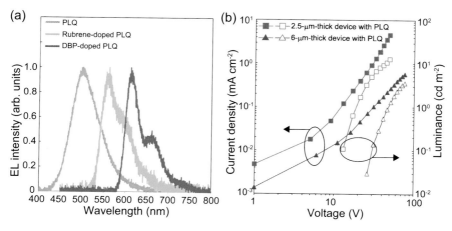

Fig. 9.18 (a) EL spectra of PLQ, rubrene-doped PLQ, and DBP-doped PLQ EMLs. (b) J–V–L characteristics of the 2.5-μm-thick and 6-μm-thick microfluidic devices with PLQ. Reprinted from ref. [37]. Copyright 2015, with permission from Elsevier

the obtained maximum EL wavelengths are identical to their PL spectra (see also Fig. 9.16b). This suggests that in the case of the device with the guest-doped PLQ, the excitons were mostly formed in the guest molecules. In addition, no significant

Fig. 9.19 Image of the EL emission degradation and recovery. The EL emission was recovered from the top edge of the light-emitting pixel by the reinjection of PLQ. Reprinted from ref. [37]. Copyright 2015, with permission from Elsevier

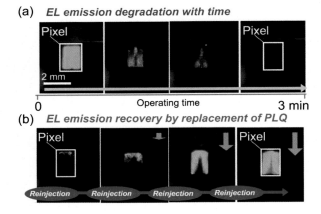

contributions from PLQ (500 nm) were observed in the EL spectra of the guest-doped PLQ. Thus, the exciton formation mechanisms are partially different between PL and EL emissions. In accordance with the energy-level diagram reported by Griffith and Forrest (2014), the LUMO levels of rubrene and DBP are located at 3.1 eV and 3.5 eV, respectively, while their HOMO levels are approximately 5.4 eV [83]. In addition, the HOMO and LUMO levels of the PLQ host are reported to be 5.8 eV and 2.6 eV, respectively [32]. Thus, the HOMO and LUMO levels of both guests are found to be inside the energy gap of the PLQ host. Therefore, the EL emission may be produced by the direct carrier recombination on guest emitters. This emission mechanism is often discussed in the literature on the solid-state OLEDs [84] and the liquid OLEDs [30]. Figure 9.18b shows the J–V–L characteristics of the 2.5-μm-thick and 6-μm-thick microfluidic devices with PLQ. It is found that the current density and luminance increase significantly with decreasing microchannel thickness. Furthermore, the turn-on voltage, which was defined at a luminance of above 0.01 cd m^{-2}, decreased with decreasing the thickness. The 2.5-μm-thick device exhibited the maximum luminance of 26.0 cd m^{-2} at the applied voltages of 61 V, and its turn-on voltage value was 13 V. However, in comparison with conventional solid-state OLEDs, the obtained luminance remained low. Therefore, for improvement of the PLQ-based device performance, the fabrication methodology of the submicron-thick electro-microchannels having functional thin films such as EIL, ETL, HIL, and HTL has to be developed.

Figure 9.19 shows the luminance recovery characteristics of the microfluidic device with PLQ. It can be clearly seen that luminance decreased with increasing operating time under the applied voltage of 70 V. This result indicates that the PLQ EML was decomposed during the carrier injection. Subsequently, a fresh PLQ was reinjected manually from the inlet nozzles into the target microchannel using a syringe under no-voltage conditions. When the constant voltage of 70 V was applied again, the recovery of the EL emission was observed at the top edge of the light-emitting pixel. This indicates the replacement of the decomposed PLQ with flesh

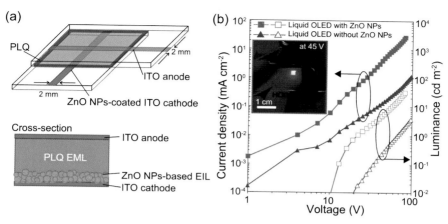

Fig. 9.20 (a) Schematic illustration of the liquid OLED having the ZnO NPs-based EIL. (b) *J–V–L* characteristics of the liquid OLEDs with and without EIL. The inset is a photograph of the liquid OLED cell with the ZnO NPs-based EIL at 45 V

one. This unique refreshable feature may be useful for future long-life display devices.

In 2020, we fabricated the liquid OLED cell having the ZnO NPs-based EIL in order to improve the EL performance of the PLQ-based device [85]. As shown in Fig. 9.20a, the ITO-coated glasses were used as both the anode and cathode substrates, and the 2-mm-wide electrode was patterned by wet etching with aqua regia. Commercial ZnO NPs (Avantama, N-10), which were dispersed in isopropyl alcohol, were spin-coated on the cathode. As shown in Fig. 9.20b, the proposed OLED [ITO anode/PLQ (EML)/ZnO NPs (EIL)/ITO cathode] exhibited the improved luminance and current density, which are more than ten times higher than those of the liquid OLED without the EIL. The maximum luminance of 32.3 cd m^{-2} at 82 V was measured. The surface area of the cathode was also increased by coating ZnO NPs on ITO. In addition, the work function of the ZnO NPs has been reported to be 4.1 eV [86], which is lower than that of ITO (4.8 eV). Thus, the electron barrier between the cathode and the LUMO level of PLQ was reduced, and the electron injections into PLQ were facilitated.

9.6.2 Flexible Electro-microfluidic Devices

The use of liquid emitters is expected to provide a new possibility for crack-free and flexible organic electronic devices. However, it is difficult to obtain flexible structure in the glass-based liquid OLED cell. Thus, it is important to development of the flexible devices that enable to keep single-μm-thick gap structures under the repeated bending. In 2014, we have developed the fabrication methodology of the flexible microfluidic OLEDs using a novel belt-transfer exposure technique [36].

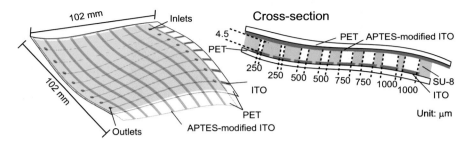

Fig. 9.21 Schematic illustration of the flexile microfluidic device. The ITO-coated PET films were used as both the anode and cathode substrates

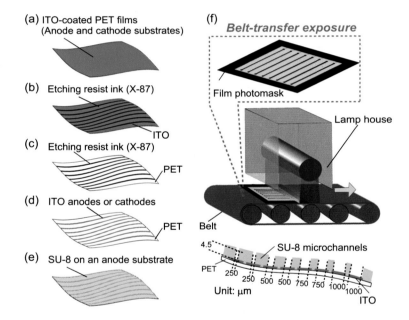

Fig. 9.22 Fabrication process of flexible SU-8 microchannels on the anode substrate by using a belt-transfer exposure technique

The device design is shown in Fig. 9.21. Here, instead of the glass substrate, the ITO-coated PET films were utilized as both anode and cathode substrates. The SU-8-based microchannels are sandwiched between an ITO anode and APTES-modified ITO cathode, and an 8 × 8 matrix of light-emitting pixels is formed in the flexible microfluidic device. The microchannel widths are designed to be 250 μm, 500 μm, 750 μm, and 1000 μm, while thickness is 4.5 μm. The dimension of the flexible microfluidic OLED was 102 mm × 102 mm. A fabrication process of flexible SU-8 microchannels is shown in Fig. 9.22. The etching resist ink (Taiyo Ink Mfg, X-87)

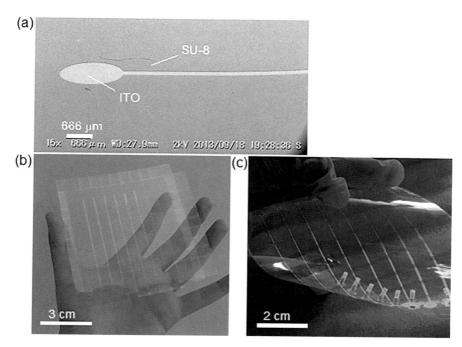

Fig. 9.23 (a) SEM image of the fabricated SU-8 microchannels on the anode substrate. (b) Fabricated flexible microfluidic device. (c) PL emission from the device with PLQ under 365 nm UV light. Reprinted from ref. [36]. Copyright 2014, with permission from Elsevier

was screen-printed on both the anode and cathode substrates (Fig. 9.22b), and then ITO was etched by a dilute aqua regia (Fig. 9.22c). The resist ink on the substrates was removed with acetone, and then the substrates were cleaned with isopropyl alcohol (Fig. 9.22d). The inlets and outlets for liquid emitters were formed on the cathode substrate. The SU-8 3005 was spin-coated on the anode substrate (Fig. 9.22e). The SU-8-coated substrate was subsequently manually aligned with a film photomask having the light interception pattern of the microchannels (Fig. 9.22e). Here, a belt-transfer exposure technique was utilized to form SU-8 microchannels, as shown in Fig. 9.22f. The exposure equipment (Japan Technology System Co., JU-C1500), which enables rapid exposure to large-area substrates, was used. The SU-8 was exposed to the UV lamp at the belt-transfer speed of 11.6 m min^{-1} and then developed. The anode and cathode substrates were bonded using APTES- and GOPTS-SAMs (see also Fig. 9.14b). Finally, PLQ was injected into the microchannels.

Figure 9.23 shows the SEM image of the fabricated SU-8 microchannels and photographs of the fabricated device. No significant bonding defects were confirmed in the microchannels and at the bonded interfaces. Furthermore, all the microchannels were filled with PLQ without leakage. The inset of Fig. 9.24 shows the photograph of the device under the voltage application. The greenish blue EL

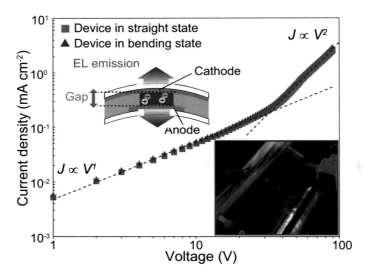

Fig. 9.24 J–V characteristics of the flexible microfluidic device with PLQ in straight and bending states. The inset is a photograph of the EL emission from the device in the bending state. Reprinted from ref. [36]. Copyright 2014, with permission from Elsevier

emission was clearly observed from the light-emitting pixels in the bending state. The J–V characteristics in straight and bending states (radius of 3 cm) are also shown in Fig. 9.24. Almost the same J–V curves were obtained in both states, indicating that the single-μm-thick microchannel structures were maintained in the bending state. In addition, the current density can be seen to increase proportionally with the increase in voltage up to 40 V ($J \propto V^1$). When the voltage higher than 40 V was applied to device, the current density increased steeply to be proportional to the square of the voltage ($J \propto V^2$). This may be due to the space-charge-limited current (SCLC). In 2011, Adachi's group has evaluated the J–V–L characteristics of the liquid OLED having the EHCz-based EML, and the SCLC behavior was also observed [31]. Based on this study, the proposed flexible microfluidic device is expected to be a highly promising technology toward future unique display applications.

9.7 Summary

This chapter focuses on the recent progress in the host-guest ECL solution, the highly luminescent rubrene solution doped with the emitting assist dopant, and functional microfluidic self-emissive devices. Novel red, green, and sky-blue ECL solutions are developed by doping DBP, TMDBQA, and TPBA molecules into the host solution. Besides, the ECL intensities of the host-guest solutions were found to be higher than those of the guest-only solution. In addition, white ECL emission can be obtained by producing yellow and sky-blue ECL emissions (TBRb and BDAVBi) simultaneously. The microfluidic ECL device with the AZO NPs-based electron injection layer, which was fabricated by the MEMS process and VUV-assisted direct bonding technique, exhibited bright orange-red emission derived from Ru(bpy)$_3^{2+}$. The fabricated microfluidic device exhibited not only ECL emissions of the prepared solutions but also EL emissions of the solvent-free molecular liquids. We expected that our finding will contribute to highly efficient solution-based and liquid-based flexible displays.

Acknowledgment The authors would like to acknowledge Dr. Ryoichi Ishimatsu (University of Fukui) for fruitful discussions. Moreover, we want to thank the former and current members of the Kasahara Lab at Hosei University, Mr. Koji Okada, Mr. Yugo Koinuma, Ms. Emiri Kato, Mr. Ryo Kawasaki, Mr. Yutaro Yamada, Mr. Shoma Hada, and Mr. Seiya Yamamoto.

References

1. Y. Seino, S. Inomata, H. Sasabe, Y.-J. Pu, J. Kido, Adv. Mater. **28**, 2638–2643 (2016)
2. K.-T. Wong, Y.-L. Liao, Y.-T. Lin, H.-C. Su, C.-c. Wu, Org. Lett. **7**, 5131–5134 (2005)
3. X.R. Wang, J.S. Chen, H. You, D.G. Ma, R.G. Sun, Jpn. J. Appl. Phys. **44**, 8480–8483 (2005)
4. H. Fukagawa, T. Shimizu, N. Ohbe, S. Tokito, K. Tokumaru, H. Fujikake, Org. Electron. **13**, 1197–1203 (2012)
5. W. Li, J. Li, D. Liu, D. Li, D. Zhang, Chem. Sci. **7**, 6706–6714 (2016)
6. H. Uoyama, K. Goushi, K. Shizu, H. Nomura, C. Adachi, Nature **492**, 234–238 (2012)
7. D.M. Hercules, Science **145**, 808–809 (1964)
8. K.N. Swanick, S. Ladouceur, E. Zysman-Colman, Z. Ding, Chem. Commun. **48**, 3179–3181 (2012)
9. R. Ishimatsu, S. Matsunami, T. Kasahara, J. Mizuno, T. Edura, C. Adachi, K. Nakano, T. Imato, Angew. Chem. Int. Ed. **126**, 7113–7116 (2014)
10. R. Ishimatsu, S. Tashiro, T. Kasahara, J. Oshima, J. Mizuno, K. Nakano, C. Adachi, T. Imato, J. Phys. Chem. B **123**, 10825–10836 (2019)
11. K. Nishimura, Y. Hamada, T. Tsujioka, K. Shibata, T. Fuyuki, Jpn. J. Appl. Phys. **40**, L945–L947 (2001)
12. K. Nishimura, Y. Hamada, T. Tsujioka, S. Matsuta, K. Shibata, T. Fuyuki, Jpn. J. Appl. Phys. **40**, L1323–L1326 (2001)
13. M. Honma, T. Horiuchi, K. Watanabe, T. Nose, Jpn. J. Appl. Phys. **53**, 112102 (2014)
14. T. Nobeshima, M. Nakakomi, K. Nakamura, N. Kobayashi, Adv. Optical. Mater. **1**, 144 (2013)
15. J.Y. Kim, S. Cheon, H. Lee, J.-Y. Oh, J.-I. Lee, H. Ryu, Y.-H. Kim, C.-S. Hwang, J. Mater. Chem. C **5**, 4214–4218 (2017)
16. H. Hwang, J.K. Kim, H.C. Moon, J. Mater. Chem. C **5**, 12513–12519 (2017)
17. J.Y. Kim, S. Cheon, D.K. Kim, S. Nam, J. Han, C.-S. Hwang, Y. Piao, J.-I. Lee, RSC Adv. **11**, 4682–4687 (2021)
18. T. Daimon, E. Nihei, J. Mater. Chem. C **1**, 2826–2833 (2013)
19. R. Nishimura, E. Nihei, Jpn. J. Appl. Phys. **55**, 042101 (2016)
20. T. Nobeshima, T. Morimoto, K. Nakamura, N. Kobayashi, J. Mater. Chem. **20**, 10630–10633 (2010)
21. T. Nobeshima, K. Nakamura, N. Kobayashi, Jpn. J. Appl. Phys. **52**, 05DC18 (2013)
22. S. Tsuneyasu, K. Ichihara, K. Nakamura, N. Kobayashi, Phys. Chem. Chem. Phys. **18**, 16317–16324 (2016)

23. S. Okamoto, K. Soeda, T. Iyoda, T. Kato, T. Kado, S. Hayase, J. Electrochem. Soc. **152**, A1677–A1681 (2005)
24. T. Kado, M. Takenouchi, S. Okamoto, W. Takashima, K. Kaneto, S. Hayase, Jpn. J. Appl. Phys. **44**, 8161–8164 (2005)
25. T. Tanaka, H. Takishita, T. Sagawa, S. Yoshikawa, S. Hayase, Chem. Lett. **38**, 742–743 (2009)
26. P. Chansri, Y.-M. Sung, Jpn. J. Appl. Phys. **55**, 02BB11 (2016)
27. P. Chansri, Y.-M. Sung, Surf. Coat. Technol. **306**, 309–312 (2016)
28. P. Chansri, J. Anuntahirunrat, J.-W. Ok, Y.-M. Sung, Vacuum **137**, 66–71 (2017)
29. S.H. Kong, J.I. Lee, S. Kim, M.S. Kang, ACS Photon **5**, 267–277 (2018)
30. D. Xu, C. Adachi, Appl. Phys. Lett. **95**, 053304 (2009)
31. S. Hirata, K. Kubota, H.H. Jung, O. Hirata, K. Goushi, M. Yahiro, C. Adachi, Adv. Mater. **23**, 889–893 (2011)
32. S. Hirata, H.J. Heo, Y. Shibano, O. Hirata, M. Yahiro, C. Adachi, Jpn. J. Appl. Phys. **51**, 041604 (2012)
33. K. Kubota, S. Hirata, Y. Shibano, O. Hirata, M. Yahiro, C. Adachi, Chem. Lett. **41**, 934936 (2012)
34. C.-H. Shim, S. Hirata, J. Oshima, T. Edura, R. Hattori, C. Adachi, Appl. Phys. Lett. **101**, 113302 (2012)
35. T. Kasahara, S. Matsunami, T. Edura, J. Oshima, C. Adachi, S. Shoji, J. Mizuno, Sens. Actuators A Phys. **195**, 219–223 (2013)
36. M. Tsuwaki, T. Kasahara, T. Edura, S. Matsunami, J. Oshima, S. Shoji, C. Adachi, J. Mizuno, Sens. Actuators A Phys. **216**, 231–236 (2014)
37. T. Kasahara, S. Matsunami, T. Edura, R. Ishimatsu, J. Oshima, M. Tsuwaki, T. Imato, S. Shoji, C. Adachi, J. Mizuno, Sens. Actuators B Chem **207**, 481–489 (2015)
38. N. Kobayashi, T. Kasahara, T. Edura, J. Oshima, R. Ishimatsu, M. Tsuwaki, T. Imato, S. Shoji, J. Mizuno, Sci. Rep. **5**, 14822 (2015)
39. N. Kobayashi, H. Kuwae, J. Oshima, R. Ishimatsu, S. Tashiro, T. Imato, C. Adachi, S. Shoji, J. Mizuno, J. Lumin. **200**, 19–23 (2018)
40. M. Kawamura, H. Kuwae, T. Kamibayashi, J. Oshima, T. Kasahara, S. Shoji, J. Mizuno, Sci. Rep. **10**, 14528 (2020)
41. S. Liu, C. Zang, J. Zhang, S. Tian, Y. Wu, D. Shen, L. Zhang, W. Xie, C.-S. Lee, Nano-Micro Lett. **14**, 14 (2022)
42. Z.-G. Wu, Y.-M. Jing, G.-Z. Lu, J. Zhou, Y.-X. Zheng, L. Zhou, Y. Wang, Y. Pan, Sci. Rep. **6**, 38478 (2016)
43. A. Perumal, H. Faber, N. Yaacobi-Gross, P. Pattanasattayavong, C. Burgess, S. Jha, M.A. McLachlan, P.N. Stavrinou, T.D. Anthopoulos, D.D.C. Bradley, Adv. Mater. **27**, 93–100 (2015)
44. T. Kasahara, S. Matsunami, T. Edura, R. Ishimatsu, J. Oshima, M. Tsuwaki, T. Imato, S. Shoji, C. Adachi, J. Mizuno, Sens. Actuators A Phys. **214**, 225–229 (2014)
45. T. Kasahara, R. Ishimatsu, H. Kuwae, S. Shoji, J. Mizuno, Jpn. J. Appl. Phys. **57**, 128001 (2018)
46. Y. Koinuma, R. Ishimatsu, H. Kuwae, K. Okada, J. Mizuno, T. Kasahara, Sens. Actuators A Phys. **306**, 111966 (2020)
47. Y. Koinuma, R. Ishimatsu, E. Kato, J. Mizuno, T. Kasahara, Electrochem. Commun. **127**, 107047 (2021)
48. K. Okada, R. Ishimatsu, J. Mizuno, T. Kasahara, Appl. Phys. Express **13**, 107001 (2020)
49. K. Okada, R. Ishimatsu, J. Mizuno, T. Kasahara, Sens. Actuators A Phys. **334**, 113329 (2022)
50. E. Kato, R. Ishimatsu, Y. Koinuma, J. Mizuno, T. Kasahara, Jpn. J. Appl. Phys. **61**, 060903 (2022)
51. N. Ichinohe, R. Ishimatsu, Jun Mizuno, T. Kasahara. Proc. 2022 Int. Conf. Electronics Packaging (ICEP). pp. 21–22 (2022)
52. S. Yamamoto, R. Ishimatsu, K. Okada, E. Kato, J. Mizuno, T. Kasahara, Proc. 2022 Int. Conf. Electronics Packaging (ICEP). pp. 23–24 (2022)

53. R. Kawasaki, R. Ishimatsu, K. Okada, J. Mizuno, T. Kasahara, Proc. 2022 Int. Conf. Electronics Packaging (ICEP). pp. 25–26 (2022)
54. E. Kato, R. Ishimatsu, J. Mizuno, T. Kasahara, Proc. 2022 Int. Conf. Electronics Packaging (ICEP). pp. 27–28 (2022)
55. E. Kato, R. Ishimatsu, J. Mizuno, T. Kasahara, Electrochemistry. **91**, 047002 (2023)
56. J. Mai, V.V. Abhyankar, M.E. Piccini, J.P. Olano, R. Willson, A.V. Hatch, Biosens. Bioelectron. **54**, 435–441 (2014)
57. Y. Li, X. Yan, X. Feng, J. Wang, W. Du, Y. Wang, P. Chen, L. Xiong, B.-F. Liu, Anal. Chem. **86**, 100653–100659 (2014)
58. P.J. Hung, P.J. Lee, P. Sabounchi, R. Lin, L.P. Lee, Biotechnol. Bioeng. **89**, 1–8 (2005)
59. L.Y. Wu, D.D. Carlo, L.P. Lee, Biomed. Microdevices **10**, 197–202 (2008)
60. J. Pihl, M. Karlsson, D.T. Chiu, Drug Discov. Today **10**, 1377–1383 (2005)
61. J. Ji, Y. Zhao, L. Guo, B. Liu, C. Ji, P. Yang, Lab Chip **12**, 1373–1377 (2012)
62. V. Srinivasan, V.K. Pamula, R.B. Fair, Lab Chip **4**, 310–315 (2004)
63. C. Priest, P.J. Gruner, E.J. Szill, S.A. Al-Bataineh, J.W. Bradley, J. Ralston, D.A. Steele, R.D. Short, Lab Chip **11**, 541–544 (2011)
64. A. Pavesi, F. Piraino, G.B. Fiore, K.M. Farino, M. Moretti, M. Rasponi, Lab Chip **11**, 1593–1595 (2011)
65. L. Tang, N.Y. Lee, Lab Chip **10**, 1274–1280 (2010)
66. H. Shinohara, T. Kasahara, S. Shoji, J. Mizuno, J. Micromech. Microeng. **21**, 085028 (2011)
67. C.W. Tang, S.A. VanSlyke, Appl. Phys. Lett. **51**, 913–915 (1987)
68. C.W. Tang, S.A. VanSlyke, C.H. Chen, J. Appl. Phys. **65**, 3610–3616 (1989)
69. K. Okumoto, H. Kanno, Y. Hamada, H. Takahashi, K. Shibata, Appl. Phys. Lett. **89**, 013502 (2006)
70. K. Okumoto, H. Kanno, Y. Hamada, H. Takahashi, K. Shibata, J. Appl. Phys. **100**, 044507 (2006)
71. Y. Zhang, S.R. Forrest, Phys. Rev. Lett. **108**, 267404 (2012)
72. Y.-Q. Zheng, J.-L. Yu, C. Wang, F. Yang, B. Wei, J.-H. Zhang, C.-H. Zeng, Y. Yang, J. Phys. D. Appl. Phys. **51**, 225302 (2018)
73. S. Izawa, M. Morimoto, S. Naka, M. Hiramoto, Adv Optical Mater **10**, 2101710 (2022)
74. H. Nakanotani, T. Higuchi, T. Furukawa, K. Masui, K. Morimoto, M. Numata, Y. Hiroyuki Tanaka, T. Sagara, C.A. Yasuda, Nat. Commun. **5**, 4016 (2014)
75. M.-S. Kim, B.-K. Choi, T.-W. Lee, D. Shin, S.K. Kang, J.M. Kim, S. Tamura, T. Noh, Appl. Phys. Lett. **91**, 251111 (2007)
76. S. Tao, Z. Hong, Z. Peng, W. Ju, X. Zhang, P. Wang, S. Wu, S. Lee, Chem. Phys. Lett. **397**, 1–4 (2004)
77. G.H. Zhang, Y.L. Hua, M.C. Petty, K.W. Wu, F.J. Zhu, X. Niu, J.L. Hui, S. Liu, X.M. Wu, S.G. Yin, J.C. Deng, Diplays **27**, 187–190 (2006)
78. W. Song, J.Y. Lee, J. Phys. D. Appl. Phys. **48**, 365106 (2015)
79. M.J. Hollamby, A.E. Danks, Z. Schnepp, S.E. Rogers, S.R. Hart, T. Nakanishi, Chem. Commun. **52**, 7344–7347 (2016)
80. A. Ghosh, T. Nakanishi, Chem. Commun. **53**, 10344–10357 (2017)
81. F. Lu, T. Nakanishi, Adv. Optical Mater. **7**, 190017 (2019)
82. T. Machida, T. Nakanishi, J. Mater. Chem. C **9**, 10661–10667 (2021)
83. O.L. Griffith, S.R. Forrest, Nano Lett. **14**, 2353–2358 (2014)
84. L.C. Picciolo, H. Murata, Z.H. Kafafi, Appl. Phys. Lett. **78**, 2378–2380 (2001)
85. Y. Yamada, H. Kuwae, T. Nomura, J. Oshima, J. Mizuno, T. Kasahara, Trans. Jpn. Inst. Electron. Packag. **13**, E20–001 (2020)
86. M. Zhang, S. Höfle, J. Czolk, A. Mertens, A. Colsmann, Nanoscale **7**, 20009–20014 (2015)